A BEGINNER'S GUIDE TO STRUCTURAL EQUATION MODELING

A Beginner's Guide to Structural Equation Modeling, fifth edition, has been redesigned with consideration of a true beginner in structural equation modeling (SEM) in mind. The book covers introductory through intermediate topics in SEM in more detail than in any previous edition.

All of the chapters that introduce models in SEM have been expanded to include easy-to-follow, step-by-step guidelines that readers can use when conducting their own SEM analyses. These chapters also include examples of tables to include in results sections that readers may use as templates when writing up the findings from their SEM analyses. The models that are illustrated in the text will allow SEM beginners to conduct, interpret, and write up analyses for observed variable path models to full structural models, up to testing higher order models as well as multiple group modeling techniques. Updated information about methodological research in relevant areas will help students and researchers be more informed readers of SEM research. The checklist of SEM considerations when conducting and reporting SEM analyses is a collective set of requirements that will help improve the rigor of SEM analyses.

This book is intended for true beginners in SEM and is designed for introductory graduate courses in SEM taught in psychology, education, business, and the social and health care sciences. This book also appeals to researchers and faculty in various disciplines. Prerequisites include correlation and regression methods.

TIFFANY A. WHITTAKER is an Associate Professor in the Department of Educational Psychology at The University of Texas at Austin, USA, where she teaches courses in structural equation modeling, statistical analysis for experimental data, and advanced statistical modeling.

RANDALL E. SCHUMACKER is a Professor of Educational Research at The University of Alabama, USA, where he teaches courses in multiple regression, multivariate statistics, and structural equation modeling.

A BEGINNER'S GUIDE TO STRUCTURAL EQUATION MODELING

Fifth Edition

Tiffany A. Whittaker and Randall E. Schumacker

Routledge
Taylor & Francis Group

NEW YORK AND LONDON

Cover image: © Getty Images

Fifth edition published 2022
by Routledge
605 Third Avenue, New York, NY 10158

and by Routledge
4 Park Square, Milton Park, Abingdon, Oxon, OX14 4RN

Routledge is an imprint of the Taylor & Francis Group, an informa business

© 2022 Taylor & Francis

First edition published by Lawrence Erlbaum Associates 1996
Fourth edition published by Routledge 2015

Library of Congress Cataloging-in-Publication Data
A catalog record has been requested for this book

ISBN: 978-0-367-49015-7 (hbk)
ISBN: 978-0-367-47796-7 (pbk)
ISBN: 978-1-003-04401-7 (ebk)

DOI: 10.4324/9781003044017

Typeset in Times New Roman
by Newgen Publishing UK

Access the Support Material: www.routledge.com/9780367477967

Tiffany would like to dedicate this book to her Mom and to Zach, for their endless support.

Randall would like to dedicate this book to Joanne: Thanks for the memories.

CONTENTS

PREFACE

APPROACH

This book presents a basic introduction to structural equation modeling (SEM). Readers will find that we have continued our tradition of keeping examples basic and easy to follow. The reader is provided with a review of correlation, covariance, multiple regression, and path analysis in order to better understand the building blocks of SEM. The book then describes the basic steps a researcher takes in conducting structural equation modeling, which include model specification, model identification, model estimation, model testing, and sometimes model modification/re-specification. The reader is subsequently provided with illustrations of factor analysis, full structural equation modeling, and multiple group modeling. The book concludes with extensive checklists that are provided for best practices when conducting analysis in SEM and reporting the results.

Our approach is to focus the researcher on the scope and application of each type of model. The book was written to provide both a conceptual and application-oriented approach. Each chapter covers basic concepts, principles, and practice while utilizing SEM software to provide a meaningful example. Chapters 3 to 9 provide the different SEM model examples with exercises to permit further practice and understanding.

The SEM model examples in the book do not require complicated programming skills, nor does the reader need an advanced understanding of statistics and matrix algebra to understand the model applications. We have provided an appendix on matrix operations for the interested reader. We encourage the understanding of the matrices used in SEM models, which is the foundation of LISREL.

GOALS AND CONTENT COVERAGE

Our main goal in this fifth edition is to provide a basic introduction to SEM model analyses, offer SEM model examples, and provide resources to better understand SEM research. Ultimately, we want readers to be able to conduct SEM analysis appropriately and be able to understand and critique published SEM research. These goals are supported by the conceptual and applied examples and exercises contained in the book chapters as well as the resources on the book's website (e.g., data sets, program codes, Internet links to important resources in SEM, and SEM-related documents).

We have organized the chapters in the book differently based on reviewer and student feedback. We begin with an introduction to SEM (what it is, some history, why conduct it, and what software is available), followed by a chapter on data entry and editing issues. Chapter 1 also introduces the drawing conventions and notation for SEM models. Chapter 2 is critical to understanding how missing data, non-normality, scale of measurement, non-linearity, outliers, and restriction of range in scores affects SEM analysis.

We proceed with correlation and multiple regression models and path models. We also learn about direct and indirect effects in path analysis. An important lesson with path models is the decomposition of the variance–covariance (correlation) matrix. This provides the essential understanding of how a theoretical model implies all, or most of, the sample relations among variables in the sample variance–covariance matrix. It also introduces the concept of model fit between the sample covariance matrix and the model implied covariance matrix.

We have set apart Chapter 5 to discuss in detail SEM principles and practice. Specifically, we detail the five SEM modeling steps. We discuss issues related to model identification and estimation. The different types of model fit assessment are presented. This is a critical chapter that should be studied in depth before proceeding with the remaining chapters in the book.

We introduce exploratory and confirmatory factor analysis in order to demonstrate how multiple measures (observed variables) share variance in defining a latent variable (construct). It introduces the concept that measurement error is modeled in the observed variables. This also permits a discussion of latent variables that are used in structural equation modeling.

Chapters 7 to 9 introduce different types of SEM models. The SEM steps of model specification, identification, estimation, testing, and modification are explained for each type of SEM model. These chapters provide a basic introduction to the rationale and purpose of each SEM model. The chapter exercises permit further application and understanding of how each SEM model is analyzed and how the results are to be interpreted.

Theoretical SEM models are present in many academic disciplines. We have tried to present a variety of different models that can be formulated and tested. This fifth edition attempts to cover SEM models and applications a researcher may use in medicine, political science, sociology, education, psychology, business, health, and the biological sciences. Our focus is to provide the basic concepts, principles, and practice necessary to test the theoretical models. We hope you become more familiar with structural equation modeling after reading the book and use SEM in your own research.

NEW TO THE FIFTH EDITION

The first edition of this book was one of the first books published on SEM, while the second edition greatly expanded knowledge of advanced SEM models. The third edition was extended to represent a more useable book for teaching SEM. The fourth edition returned to a basic introductory book with concepts, principles, and practice related to many different SEM models, including advanced SEM models (e.g., multilevel SEM models, latent growth models, and latent interaction models). The fifth edition now devotes more space to cover the important introductory through intermediate topics in SEM.

Our fifth edition reorganized chapters to represent the progression of topics taught in a standard introduction to structural equation modeling course. The fifth edition offers several changes and additions not in the previous editions. Chapter 2 was expanded to include step-by-step guides to examine outliers and assumptions in SEM. More information is provided with respect to handling outliers and violations of different assumptions. The missing data section was developed to include more information regarding missing data mechanisms and how to handle missing data depending upon the missing data mechanisms.

Chapter 3 now combines Chapters 3 and 4 from the previous edition. This chapter was expanded to include step-by-step instructions to perform multiple regression analysis using SEM software. The content of Chapter 4 was updated to include more information about testing indirect effects. An example model is used as a guide to help readers analyze and interpret findings in a path model with indirect effects. Chapter 5 was developed to include more information concerning model specification, identification, estimation, testing, and modification/re-specification in SEM. More information is provided about when to use different estimators that are available in SEM software, assessing model fit, and model modification. Chapter 5 also includes a discussion about estimation problems that may be encountered and how to remedy certain estimation problems.

Chapter 6 now includes information about parallel analysis to help with the selection of the number of factors in exploratory factor analysis. An example model is used as a guide to help readers analyze and interpret findings in an exploratory factor analysis and a confirmatory factor analysis. Chapter 7 is a new chapter that focuses on a full SEM. Some of the information from Chapter 7 in the previous edition will remain. Specifically, the two-step SEM approach is still discussed where it is recommended to fit the measurement model followed by the structural model. The information about model fit has moved to Chapter 5. Again, an example model is used as a guide to help readers analyze and interpret findings in a full SEM.

Chapter 8 content was updated to include information about the bifactor model which is another parameterization of a general factor as in a higher-order/second-order model. Example models are illustrated to help readers to be able to analyze and interpret the results for second-order and bifactor models. Chapter 9 includes detailed illustrations of a multiple group observed variable path model and a multiple group confirmatory factor analysis (CFA)/measurement model. This chapter was updated with more discussion about invariance testing methods and considerations. Again, example models are used to highlight the use of multiple group modeling.

Chapter 10 was Chapter 16 in the previous edition. Much of the content from Chapter 16 in the previous edition is referenced in chapters that introduce different models, including Chapter 4 and Chapters 6–9. This chapter still includes the checklist for SEM, but there is more elaboration on certain items mentioned in the checklist. For instance, more information concerning adequate sample size and power analyses is provided. In addition to these changes, LISREL (version 11) and M*plus* (version 8.6) are the only two software programs illustrated in the book and on the website.

SEM methods, applications, techniques, issues, and trends have advanced over the past several decades. The fifth edition updates the chapter exercises as well as the relevant references in SEM. We also continue to provide key references for certain topics in SEM.

This fifth edition is intended to provide introductory information to get a beginner to start understanding and using SEM. We hope you find that this fifth edition provides the updated information a beginner would need regarding selection of software, trends, and topics in SEM today.

ACKNOWLEDGMENTS

The fifth edition of this book represents many years of interacting with our colleagues and students who use SEM. As before, we are most grateful to the pioneers in the field of structural equation modeling, particularly to Karl Jöreskog, Dag Sörbom, Peter Bentler, James Arbuckle, and Linda and Bengt Muthén. These individuals have developed and shaped the new advances in the SEM field as well as the content of this book, plus provided SEM researchers with software programs, data sets, and examples. We are grateful to their companies (IBM-Amos, EQS, LISREL, and M*plus*) for providing software to use for SEM analyses.

We wish to extend many thanks to both of the wonderful LISREL and M*plus* support teams for quickly and patiently responding to many emailed questions during the writing of this book. We wish to thank Danielle Dyal, Matt Bickerton, and the production department at Routledge for coordinating all of the activity required to get this book into print. We also want to thank our students for all the questions that keep us continuing to learn as we search for the answers.

ABOUT THE AUTHORS

TIFFANY A. WHITTAKER is an Associate Professor in the Department of Educational Psychology at The University of Texas at Austin, USA, where she teaches courses in fundamental statistics, statistical analysis for experimental data, structural equation modeling, and advanced statistical modeling. She regularly teaches a workshop on structural equation modeling at the Summer Statistics Institute hosted by the Department of Statistics and Data Sciences at The University of Texas at Austin.

Her principal methodological research interest deals with the various facets of model specification, including, but not limited to, model comparison/selection and model modification methods. With the use of simulation techniques, she examines the performance of these different model specification approaches under manipulated conditions. She employs these models within the structural equation modeling (SEM), multilevel modeling (MLM), and item response theory (IRT) arenas. Her research has been published in numerous journals, such as *Structural Equation Modeling*, *Multivariate Behavioral Research*, *American Journal of Evaluation*, *Applied Psychological Measurement*, and the *Journal of Experimental Education*.

She serves on the editorial boards of the *Structural Equation Modeling* and *AERA Open* journals. She was awarded the Excellence in Teaching Award in the Department of Educational Psychology for the 2015–2016 academic year. She has been the recipient of the *Elizabeth Glenadine Gibb Teaching Fellowship in Education* (2014–2015), the *Margie Gurley Seay Centennial Professorship in Education* (2017–2018), and the *Judy Spence Tate Fellowship for Excellence* (2018–2021) in the College of Education at The University of Texas at Austin.

She is an active member of the Structural Equation Modeling Special Interest Group (SIG) of the American Educational Research Association (AERA). She was elected as the Co-Chair of the Structural Equation Modeling SIG in 2009 and 2021 to serve as Chair of the Structural Equation Modeling SIG in the following year (2010 and 2022, respectively). She can be contacted at The University of Texas at Austin, Department of Educational Psychology, 1912 Speedway; STE 5.708, Mail Code: D5800, Austin, TX 78712 or by email at t.whittaker@austin.utexas.edu.

RANDALL E. SCHUMACKER is a Professor of Educational Research at The University of Alabama, where he teaches courses in multiple regression, multivariate statistics, and structural equation modeling. His research interests are varied, including applications in SEM, psychometrics, and meta-analysis. He has taught several international and national workshops on structural equation modeling.

Randall has written and co-edited several books, including *A Beginner's Guide to Structural Equation Modeling (4th edition)*, *Advanced Structural Equation Modeling*: *Issues and Techniques*, *Interaction and Non-Linear Effects in Structural Equation Modeling*, *New Developments and Techniques in Structural Equation Modeling*, *Understanding Statistical Concepts Using S-PLUS*, *Understanding Statistics Using R*, *Learning Statistics using R*, and *Using R with Multivariate Statistics*.

Randall has published in several journals including *Academic Medicine*, *Educational and Psychological Measurement*, *Journal of Applied Measurement*, *Journal of Educational and Behavioral Statistics*, *Journal of Research Methodology*, *Multiple Linear Regression*, and *Structural Equation Modeling*. He has served on the editorial boards of numerous journals and is a member of the American Educational Research Association, Past-President of the Southwest Educational Research Association, and Emeritus Editor of the *Structural Equation Modeling* journal.

Dr. Schumacker was the 1996 recipient of the *Outstanding Scholar Award*, and the 1998 recipient of the *Charn Oswachoke International Award*. In 2010, he launched the *DecisionKit* App for the iPhone and iPad. In 2011, he received the Apple iPad Award and in 2012, he received the *CIT Faculty Technology Award*. In 2013, he received the *McCrory Faculty Excellence in Research Award* from the College of Education at The University of Alabama. In 2014, Dr. Schumacker was the recipient of the *Structural Equation Modeling Service Award* at the American Educational Research Association, where he founded the Structural Equation Modeling Special Interest Group. He can be contacted at The University of Alabama, College of Education, P.O. Box 870231, 316 Carmichael Hall, Tuscaloosa, AL 35487–0231, USA or by email at rschumacker@ua.edu.

Chapter 1

INTRODUCTION

WHAT IS STRUCTURAL EQUATION MODELING?

Structural equation modeling (SEM) depicts relations among observed and latent variables in various types of theoretical models, providing quantitative tests of hypotheses by the researcher. Basically, various theoretical models may be hypothesized and tested with SEM. The SEM models hypothesize how sets of observed variables are interrelated, how sets of variables define constructs, and/or how different constructs are related to each other. For example, an educational researcher might hypothesize that a student's home environment influences their subsequent achievement in school. A marketing researcher may hypothesize that consumer trust in a corporation leads to increased financial performance for that corporation. A health care professional might believe that increased stress levels can result in increased health risks.

In each example, based on theory and empirical research, the researcher wants to test whether a set of observed variables are related or define the constructs that are hypothesized to be related in a certain way. The goal of SEM is to test whether the theoretical model is supported by sample data. If the sample data support the theoretical model, then the hypothesized relations among the constructs are strengthened. If the sample data do not support the theoretical model, then an alternative theoretical model may need to be specified and tested. Consequently, SEM allows researchers to test theoretical models in order to advance our understanding of the complex relations among constructs.

SEM can test various types of theoretical models. The first types discussed in this book include linear regression, path, and confirmatory factor analysis (CFA) models, which form the basis for understanding the many different types of SEM models. The regression models use observed variables, while path models can use either observed or latent variables. CFA models, by definition, use observed

DOI: 10.4324/9781003044017-1

variables to define latent variables. Thus, these two types of variables, *observed variables* and *latent variables*, are used depending upon the type of SEM model.

NOTATION AND TERMINOLOGY

Structural equation models are commonly demonstrated in diagrams with various symbols representing observed and latent variables and their interrelationships, such as squares, circles, one-headed arrows, and two-headed arrows. Figure 1.1 includes a summary of the common diagrams used in SEM. In SEM diagrams, squares symbolize *observed* variables whereas circles symbolize *unobserved* or *latent* variables.

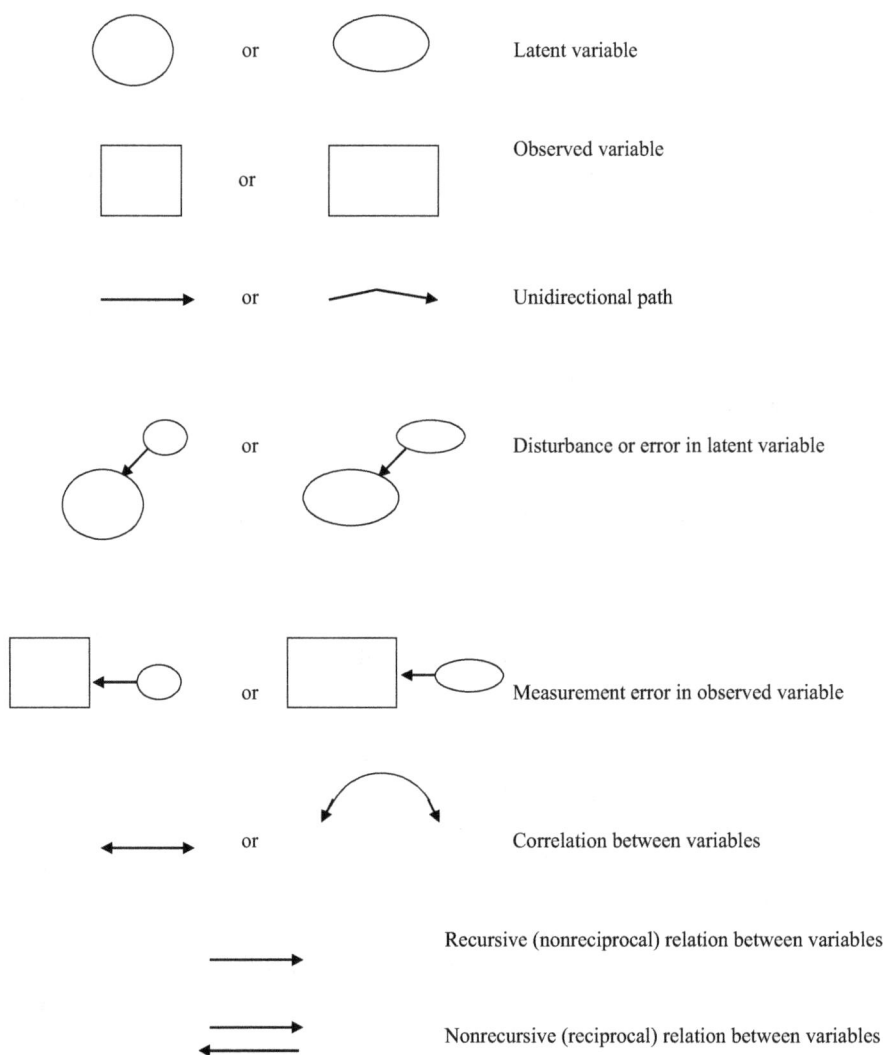

Figure 1.1:   COMMON SEM DIAGRAM SYMBOLS

Latent variables, also called constructs or factors, are variables that are not directly observed or measured. Latent variables are not directly observed or measured, but are inferred from a set of observed variables that we actually measure using tests, surveys, scales, and so on. For example, intelligence is a latent variable that represents a psychological construct. The confidence of consumers in American business is another latent variable, one representing an economic construct. Stress is a third latent variable, one representing a health-related construct.

The observed variables, also called measured, manifest, or indicator variables, are a set of variables that we use to define or infer the latent variable or construct. Researchers use several indicator variables to define a latent variable. For example, the Wechsler Intelligence Scale for Children-Revised (WISC-R) is an instrument that produces several measured composite variables (or observed scores), which are used to infer different components of the construct of a child's intelligence. These variables could be used as indicator variables of intelligence to indicate or define the construct of intelligence (a latent variable). The Dow Jones index is a standard measure of the American corporate economy strength construct. Other indicator variables that could be used in addition to the Dow Jones index to define economy strength (a latent variable) include gross national product, retail sales, and export sales. The number of times a person exercises per week is an indicator of a health-related latent variable that could be defined as *Health Risk*. Other indicator variables that could also be used as an indicator for this *Health Risk* factor include diet and body mass index (BMI). These models are commonly referred to as confirmatory factor analysis (CFA) or measurement models, which test whether the relationships among indicator variables are explained well by the latent variable. These measurement models can then be used in larger structural models which test the hypothesized relations among the latent variables.

Observed and latent variables are defined as either *independent variables* or *dependent variables*. An independent variable, also referred to as an *exogenous variable*, is a variable that is not influenced or directly affected by any other variable in the model. A dependent variable, also called an *endogenous variable*, is a variable that is influenced or directly affected by other variables in the model. Figure 1.1 illustrates other symbols used in SEM. One-headed arrows represent direct effects while two-headed arrows represent that two variables are simply related. The researcher specifies the independent (exogenous) and dependent (endogenous) variables and other interrelationships in a model using one-headed and two-headed arrows.

For instance, the educational researcher hypothesizes that a student's home environment (independent/exogenous latent variable) influences school achievement (dependent/endogenous latent variable). The marketing researcher believes that consumer trust in a corporation (independent/exogenous latent variable) leads to increased financial performance for that corporation (dependent/endogenous latent variable). The health care professional wants to determine whether stress (independent/exogenous latent variable) influences health risk (dependent/ endogenous latent variable).

The basic SEM models (regression, path, and CFA) illustrate the use of observed variables and latent variables which are defined as independent or dependent in the model. A regression model consists solely of observed variables where a single observed dependent variable is predicted or explained by one or more observed independent variables. For example, a child's intelligence and their time spent on homework (observed independent variables) are used to predict the child's achievement score (observed dependent variable). A depiction of this model is shown in Figure 1.2, wherein the intelligence (IQ) variable and the homework time (HW) variable are hypothesized to directly impact achievement (ACH) while covarying. You will notice that variances (represented by a circular two-headed arrow) are associated with both IQ and HW because they are independent (exogenous) variables. This is because these variables are free to vary with freely estimated variances given that they are not directly impacted by other variables in the model. You will also notice that an error variance is associated with ACH because it is a dependent (endogenous) variable. Unless both IQ and HW explain 100% of the variance in ACH, there will be unexplained variability in ACH that is not explained by both IQ and HW. Thus, dependent (endogenous) variables in SEM have error terms associated with them. These are typically represented via circles in diagrams because they are also unobserved or latent variables. They also have direct effects on their respective dependent (endogenous) variables. Thus, errors are usually considered independent (exogenous) variables because they have direct effects on dependent (endogenous) variables. As such, errors also have variances (represented by a circular two-headed arrow) associated with them.

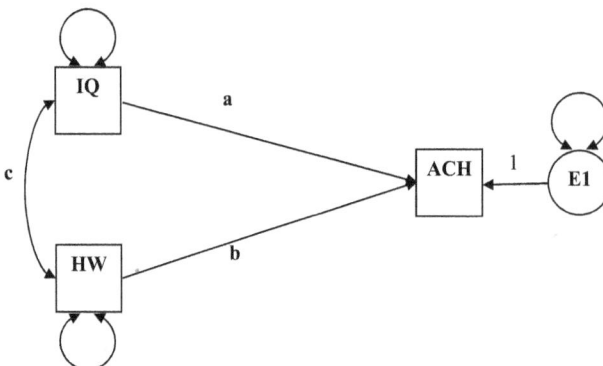

▪ **Figure 1.2:**   ACHIEVEMENT MULTIPLE REGRESSION MODEL

A path model can also be specified entirely with observed variables, but the flexibility allows for multiple observed independent variables and multiple observed dependent variables – for example, export sales, gross national product, and NASDAQ index (observed independent/exogenous variables) influence consumer trust and consumer spending (observed dependent/endogenous variables). Path models are generally more complex models than regression models, include direct and indirect effects, and can include latent variables. Illustrated in Figure 1.3 is an observed variable path model wherein home environment, school environment, and relations with peers (covarying independent/exogenous variables) explain student–parent relations and student–teacher relations (dependent/endogenous variables), which subsequently directly impact student achievement (dependent/ endogenous variable).

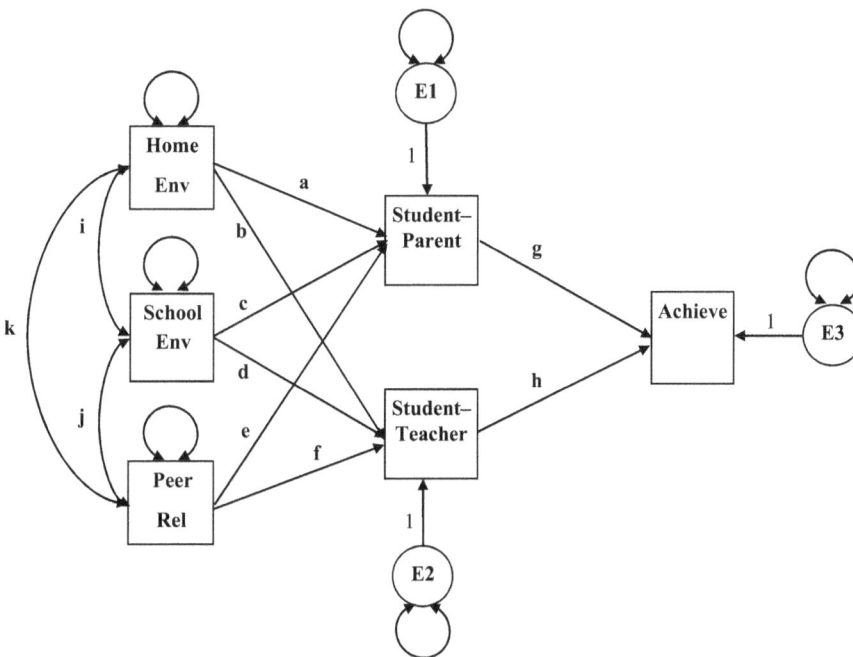

■ **Figure 1.3:** ACHIEVEMENT OBSERVED VARIABLE PATH MODEL

Confirmatory factor models consist of observed variables that are hypothesized to measure both independent and dependent latent variables. Figure 1.4 illustrates an SEM model wherein one factor directly impacts another factor. As shown in Figure 1.4, responses to items asking how often participants have felt nervous, felt not in control of life events, and felt disorganized in the last week are observed measures of the independent (exogenous) latent variable, *Stress*, while blood pressure, cholesterol, and restless sleep are observed measures of the dependent (endogenous) latent variable, *Health Risk*. The dependent (endogenous) *Health*

Risk factor has error associated with it, representing unexplained variability. You will notice in Figure 1.4 that the indicator variables are dependent (endogenous) variables in measurement models with errors (E1–E6). Errors associated with latent dependent (endogenous) factors are commonly differentiated from the errors associated with observed dependent (endogenous) variables and typically called disturbances (e.g., D1; see Figure 1.4).

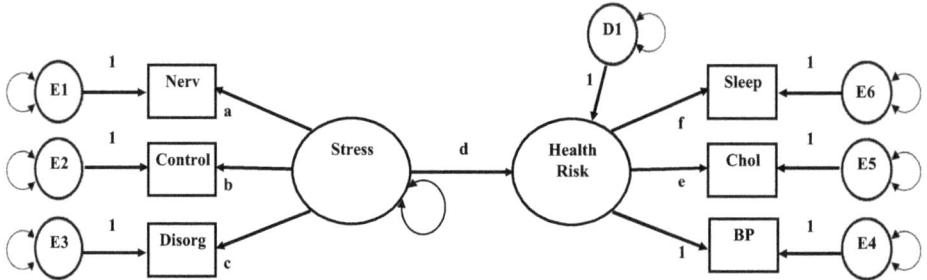

■ **Figure 1.4:**   STRESS AND HEALTH RISK SEM MODEL

HISTORY OF STRUCTURAL EQUATION MODELING

To discuss the history of structural equation modeling, we explain the following four types of related models and their chronological order of development: regression, path, confirmatory factor, and structural equation modeling (see Matsueda, 2012 for additional coverage of the history of SEM). The first models involve linear regression models which use a correlation coefficient and the least squares criterion to compute regression weights. Regression models were made possible because Karl Pearson created a formula for the correlation coefficient in 1896 that provided an index for the relation between two variables (Pearson, 1938). The regression model permits the prediction of observed dependent variable scores (*Y scores*), given a linear weighting of a set of observed independent scores (*X scores*). The linear weighting of the independent variables is done using regression coefficients, which are determined based on minimizing the sum of squared residual error values. The selection of the regression weights is therefore based on the *Least Squares Criterion*. The mathematical basis for the linear regression model is found in basic algebra. Regression analysis provides a test of a theoretical model that may be useful for prediction. For example, Delucchi (2006) used regression analysis to predict student exam scores in statistics (dependent variable) from a series of collaborative learning group assignments (independent variables). The results provided some support for collaborative learning groups improving statistics exam performance, although not for all tasks.

Some years later, Charles Spearman (1904, 1927) used the correlation coefficient to determine which items correlated or went together to create a measure of general intelligence. His basic idea was that if a set of items correlated or went together, individual responses to the set of items could be summed to yield a score that would measure or define a construct. Spearman was the first to use the term *factor analysis* in defining a two-factor construct for a theory of intelligence. In 1940, D. N. Lawley and L. L. Thurstone further developed applications of factor models, and proposed instruments (sets of items) that yielded observed scores from which constructs could be inferred. Most of the aptitude, achievement, and diagnostic tests, surveys, and inventories in use today were created using factor analytic techniques. The term *confirmatory factor analysis* (CFA) used today is based in part on earlier work by Howe (1955), Anderson and Rubin (1956), and Lawley (1958). The CFA method was more fully developed by Karl Jöreskog in the 1960s to test whether a set of items defined a construct. Jöreskog completed his dissertation in 1963, published the first article on CFA in 1969, and subsequently helped develop the first CFA software program. Factor analysis has been used for more than 100 years to create measurement instruments in many academic disciplines. CFA today uses observed variables derived from measurement instruments to test the existence of a theoretical construct. For instance, Goldberg (1990) used CFA to test the *Big Five* model of personality. His five-factor model of extraversion, agreeableness, conscientiousness, neuroticism, and intellect was tested through the use of multiple indicator variables for each of the five hypothesized constructs.

The path model was developed by Sewell Wright (1918, 1921, 1934), a biologist. Path models use correlation coefficients and multiple regression equations to model more complex relations among observed variables. The first application of path models dealt with animal behavior. Unfortunately, path analysis was largely overlooked until econometricians reconsidered it in the 1950s as a form of simultaneous equation modeling (Wold, 1954) and sociologists rediscovered it in the 1960s (Duncan, 1966) and 1970s (Blalock, 1972). A path analysis involves solving a set of simultaneous regression equations that theoretically establish the relations among the observed variables in the path model. Parkerson, Lomax, Schiller, and Walberg (1984) conducted a path analysis to test Walberg's theoretical model of educational productivity for fifth- through eighth-grade students. The relations among the following variables were analyzed in a single model: home environment, peer group, media, ability, social environment, time on task, motivation, and instructional strategies. All of the hypothesized paths among those variables were shown to be statistically significant, providing support for the educational productivity path model.

Structural equation models (SEMs) combine path models and confirmatory factor models when establishing hypothesized relations among latent variables. The early development of SEM models was due to Karl Jöreskog (1969, 1973), Ward Keesling (1972), and David Wiley (1973). The approach was initially known as the JKW model, but became known as the *li*near *s*tructural *rel*ations (LISREL) model with the development of the first software program in 1973, which was called LISREL.

Jöreskog and van Thillo originally developed the LISREL software program at the Educational Testing Service (ETS) using a matrix command language that used Greek and matrix notation. The first publicly available version, LISREL III, was released in 1976. By 1993, LISREL 8 was released, introducing the SIMPLIS (SIMPle LISrel) command language in which equations were written instead using variable names. In 1999, the first interactive version of LISREL was released. LISREL 8 introduced the dialog box interface using pull-down menus and point-and-click features to develop models. The path diagram mode permitted drawing a program to develop models. LISREL 9 was released with new features to address categorical and continuous variables. LISREL 10.3 introduced combined functionality of LISREL and PRELIS. LISREL 11 was substantially updated, including self-contained path diagram functionality and run button enabling of appropriate LISREL or PRELIS syntax files.

Karl Jöreskog was recognized by Cudeck, du Toit, and Sörbom (2001) who edited a Festschrift in honor of his contributions to the field of structural equation modeling. Their volume contains chapters by scholars who addressed the many topics, concerns, and applications in the field of structural equation modeling today, including milestones in factor analysis; measurement models; robustness, reliability, and fit assessment; repeated measurement designs; ordinal data; and interaction models.

The field of structural equation modeling across all disciplines has expanded since 1994. Hershberger (2003) found that between 1994 and 2001, the number of journal articles concerned with SEM increased; the number of journals publishing SEM research increased; SEM became a popular choice among multivariate methods; and the *Structural Equation Modeling* journal became the primary source for technical developments in structural equation modeling, and continues so today. SEM research articles are now more prevalent than ever in professional journals of several different academic disciplines (medicine, psychology, business, education, etc.).

WHY CONDUCT STRUCTURAL EQUATION MODELING?

Why is structural equation modeling popular? There are at least four major reasons for the popularity of SEM. First, researchers are becoming more aware of the need to use multiple observed variables to investigate their area of scientific inquiry. Basic statistical methods only utilize a limited number of independent and dependent variables, restricting the theoretical hypotheses that may be tested among multiple variables. The use of a small number of variables to understand complex phenomena is restricted in focus. For instance, the bivariate correlation is not sufficient for examining prediction when using multiple variables in a regression equation. In contrast, structural equation modeling permits relations among multiple variables to be modeled and statistically tested. SEM techniques are therefore a more preferred method to support (or disconfirm) theoretical models.

Second, an increased concern has been given to the validity and, particularly, the reliability of observed scores from measurement instruments. Specifically, the negative impact of measurement error in observed variable path models (i.e., biased estimates and increased Type I or Type II errors) has become a major issue in many disciplines (Cole & Preacher, 2014). Measurement or CFA models explicitly take the measurement error associated with the observed indicator variables into account when statistically analyzing data. Thus, including measurement models in SEM analyses can result in more confidence associated with the inferences drawn from the analyses.

Third, structural equation modeling has matured over the past 45 years, especially the software programs and the ability to analyze more advanced theoretical SEM models. For example, group differences in theoretical models can be tested with multiple-group SEM models. The analysis of educational data collected at different nested levels (classes, schools, and school districts) with student data is now possible using multilevel SEM. SEM models are no longer limited to linear relations. Interaction terms can now be included in a SEM model so that main effects and interaction effects can be tested. The improvement in SEM software has led to advanced SEM models and techniques, which have provided researchers with an increased capability to analyze sophisticated theoretical models of complex phenomena.

Fourth, the SEM software programs have become increasingly user-friendly. Until 1993, SEM modelers had to input the program syntax for their models using Greek and matrix notation. At that time, many researchers sought help because of the complex programming and knowledge of the SEM syntax that was required. Today, SEM software programs are easier to use and contain features similar to other Windows-based software packages, such as pull-down menus, data spreadsheets, and a simple set of commands. Nonetheless, user-friendly SEM software, which has permitted more access by researchers, comes with concerns about proper usage. Researchers need the prerequisite training in statistics and, specifically, in SEM. Fortunately, there are courses, workshops, and textbooks for SEM that are widely available, which can help researchers avoid mistakes and errors in analyzing sophisticated theoretical models using SEM. Which SEM software program you choose may also influence your level of expertise and what type of models can be analyzed.

BRIEF CONSIDERATION OF CAUSALITY IN SEM

With the various types of relationships that may be hypothesized in SEM models, it is worthy to briefly consider the issue of causal inference. By no means is the coverage of this topic exhaustive. Interested readers should consult formative readings in the area (viz., Davis, 1985; Sobel, 1995; Pearl, 2000, 2012; Mulaik, 2009; Bollen & Pearl, 2013). As illustrated in Figure 1.1 and in the models described in this chapter, one-headed arrows represent hypotheses concerning causal directions. For instance, in Figure 1.2, IQ is hypothesized to directly affect student achievement. Based on a theoretical basis, the implication is that IQ causes student achievement outcomes. In SEM, three conditions are generally needed to deduce causality: association, isolation, and temporal precedence.

Association basically means that the cause (X) and effect (Y) are shown to covary with one another. Isolation denotes that the cause (X) and effect (Y) still covary when they are isolated from other influential variables. This condition is commonly the most challenging to satisfy in its entirety, but it could be closer to attainment if data are collected for those variables that may well impact the relationship between the cause and effect (e.g., motivation) and controlled for statistically. Design considerations (e.g., collecting data only on eighth grade students to avoid the influence of grade differences) may also be implemented to help meet this condition. Temporal precedence denotes that the theorized cause happens before the theorized effect in time. Temporal precedence may be achieved by way of data collection methods used in experimental designs with random assignment or in quasi-experimental designs with manipulation of treatment exposure. Temporal precedence may also be accomplished by collecting data for the outcome variables following the collection of data on the causal variables.

STRUCTURAL EQUATION MODELING SOFTWARE

Structural equation modeling can be easily understood if the researcher has a grounding in basic statistics, correlation, regression, and path analysis. Some SEM software programs provide a pull-down menu with these capabilities, while others come included in a statistics package where they can be computed. Although the LISREL program was the first SEM software program, other software programs have subsequently been developed since the mid–1980s. These include EQS, M*plus*, R, PROC CALIS (SAS), AMOS (SPSS), LISREL, and SEM (STATA), to name a few. These software programs are each unique in their own way, with some offering specialized features for conducting different SEM applications. Many of these SEM software programs provide statistical analysis of raw data (means, correlations, missing data conventions), provide routines for handling missing data and detecting outliers, generate the program's syntax, diagram the model, and provide for the importing and exporting of data and figures of a theoretical model. Also, many of the SEM software programs come with sets of data and program examples that are clearly explained in their user guides. Many of these software programs have been reviewed in the *Structural Equation Modeling* journal.

The pricing information for SEM software varies, depending on individual, group, or site license arrangements; corporate versus educational settings; and even whether one is a student or faculty member. Furthermore, newer versions and updates necessitate changes in pricing. We are often asked to recommend a software package to a beginning SEM researcher; however, given the different individual needs of researchers and the multitude of different features available in these programs, we are not able to make such a recommendation. Ultimately, the decision depends upon the researcher's needs and preferences.

The many different SEM software programs are either stand-alone or part of a statistical package. Most SEM programs run in the Windows environment, but not all. Some have demo or trial versions, while others are free. Several SEM software programs are described below in alphabetical order with links to their websites. These links are also on the book website.

AMOS (SPSS)

AMOS is an add-on purchase in the IBM SPSS statistical package. It uses diagramming tools to draw the SEM model, then links the model and variables to an SPSS data set. Some of its features are the ability to drag and drop the variable names into the model from the SPSS data set, the ability to export the SEM model diagram to Word via a clipboard tool, the analysis of multiple-group models, and the ability to conduct an automated specification search for a better model. More information about AMOS can be found at: www.ibm.com/us-en/marketplace/structural-equation-modeling-sem

CALIS Procedure (in SAS)

PROC CALIS is a procedure in the SAS statistical package. It uses SAS data sets and specific SAS syntax code. Statisticians that frequently use SAS also use this program to conduct a test of their SEM model. Some of its features are the flexibility to describe a model with structural equation statements similar to EQS using the LINEQS option or specify complex matrix models using MATRIX statements. The CALIS procedure also offers several optimization algorithms (NLOPTIONS) to obtain parameter estimates and standard errors when problems are encountered in the data due to non-normality and non-linearity. More information on this procedure can be found at: http://support. sas.com/documentation/cdl/en/statug/63033/HTML/default/viewer.htm#calis_toc.htm

EQS: Stand-alone Software

EQS software comes with a program manual, user's guide, and free technical support. Some of its features are new and improved missing data procedures, reliability coefficients for factor models, Satorra-Bentler robust corrected standard errors in the presence of non-normal and missing data, multilevel modeling options, bootstrap simulation capabilities, case weighting, and a simple set of syntax commands. More information about EQS can be found at: https://mvsoft.com/

JMP: Stand-alone Software / SAS Interface

JMP is a graphical user interface software product. It is a statistical, graphical, and spreadsheet-based software product that does not necessarily require writing program syntax. SEM runs in JMP version 15 for Mac and Windows operating systems. SAS structural equation modeling (SEM) for JMP is an application that enables researchers to use SAS and JMP to draw models by using an interface that is built on the SAS/STAT® CALIS procedure. JMP can be used to analyze path models, CFA models, full SEM models, and latent growth curve models using full information maximum likelihood estimation. SAS created JMP in 1989 to allow data to be explored visually by scientists and engineers. JMP, once a single product, has grown into a collection of statistical tools for data cleaning, analysis, and sharing. All of the software is visual, interactive, comprehensive, and extensible. More information about JMP is available at: www.jmp.com/software/

LISREL: Stand-alone Software

LISREL for Windows has evolved from a matrix only software (in older versions of LISREL) to a simple set of syntax commands (SIMPLIS). You can obtain output in matrix format or obtain the basic results. The software comes with a free two-week trial of the fully functional software. User guides, sets of data, and examples for the many different types of SEM models with a help menu to search for information and explanations of analysis and procedures are available.

PRELIS is a pre-processor for the data to check for missing data, non-normality (skewness and kurtosis), and data type (nominal, ordinal, interval/ratio). The pull-down menu in LISREL also provides for data imputation in the presence of missing data; classical and ordinal factor analysis; censored, logistic, and probit regression; bootstrapping; and output options to save files, such as the asymptotic covariance matrix. Other features on the pull-down menu include multilevel modeling and survey data analysis using the generalized linear model. Students taking a course or workshop from an instructor with a current license of LISREL will be allowed free access to LISREL software for the duration of the course or workshop through their free educator benefit program. More information about LISREL can be found at: www.ssicentral.com/index.php/product/lisrel

M*plus*: Stand-alone Software

M*plus* comes with a demo version (with certain restrictions), user's guide, and examples. It runs on Windows, Mac, and Linux operating systems. M*plus* uses a simple set of commands to specify and test many different types of SEM models. In addition to basic SEM analyses, some other features in M*plus* include Bayesian SEM, exploratory SEM, IRT model specifications, mixture modeling, Monte Carlo simulation, and intensive longitudinal data modeling. The company provides a variety of short courses and presentations. Many examples, white papers, videos, and references are also provided for SEM researchers on the M*plus* website. More information about M*plus* can be found at: www.statmodel.com/

OpenMx: Stand-alone Software (Free); R Interface

OpenMx is a free and open-source software that can be used with R to estimate various advanced multivariate statistical models. It consists of a library of functions and optimizers allow users to quickly define an SEM model and estimate model parameters given observed variables. It runs on Mac, Windows, and Linux operating systems. The same script programs in Windows will run in Mac or Linux. OpenMx can be used by drawing path models or specifying models in terms of matrix algebra. It is extremely functional because it takes advantage of the R programming environment. More information about OpenMx can be found at: https://openmx.ssri.psu.edu/

R: Stand-alone Software (Free)

R is a library of packages with functions that permit flexible writing of script files to conduct statistical analysis in Windows, Mac, or Linux operating systems. Two structural equation modeling packages available in R are *sem* and *lavaan*. Fox (2006) published an article about the *sem* package in the *Structural Equation*

Modeling journal. Yves (2012) published an article about the *lavaan* package in the *Journal of Statistical Software*. More information about R, R installation, and R packages can be found at: www.cran.r-project.org

SEM (in STATA)

SEM is included in the STATA statistical package as a collection of procedures with data sets and examples. The company provides video tutorials, training sessions, and short courses. The software is available on Windows, Mac, and UNIX (LINUX) operating systems. SEM uses either the *SEM Builder* option or command language to create path diagrams of models. *SEM Builder* uses the drag-n-drop approach to create the path diagrams, run the model, and display the results on the path diagram. Some features of interest are the mediation analysis, MIMIC models, latent growth curve models, multilevel models, and generalized linear models with binary, count, ordinal, and nominal outcomes. More information about SEM in STATA can be found at: www.stata.com/features/structural-equation-modeling/

SEPATH (in Statistica)

SEPATH is included in Statistica (now part of the TIBCO Data Science Platform) as a collection of procedures with data sets and examples. The SEM model is specified and the data set input into SEPATH via a dialog box in Statistica. Dialog boxes also permit the selection of analysis parameters (type of matrix to analyze, estimation method, constraints, etc.) and report the results (parameter estimates, standard errors, etc.). A path diagram of the SEM model can also be generated. Some features of this software are the Wizards used to create CFA and SEM models; Monte Carlo analysis; and explanation of important concepts in SEM (iteration, estimation, model identification, non-centrality based fit indices, standard error differences between correlation versus covariance matrix input). You can find more information about SEPATH at: www.tibco.com/products/data-science

SOFTWARE CONSIDERATIONS

We have listed many of the different SEM software programs available today for the SEM modeler and explained some of the features available in each of the software programs. If you are new to structural equation modeling, check what software is available at your company or university. You may wish to learn the software program that comes with a statistics package (IBM – AMOS, SAS – PROC CALIS, STATA – SEM, Statistica – SEPATH). Alternatively, you can purchase a stand-alone software package (EQS, LISREL, M*plus*) or download software that is free (OpenMx, R – sem, R – lavaan). Many researchers choose SEM software based on whether it runs in a Windows, Mac, or UNIX (Linux) environment on their computer. We will demonstrate regression models, path models, confirmatory factor models, and the various SEM models in the book using LISREL (version 11) and M*plus* (version 8.6) software programs. We understand that the reader may not have access to these software programs, however, the data sets and examples in the book can be run with the SEM software available to you. The data and programs on the book website were saved as ASCII files, which permit easy copying and use in any SEM software program. The choice and installation of software to use with the data sets and examples is left up to the reader.

BOOK WEBSITE

This book has a website on the Internet located at www.routledge.com/cw/ whittaker. From the website you will be able to access or download the following resources:

- Answers to all of the chapter exercises.
- Computer files and/or diagrams for the SEM modeling types presented in the book, which include LISREL and M*plus* examples.
- Data sets used in the SEM model examples.
- Links to web pages that have more information about SEM.
- Suggested readings or journal articles in PDF format for many of the SEM topics covered in the book.

The purpose of the website is to support learning how the different SEM models are created and tested using LISREL and M*plus* software programs. It also allows for the expansion of the website to include other data sets, examples, and software usage in the future. The availability of data and syntax files will assist in the analysis and interpretation of the examples in the book using the corresponding software program.

You will be able to open the syntax files from the different software programs in Notepad because they will be in ASCII text format. This provides a visual inspection of how each software program has unique syntax commands with the same model specifications. Some of the exercises involve real data sets, which will provide a practical guide for programming, analysis, and interpretation. Some of the SEM model examples in the book may only contain the solution and interpretation of results with the data and program on the website.

SUMMARY

In this chapter, we briefly described structural equation modeling, presented the history of structural equation modeling, and explained four reasons why researchers would choose to conduct structural equation modeling. We introduced structural equation modeling by describing basic types of variables – that is, latent, observed, independent, and dependent – and basic types of SEM models – that is, regression, path, confirmatory factor analysis, and structural equation models. Causality in SEM was briefly considered. The chapter concluded with a brief listing of the different structural equation modeling software programs and where to obtain them. The website contains answers to the chapter exercises, data sets, programs, links, and other resources referenced in the book.

EXERCISES

1. Define the following terms:
   a. Latent variable
   b. Observed variable
   c. Dependent/endogenous variable
   d. Independent/exogenous variable
2. Explain the difference between a dependent/endogenous latent variable and a dependent observed variable.
3. Explain the difference between an independent/exogenous latent variable and an independent observed variable.
4. List the four reasons why a researcher would conduct structural equation modeling.
5. Figure 1.5 is adapted from Howard and Maxwell (1982). Based on this figure, answer the following questions:
   a. Which variables in the model, either observed or latent, are serving as independent/exogenous variables and dependent/endogenous variables?
   b. What do the errors (E1–E3) represent in the model?
   c. What do the circular two-headed arrows represent in the model?

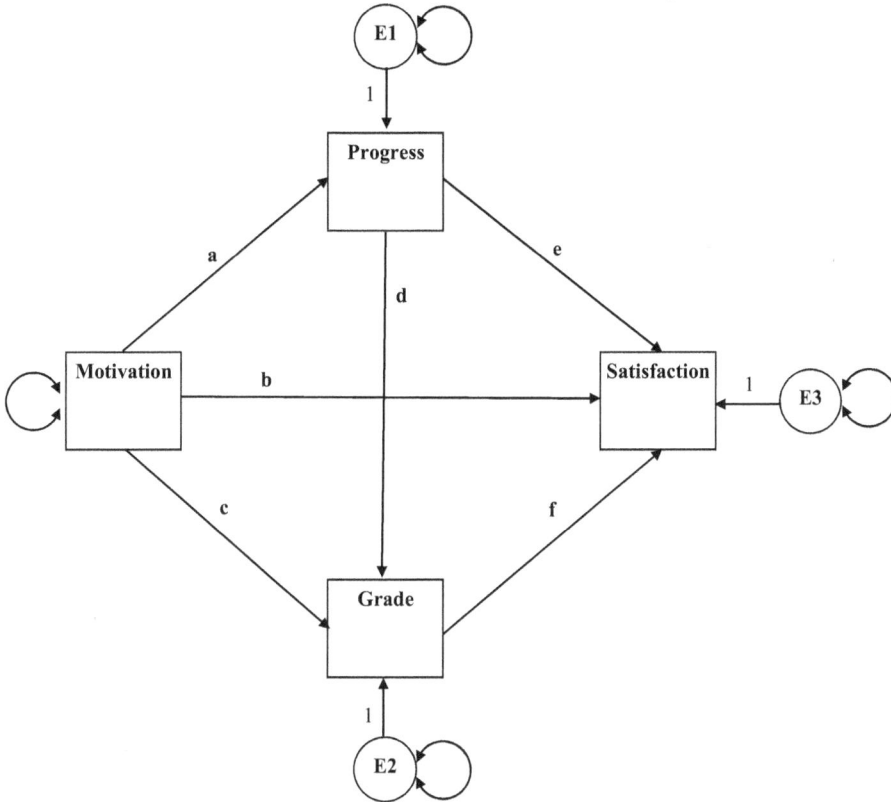

■ **Figure 1.5:** HOWARD & MAXWELL PATH MODEL

6. Figure 1.6 is a one-factor CFA or measurement model. Based on this figure, answer the following questions:
   a. Which variables in the model, either observed or latent, are serving as independent/exogenous variables and dependent/endogenous variables?
   b. What do the errors (E1–E4) represent in the model?
   c. What do the circular two-headed arrows represent in the model?

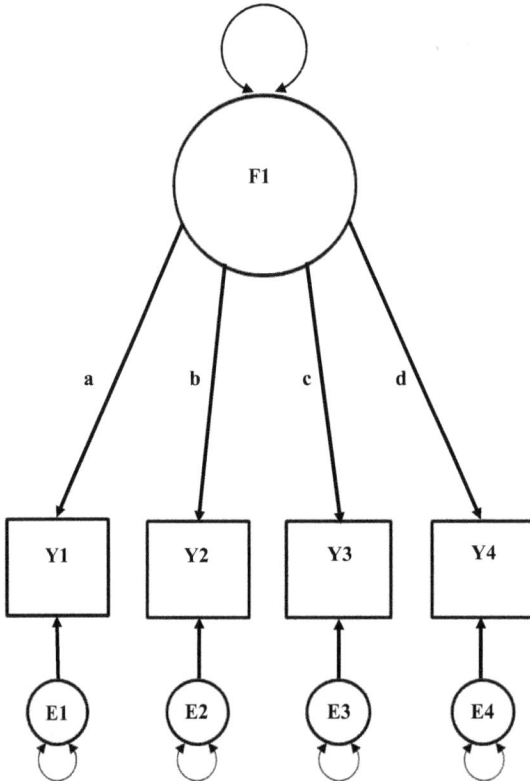

■ **Figure 1.6:**   ONE-FACTOR CFA MODEL WITH FOUR INDICATORS

REFERENCES

Anderson, T. W., & Rubin, H. (1956). Statistical inference in factor analysis. In J. Neyman (Ed.), *Proceedings of the third Berkeley symposium on mathematical statistics and probability, Vol. V* (pp. 111–150). Berkeley, CA: University of California Press.

Blalock, H. M., Jr. (Ed.) (1972). *Causal models in the social sciences.* London: Macmillan.

Bollen, K. A., & Pearl, J. (2013). Eight myths about causality and structural equation models. In S. L. Morgan (Ed.), *Handbook of causal analysis for social research* (pp. 301–328). New York: Springer.

Cole, D. A., & Preacher, K. J. (2014). Manifest variable path analysis: Potentially serious and misleading consequences due to uncorrected measurement error. *Psychological Methods*, 19(2), 300–315.

Cudeck, R., du Toit, S., & Sörbom, D. (Eds). (2001). *Structural equation modeling: Present and future. A Festschrift in honor of Karl Jöreskog.* Lincolnwood, IL: Scientific Software International.

Davis, J. A. (1985). *The logic of causal order*. Beverly Hills, CA: Sage Publications.

Delucchi, M. (2006). The efficacy of collaborative learning groups in an under-graduate statistics course. *College Teaching*, 54, 244–248.

Duncan, O. D. (1966). Path analysis: Sociological examples. *The American Journal of Sociology*, 72(1), 1–16.

Fox, J. (2006). Structural equation modeling with the sem package in R. *Structural Equation Modeling*, 13, 465–486.

Goldberg, L. (1990). An alternative "description of personality": Big Five factor structure. *Journal of Personality and Social Psychology*, 59, 1216–1229.

Hershberger, S. L. (2003). The growth of structural equation modeling: 1994–2001. *Structural Equation Modeling*, 10(1), 35–46.

Howard, G. S., & Maxwell, S. E. (1982). Do grades contaminate student evaluations of instruction? *Research in Higher Education*, 16(2), 175–188.

Howe, W. G. (1955). *Some contributions to factor analysis* (Report No. ORNL–1919). Oak Ridge National Laboratory, Oak Ridge, TN.

Jöreskog, K. G. (1963). *Statistical estimation in factor analysis: A new technique and its foundation*. Stockholm: Almqvist & Wiksell.

Jöreskog, K. G. (1969). A general approach to confirmatory maximum likelihood factor analysis. *Psychometrika*, 34, 183–202.

Jöreskog, K. G. (1973). A general method for estimating a linear structural equation system. In A. S. Goldberger & O. D. Duncan (Eds.), *Structural equation models in the social sciences* (pp. 85–112). New York: Seminar.

Keesling, J. W. (1972). *Maximum likelihood approaches to causal flow analysis*. Unpublished doctoral dissertation. Chicago, IL: University of Chicago.

Lawley, D. N. (1958). Estimation in factor analysis under various initial assumptions. *British Journal of Statistical Psychology*, 11, 1–12.

Matsueda, R. L. (2012). Key advances in structural equation modeling. In R. H. Hoyle (Ed.), *Handbook of structural equation modeling* (pp. 3–16). New York: Guilford Press.

Mulaik, S. A. (2009). *Linear causal modeling with structural equations*. Boca Raton, FL: Chapman & Hall/CRC.

Parkerson, J. A., Lomax, R. G., Schiller, D. P., & Walberg, H. J. (1984). Exploring causal models of educational achievement. *Journal of Educational Psychology*, 76, 638–646.

Pearl, J. (2000). *Causality: Models, reasoning, and inference*. New York: Cambridge University Press.

Pearl, J. (2012). The causal foundations of structural equation modeling. In R. H. Hoyle (Ed.), *Handbook of structural equation modeling* (pp. 68–91). New York: Guilford Press.

Pearson, E. S. (1938). *Karl Pearson. An appreciation of some aspects of his life and work*. Cambridge: Cambridge University Press.

Sobel, M. E. (1995). Causal inference in the social and behavioral sciences. In G. Arminger, C. C. Clogg, & M. E. Sobel (Eds.), *Handbook of statistical modeling for the social and behavioral sciences* (pp. 1–35). New York: Plenum.

Spearman, C. (1904). The proof and measurement of association between two things. *American Journal of Psychology*, 15, 72–101.

Spearman, C. (1927). *The abilities of man.* New York: Macmillan.

Wiley, D. E. (1973). The identification problem for structural equation models with unmeasured variables. In A. S. Goldberger & O. D. Duncan (Eds.), *Structural equation models in the social sciences* (pp. 69–83). New York: Seminar.

Wright, S. (1918). On the nature of size factors. *Genetics*, 3, 367–374.

Wright, S. (1921). Correlation and causation. *Journal of Agricultural Research*, 20, 557–585.

Wright, S. (1934). The method of path coefficients. *Annals of Mathematical Statistics*, 5, 161–215.

Wold, H. O. A. (1954). Causality and Econometrics. *Econometrica*, 22(2), 162–177.

Yves, R. (2012). lavaan: An R package for structural equation modeling. *Journal of Statistical Software*, 48(2), 1–36.

Chapter 2

DATA ENTRY AND EDITING ISSUES

DATA SET FORMATS

An important first step in using an SEM software program is to be able to prepare either a raw data file or a summary data file to use when fitting specific models. Many of the software programs permit options to read in text files (ASCII), comma separated files (CSV), data from other programs, such as SPSS, SAS, or Excel, enter correlation or covariance matrices directly into a program, or read in raw data and save as a system file. A system file is a file type unique to the software program. For example, *.LSF* files are system files in LISREL that appear in spreadsheet format. The spreadsheet format is similar to that found in SPSS. The LISREL system file activates a pull-down menu that permits data editing features, data transformations, statistical analysis of data, graphical displays of data, multi-level modeling, and other related features.

We present an example using data on the cholesterol levels of 28 patients treated for heart attacks. The raw data are from the file, ***chollev.dat***, which is located on the book website and was taken from previous LISREL Student version examples.

DOI: 10.4324/9781003044017-2

```
 chollev - Notepad
File  Edit  Format  View  Help
270.000    218.000    156.000
236.000    234.000     -9.000
210.000    214.000    242.000
142.000    116.000     -9.000
280.000    200.000     -9.000
272.000    276.000    256.000
160.000    146.000    142.000
220.000    182.000    216.000
226.000    238.000    248.000
242.000    288.000     -9.000
186.000    190.000    168.000
266.000    236.000    236.000
206.000    244.000     -9.000
318.000    258.000    200.000
294.000    240.000    264.000
282.000    294.000     -9.000
234.000    220.000    264.000
224.000    200.000     -9.000
276.000    220.000    188.000
282.000    186.000    182.000
```

■ **Figure 2.1:**   CHOLLEV.DAT – SPACE DELIMITED FILE

Cholesterol levels were measured after 2 days (column 1), after 4 days (column 2), and after 14 days (column 3). Each row contains data for one participant, and only 19 of the 28 patients had complete data. Missing data were denoted with a value of –9.000 (more discussion about missing data is forthcoming). Both LISREL and M*plus* can read in external raw data files which may be saved as a tab or space delimited (.dat or .txt) file (as shown in Figure 2.1), as well as a comma separated file (.csv; see *chollev.csv* in Figure 2.2).

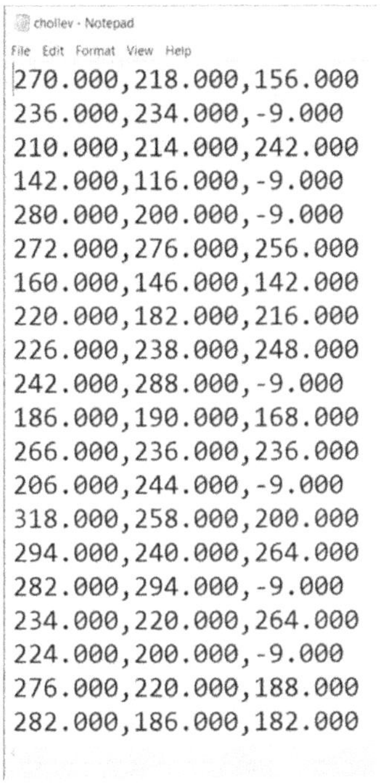

```
choliev - Notepad
File  Edit  Format  View  Help
270.000,218.000,156.000
236.000,234.000,-9.000
210.000,214.000,242.000
142.000,116.000,-9.000
280.000,200.000,-9.000
272.000,276.000,256.000
160.000,146.000,142.000
220.000,182.000,216.000
226.000,238.000,248.000
242.000,288.000,-9.000
186.000,190.000,168.000
266.000,236.000,236.000
206.000,244.000,-9.000
318.000,258.000,200.000
294.000,240.000,264.000
282.000,294.000,-9.000
234.000,220.000,264.000
224.000,200.000,-9.000
276.000,220.000,188.000
282.000,186.000,182.000
```

■ **Figure 2.2:**   CHOLLEV.CSV – COMMA SEPARATED FILE

In M*plus*, the user must specify the location and name of the ASCII file in which the data are contained. Figure 2.3 presents an excerpt from an M*plus* input file with commands necessary to read in the cholesterol data (M*plus* commands will be described in more detail subsequently).

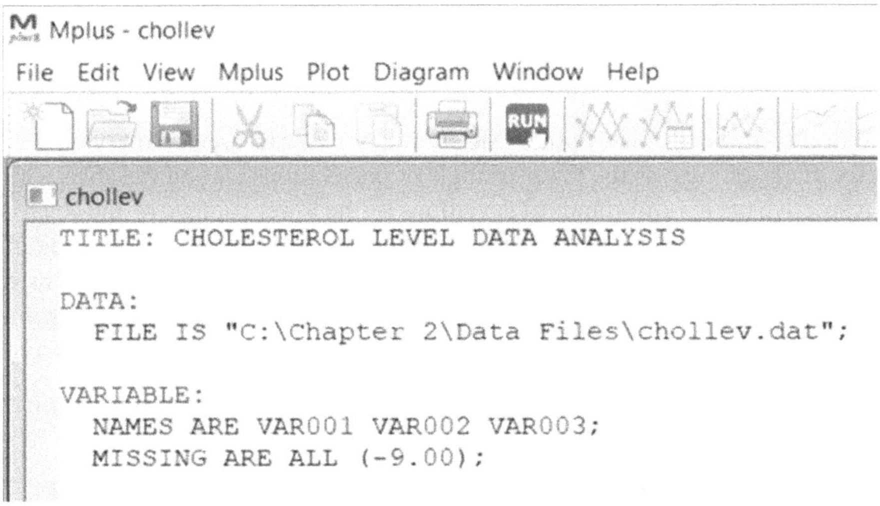

```
M  Mplus - chollev

File  Edit  View  Mplus  Plot  Diagram  Window  Help

■ chollev

   TITLE: CHOLESTEROL LEVEL DATA ANALYSIS

   DATA:
     FILE IS "C:\Chapter 2\Data Files\chollev.dat";

   VARIABLE:
     NAMES ARE VAR001 VAR002 VAR003;
     MISSING ARE ALL (-9.00);
```

■ **Figure 2.3:**   EXCERPT FROM M*PLUS* INPUT FILE FOR CHOLLEV.DAT

We used the import data feature in LISREL (for List-Directed Data), which prompted us to save the raw data file as a LISREL system file type (*chollev.LSF*). The LISREL main menu should look like the image in Figure 2.4.

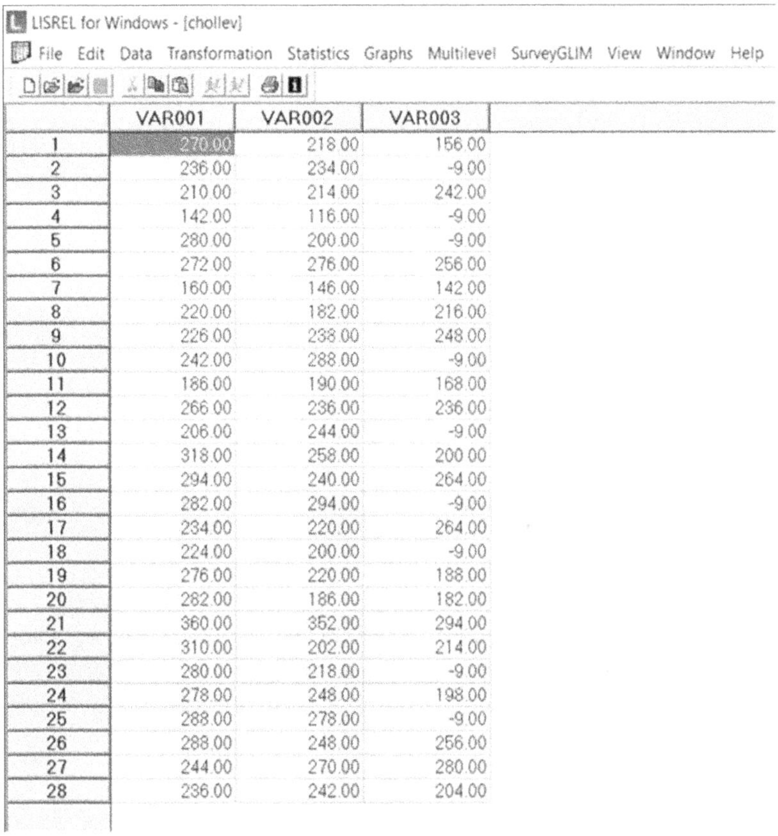

	LISREL for Windows - [chollev]

File Edit Data Transformation Statistics Graphs Multilevel SurveyGLIM View Window Help

	VAR001	VAR002	VAR003	
1	270.00	218.00	156.00	
2	236.00	234.00	-9.00	
3	210.00	214.00	242.00	
4	142.00	116.00	-9.00	
5	280.00	200.00	-9.00	
6	272.00	276.00	256.00	
7	160.00	146.00	142.00	
8	220.00	182.00	216.00	
9	226.00	238.00	248.00	
10	242.00	288.00	-9.00	
11	186.00	190.00	168.00	
12	266.00	236.00	236.00	
13	206.00	244.00	-9.00	
14	318.00	258.00	200.00	
15	294.00	240.00	264.00	
16	282.00	294.00	-9.00	
17	234.00	220.00	264.00	
18	224.00	200.00	-9.00	
19	276.00	220.00	188.00	
20	282.00	186.00	182.00	
21	360.00	352.00	294.00	
22	310.00	202.00	214.00	
23	280.00	218.00	-9.00	
24	278.00	248.00	198.00	
25	288.00	278.00	-9.00	
26	288.00	248.00	256.00	
27	244.00	270.00	280.00	
28	236.00	242.00	204.00	

▪ **Figure 2.4:** LISREL MAIN MENU SCREEN

In addition to raw data files, both LISREL and M*plus* can use summary data as input for analyses, such as the lower triangle of correlations among variables with their respective means and standard deviations or the lower triangle of the covariance matrix with respective variable means. Figure 2.5 illustrates the contents of an external ASCII file that can be used in M*plus*. In the summary data file (named *chollev_summary.txt*), means are on the first row and standard deviations are on the second row, which are then stacked on top of the lower triangle of the correlation matrix (assuming non-missing data).

```
chollev_summary - Notepad
File   Edit   Format   View   Help
253.929        230.643        222.237
47.710         46.967         39.207
1.000
0.673          1.000
0.435          0.801          1.000
```

■ **Figure 2.5:** CONTENTS IN EXTERNAL ASCII FILE WITH SUMMARY DATA FOR M*PLUS*

Figure 2.6 shows an excerpt from a SIMPLIS syntax file with the same summary data directly included in the program.

```
Correlation Matrix
  1.000
  0.673        1.000
  0.435        0.801        1.000
Means
  253.929      230.643      222.237
Standard Deviations
  47.710       46.967       39.207
```

■ **Figure 2.6:** EXCERPT FROM SIMPLIS SYNTAX FILE WITH SUMMARY DATA INCLUDED

You may have noticed that the data files for LISREL and for M*plus* do not include the actual variable names. Instead, the user must define the variable names either when importing the data into LISREL or when writing the LISREL and M*plus* programs. Thus, the statistical analysis of data in SEM requires you to become familiar with the software input and output data features. For input, the most common approaches are inputting raw data from ASCII data files (e.g., .txt or .dat), comma separated data files (.csv), covariance matrices (with or without means), or correlation matrices with standard deviations (with or without means). SEM software will convert raw data or the correlation matrix with standard deviations to a covariance matrix in modeling applications. As such, the sample covariance matrix is the unit of analysis in SEM applications because it includes information regarding the variability of all of the variables, which is important for standard error estimation.

All SEM software programs can read in raw or summary data files. The type of data (raw or summary) to use as input for analysis in SEM software depends upon certain considerations, such as data availability, the types of variables to be analyzed, the normality of variables, the presence of missing data, and model complexity. For instance, the convenient feature of being able to use summary data is that researchers can duplicate analyses done in published articles, given that the authors included the correlation matrix with means and standard deviations in the publication (which is generally standard practice). Thus, researchers duplicating previous analyses conducted in published papers would only have access to the summary data provided in the publication. If ordered categorical variables to be modeled are endogenous (dependent) variables, raw data must be available to use estimators that account for the ordinal nature of the data. Similarly, if data are not normally distributed, alternative estimators may need to be used to more accurately estimate standard errors and fit indices (estimation methods will be discussed later in Chapter 5). If missing data are present, raw data will need to be used in which missing data points are identified so that missing data methods may be used to yield more accurate parameter estimates. Another consideration is model complexity. For example, if a researcher wants to estimate model parameters while controlling for several variables (e.g., more than 3), it will require that all control variables covary with all exogenous (independent) variables and directly affect all endogenous (dependent) variables in the model. You can see how this will exponentially increase the number of parameters to estimate in a model as the number of control variables increases. Thus, a researcher could use a partial correlation matrix in which the relationships between the control variables and both the independent and dependent variables has been partialled out (with respective standard deviations) as input for the analyses.

SEM software output also varies, with some providing graphs and descriptive statistics, and others having options to output scores, different types of correlation matrices, or saving an asymptotic covariance matrix. Input and output options in both LISREL and M*plus* will be demonstrated in examples in subsequent chapters. Prior to conducting SEM analyses, however, a researcher should also be aware of important data editing features needed, whether available in a statistics package or the SEM software.

DATA EDITING ISSUES

Statistical packages that provide SEM analyses permit the researcher to conduct much needed data screening and editing options, including SPSS – AMOS, SAS – PROC CALIS, Statistica – SEPATH, R – sem or lavaan, and STATA – SEM. Nothing prevents a researcher from using a statistical package to edit data prior to using it in a stand-alone SEM software program (e.g., LISREL, M*plus*). There are several key issues related to data in the field of statistics that impact our analyses once data have been imported into a software program. These data issues commonly concern the measurement scale of variables, restriction in the range of data, outliers, non-normality, linearity, and missing data. Each of these data issues will be discussed because they not only affect traditional statistics, but present additional problems and concerns in structural equation modeling.

Measurement Scale

How variables are measured or scaled influences the type of statistical analyses we perform (Stevens, 1946; Anderson, 1961). Properties of the scale also guide our understanding of permissible mathematical operations. For example, a nominal variable implies mutually exclusive groups. Biological blood type, using the ABO/ RhD classification system, has 8 mutually exclusive groups: A+, A–, B+, B–, AB+, AB–, O+, and O–. An individual can only be in one of the groups that define the levels of the variable. In addition, it would not be meaningful to calculate a mean and a standard deviation on the variable *blood type*. Consequently, the number or percentage of individuals in each blood type category is the only mathematical property of scale that makes sense. Other examples of nominal variables include occupation, social security number, and type of smartphone one owns.

An ordinal variable, for example, attitude toward school, that is scaled *strongly agree*, *agree*, *neutral*, *disagree*, and *strongly disagree*, implies mutually exclusive categories that are ordered or ranked. When levels of a variable have properties of scale that involve mutually exclusive groups that are ordered, only certain mathematical operations are permitted. A comparison of ranks between groups is a permitted mathematical operation, while calculating a mean and standard deviation is not. Undergraduate class level, position finishing in a race, and college football team rankings are examples of ordinal scale variables.

Final exam scores, an example of an interval variable, possess the property of scale that implies equal intervals between the data points, but no true zero point. This property of scale permits the mathematical operation of computing a mean and a standard deviation. Similarly, a ratio variable, for example, weight, has the property of scale that implies equal intervals and a true zero point (weightlessness). Therefore, ratio variables also permit mathematical operations of computing a mean and a standard deviation. The difference between interval and ratio level data is whether a true zero point exists. Other examples of interval scale variables are grade point average, temperature in Celsius or Fahrenheit, and IQ score. Height, count data (e.g., number of pizzas ordered in a month), and salary are other examples of ratio scale variables.

A question often asked by students is why certain interval scale variables, particularly grade on an exam or an IQ test score, are not considered ratio scale variables because a zero could potentially be attained on these types of measures. The best way to differentiate interval versus ratio scale variables is that a value of zero is the total lack of the attribute being measured. It may be tempting for a student to indicate that they scored twice as many points on an exam than another student, but the zero on the interval scale exam score variable does not reflect a lack of that attribute being measured. For instance, if someone scored a zero on an English exam, that does not substantiate that the person has no knowledge of any kind concerning what is being measured on the English exam. In contrast, it would be appropriate for someone to indicate that they ordered twice as many pizzas in a month than their friend. Thus, ratios can be calculated with ratio scale variables. In the end, achievement outcomes and attributes measured in the social and behavioral sciences are generally treated as interval scale variables.

Our use of different variables requires us to be aware of their properties of scale and what mathematical operations are possible and meaningful, especially in SEM where variance–covariance matrices are used. Different types of correlation coefficients (Pearson, polychoric, polyserial) will produce different types of covariance matrices among variables depending upon the level of measurement, and this creates a unique consideration in SEM. It is very important to understand how continuous variables, ordinal variables, and group or categorical variables are to be used appropriately in SEM. We will address the use of different variable types and correlation matrices in subsequent chapters of the book.

Restriction of Range

Data values at the interval or ratio level of measurement can be further defined as being discrete or continuous. For example, final exam scores could be reported in whole numbers or discrete values (e.g., 70, 82, 94). Similarly, the number of children in a family would be considered discrete (i.e., 3 children). In contrast, a continuous variable can, in theory, be any value within the range of values for the variable and may be reported using decimals. For example, a student's grade point average could be reported as 3.78 on a 5-point scale.

Karl Jöreskog and Sörbom (1996) provided a criterion based on his research that defines whether a variable is ordinal or interval. If a variable has fewer than 15 levels (e.g., minimum value = 2 and maximum value = 10 and all values are discrete), it should be considered an ordinal variable. In contrast, a variable with 15 or more levels (e.g., minimum value = 2 and maximum value = 20 and all values are discrete) is considered a continuous variable. This 15-point range criterion permits the Pearson correlation coefficient values to vary between +/– 1.0.

Variables with fewer distinct levels restrict the value of the Pearson correlation coefficient such that it may only vary between +/– 0.5. SEM software requires you to understand the types of variables being used in the model.

Outliers

Outliers or influential data points can be defined as data values that are extreme or atypical on either the independent (X variables) or dependent (Y variables) variables or both (Ho & Naugher, 2000). Outliers can occur as a result of observation errors, data entry errors, instrument errors based on layout or instructions, or actual extreme values from self-report data. Because outliers affect the mean, the standard deviation, and correlation coefficient values, they must be explained, deleted, adjusted, or accommodated by using robust statistics.

Visually inspecting the distributions of quantitative variables in a data set can be a helpful tool when trying to detect potential outlying cases. For instance, boxplots, scatterplots, histograms, stem and leaf plots, and frequency distributions can all be useful visual outlier detection tools. Using the data from *chollev.dat* (see Figure 2.4), boxplots are displayed in Figure 2.7.

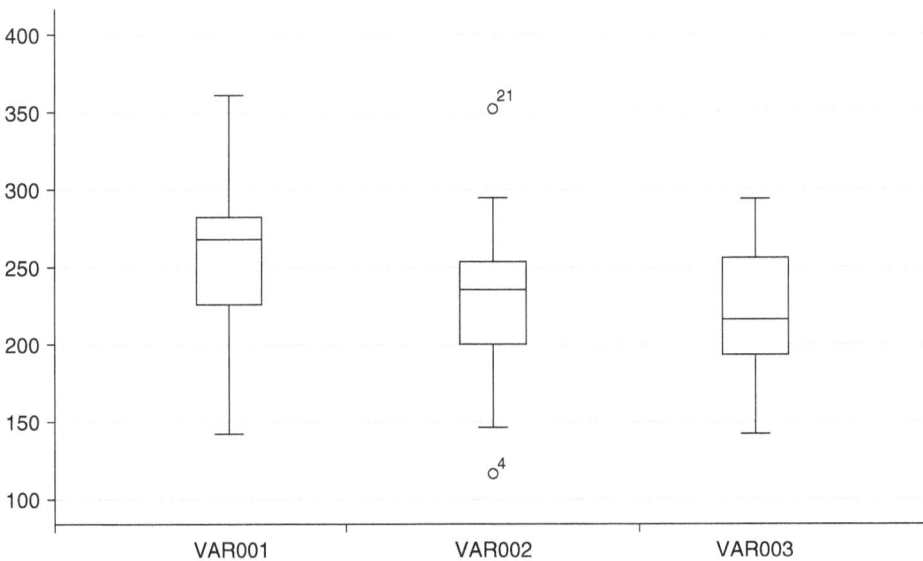

■ **Figure 2.7:**   BOXPLOTS FOR DATA IN CHOLLEV.DAT

These boxplots were created using SPSS, but boxplots can be created using other statistical packages (e.g., R, SAS, STATA). The horizontal line in each box represents the median score for each variable. The top of the box denotes the 3rd quartile (Q3; or the 75th percentile) and the bottom of the box denotes the 1st quartile (Q1; or the 25th percentile). Thus, the box encompasses the interquartile range (Q3 – Q1) or the middle 50% of scores. The bars that extend vertically from the top and the bottom of the box are called whiskers and are demarcated in SPSS by smaller horizontal bars relative to the box. Thus, whiskers are sometimes referred to as T-bars or inner fences which mark the minimum and maximum scores with the exception of outlying scores, which are classified as 1.5 times the interquartile range (IQR). If any scores appear outside of the whiskers, they are regarded as outliers. In Figure 2.7, you will notice that two scores fall beyond the whiskers for VAR002 and are denoted with circles and their respective observation number, indicating that they are beyond 1.5 times the IQR. These outlying scores belong to observations 4 and 21 who have cholesterol levels of 116 and 352, respectively (see Figure 2.4).

A scatterplot between two variables is another useful visualization tool that can highlight outlying cases. A scatterplot between VAR001 and VAR002 from *chollev. dat* is illustrated in Figure 2.8.

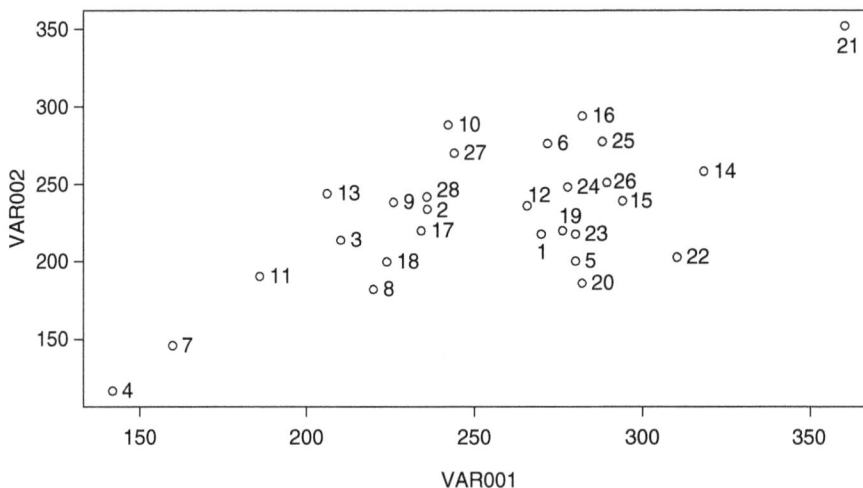

■ **Figure 2.8:**   SCATTERPLOT FOR DATA IN CHOLLEV.DAT

Similar to the information provided by the boxplots, you can see that observations 4 and 21 are farthest from the majority of points on the scatterplot. Although observation 7 did not appear as an outlier in the boxplots, they also appear to be far away from the majority of the points in the scatterplot, which requires further exploration. Scatterplots can be created using stand-alone statistical software packages (e.g., R, SAS, SPSS, STATA) as well as with LISREL and M*plus*.

In addition to visual outlier detection methods, there are statistical calculations that may be used to help detect outliers. A commonly used method is examining the standardized (z) scores of each quantitative variable: $\dfrac{X - Mean}{SD}$, where X is a participant's score for a variable, $Mean$ is the mean of the values of X, and SD is the standard deviation of the values of X. z scores greater than $\pm\, 3$ are typically considered to be atypical or outlying scores. For example, the z scores associated with VAR002 for observations 4 and 21 are –2.44 and 2.58, respectively. Thus, these observations would not be classified as outliers using the $\pm\, 3$ z score cutoff. A criticism of z scores is that they are calculated using the mean and the standard deviation, which are heavily influenced by outliers. In addition, z scores greater than the $\pm\, 3$ cutoff can also occur more frequently as sample size increases. Standardized z scores are easily calculated in basic statistical software packages (e.g., R, SAS, SPSS, STATA).

A more robust estimate to help detect outliers as compared to z scores is the median absolute deviation (MAD; Hampel, 1974): $b * Median_{md}\left(|X - Median|\right)$, where X is a participant's score for a variable, $Median$ is the median of the X values, $Median_{md}$ is the median of the absolute deviations around the median $\left(|X - Median|\right)$, and b is a constant equal to 1.4826. Thus, the MAD is calculated by taking the median of the absolute deviations around the median $\left(|X - Median|\right)$ and multiplying it by the constant (1.4826). Next, a participant's deviation score around the median is divided by the MAD. If it is greater than a $+ 3$ cutoff, that score would be considered atypical. The median for VAR002 is 235 and the MAD for VAR002 is 41.5128. Dividing the absolute deviation scores around the median $\left(|X - Median|\right)$ by the MAD results in values of 2.87 and 2.82 for observations 4 and 21, respectively. These values are not greater than the $+ 3$ cutoff value, suggesting that these scores are not outliers. Researchers can also adjust the cutoff values (as they could with z scores) to be more likely to detect outliers by using cutoffs equal to $+ 2.5$ or $+ 2$. MAD can be calculated using functions in R and SAS, using a module in STATA, and can be easily computed in Excel and SPSS.

The methods previously described help detect univariate outliers, which are atypical scores on one variable. Multivariate outliers are atypical scores on two or more variables. The most commonly used method to detect multivariate outliers is to calculate the Mahalanobis distance (MD) for each participant. Simply, MD signifies the difference between the set of observed scores and their respective means while taking the intercorrelations among the variables into account. The squared values of MD are chi-square $\left(\chi^2\right)$ distributed with degrees of freedom (df) that are equal to the number of variables. Cases with squared MD values that exceed the χ^2 critical with given df at an alpha of .001 have been suggested as multivariate outlying cases. Thus, cases with squared MD values associated with p-values less than .001 are considered outliers. MD values with corresponding p-values can be saved in M*plus* and they can be calculated in other software programs, including

R, SAS, and SPSS. It is important to note that MD is calculated using observed means and standard deviations, which may be impacted by outlying cases. Thus, using the median as the central tendency measure instead of the mean may be more appropriate.

Once outliers have been detected, the next step is to determine the potential cause of the atypical data points. If the cause of the outlying scores can be attributed to errors (e.g., data entry or observation errors), the scores could either be corrected or omitted from the analyses. However, when the cause of outlying scores cannot be attributed to errors, the decision about the treatment of the outliers becomes more challenging. If it is determined that a participant is not from the target population of interest, that participant's data may be deleted from further analyses. For example, if the target population consists of college students required to take remedial courses, college students who are not required to take remedial courses, but are taking them as a "refresher," would clearly not be from the same population. As such, the students not required to take remedial courses could be deleted from the data analyses.

If the outlying scores are valid (yet extreme) responses, it is recommended that the analysis be conducted with and without the outlying cases to determine whether the findings change substantially. In SEM, researchers should check the fit of the theoretical model (discussed in Chapter 5) to the data with and without outliers included. If model fit does not change substantially, the outlying cases can remain in the analysis. In contrast, if model fit does change considerably with and without outliers included in the analysis, there are a few options available that may help mitigate the impact of the outlying cases. Transformations can be applied to the variable with outlying cases. For instance, computing the natural log of a variable can eradicate outliers. However, transformations may not always fix the problem and the results for the transformed variable must be interpreted for the log of the variable (not for the original metric of the variable). It has been suggested that case-robust estimation procedures be implemented wherein cases are weighted based on their distance from the center of the distribution (as defined by MD), with more extreme outlying cases being down-weighted. Yuan and Bentler (1998) described a two-stage estimation procedure and Yuan and Zhong (2008) described an iteratively reweighting estimation procedure. These two estimation procedures have demonstrated promising results compared to non-robust estimation procedures (e.g., maximum likelihood estimation) with outliers (Zhong & Yuan, 2011). Nonetheless, these case-robust estimation procedures are not widely available in SEM software, with the exception of EQS software. Researchers could implement estimation procedures that provide robust standard errors and test statistics (e.g., maximum likelihood robust in LISREL, M*plus*, and R), acknowledging that estimates may be biased. The deletion of outlying cases has also been suggested because the impact of a few outlying cases should not be able to distort the findings based on the majority of participants (Judd, McClelland, & Ryan, 2009).

In the end, a discussion of the outlier detection process and how outliers were treated should be detailed in research papers. Also, researchers should report the results with and without outlying cases included as part of the analysis. Although outlying cases will result in additional efforts relating to the data analysis tasks, they can still provide theoretically meaningful information about potential subpopulations of interest that are underrepresented.

Normality

An assumption in SEM is that the endogenous (dependent) variables are from a multivariate normal distribution. Normality is important because the default estimator in SEM software programs is maximum likelihood (ML), which assumes normality. Non-normality in the endogenous variables can lead to inaccurate model fit statistics and parameter estimates (as well as their respective standard errors) when using ML estimation (or other estimators that assume normality).

In practice, researchers commonly examine the univariate distributions of the endogenous (dependent) variables when assessing the normality assumption. Two characteristics that can impact the normality of a distribution of scores are skewness and kurtosis. Skewness refers to the symmetry of a distribution and kurtosis refers to the heaviness of the tails of the distribution as well as the peak of the distribution. The normal distribution is symmetrical and has moderate tails with a moderately peaked distribution. Normally skewed distributions contain a majority of scores in the center of the distribution. Positively skewed distributions contain a majority of scores in the lower region of the distribution whereas negatively skewed distributions contain a majority of scores in the higher region of the distribution. Figure 2.9 demonstrates normally distributed scores as well as both positively and negative skewed distributions.

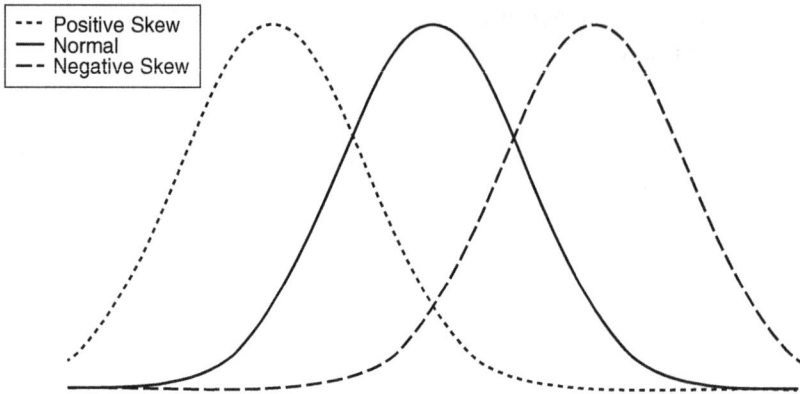

■ **Figure 2.9:** NORMAL AND SKEWED DISTRIBUTIONS

The positive and negative terms for skew refer to the location of the longer tail of the distribution and not where the majority of scores are located within the distribution. As seen in Figure 2.9, the positively skewed distribution has a longer tail falling in the right or positive portion of the distribution whereas the negatively skewed distribution has a longer tail falling in the left or negative portion of the distribution. Distributions that are more peaked with heavy tails are designated as leptokurtic and distributions that are less peaked with light tails are designated as platykurtic. Figure 2.10 illustrates normally distributed scores as well as both leptokurtotic and platykurtic distributions.

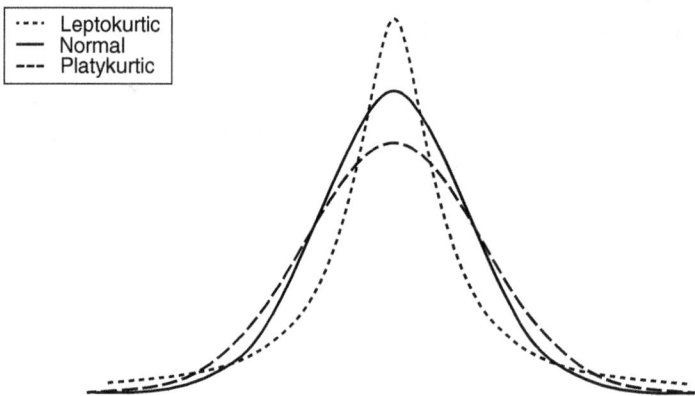

■ **Figure 2.10:** NORMAL AND KURTOTIC DISTRIBUTIONS

As with outliers, visual inspections of the distributions of quantitative variables in a data set are useful when examining deviations from normality. Histograms and stem and leaf plots can be beneficial visual aids. Using the data from *chollev.dat* (see Figure 2.4), a histogram of scores for VAR001 with a superimposed normal curve is shown in Figure 2.11. From inspecting the histogram, the distribution of VAR001 is fairly normally distributed.

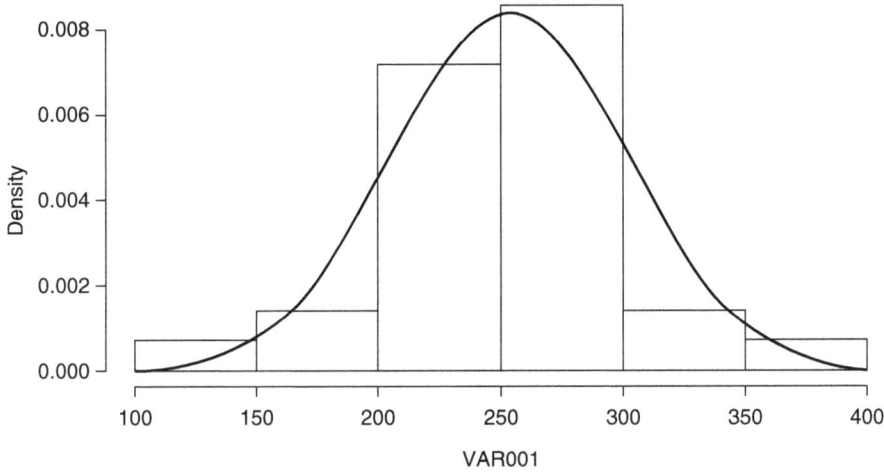

Figure 2.11: HISTOGRAM WITH NORMAL CURVE FOR VAR001 IN CHOLLEV.DAT

Normal quantile-quantile (Q–Q) plots may also be examined to detect departures from normality. Figure 2.12 illustrates a normal Q–Q plot for VAR001 from *chollev.dat*. As seen in Figure 2.12, normal Q–Q plots plot the observed quantiles of scores (shown as dots) against the quantiles of scores that would be expected if the scores were distributed normally (shown as a solid line). If points deviate considerably from the straight line, the plot is suggesting non-normality of the scores.

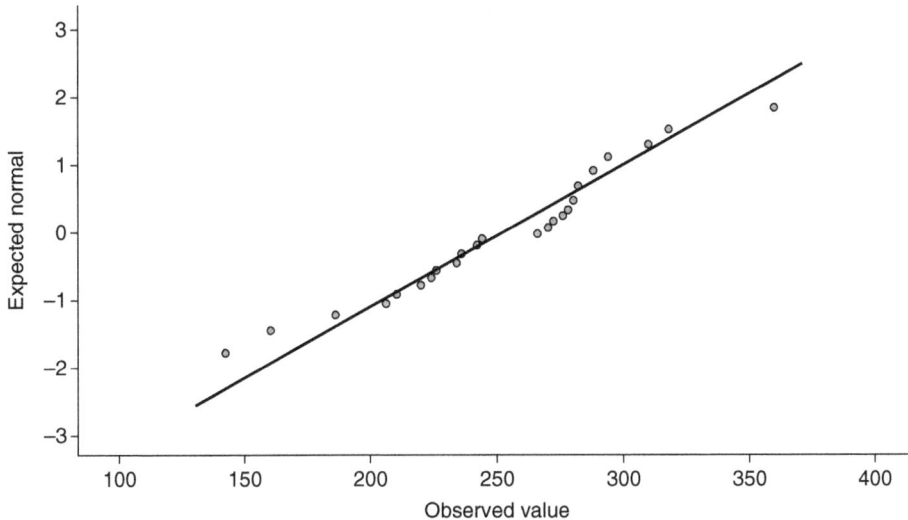

Figure 2.12: NORMAL QUANTILE-QUANTILE PLOT FOR VAR001 IN CHOLLEV.DAT

It must be noted that distributions of scores based on small sample sizes will not necessarily appear to be normally distributed. Thus, visual inspections of the data for departures from normality with small samples (e.g., *n* < 20) may not provide useful information pertaining to the normality assumption. Histograms, stem and leaf plots, and normal Q–Q plots are available in standard statistical software packages (e.g., R, SAS, SPSS, STATA). Histograms may be obtained in LISREL and M*plus* and Q–Q plots are also available in M*plus*.

In addition to visually inspecting the distribution of scores, researchers can also evaluate descriptive statistics associated with the quantitative variables. Specifically, skewness and kurtosis statistics can be computed for quantitative variables in a data set, which provide useful information. Skewness and kurtosis values close to zero are indicative of a normal distribution. Positive skewness values indicate positively skewed distributions whereas negative skewness values indicate negatively skewed distributions. Positive kurtosis values indicate leptokurtic distributions whereas negative kurtosis values indicate platykurtic distributions. Research in SEM has suggested that univariate skew and kurtosis values greater than 2 and 7, respectively, can begin to introduce problems when using maximum likelihood estimation (West, Finch, & Curran, 1995).

In addition to using cutoff values, *z* scores can be calculated by dividing the skewness and kurtosis statistics by their respective standard errors. Absolute values of the *z* scores associated with skewness or kurtosis less than or equal to 1.96 (which assumes an alpha = .05) would suggest that the distribution is normally distributed. For example, the skewness associated with VAR001 from ***chollev.dat*** is –.351 (*SE* = .441) and the kurtosis is equal to .473 (*SE* = .858). Thus, the distribution of scores for VAR001 is marginally negatively skewed and leptokurtic. The *z* scores associated with skewness and kurtosis are –.796 (–.351/.441) and .551 (.473/.858), respectively. Because the *z* scores are not greater than ± 1.96, the distribution of scores for VAR001 would be considered normally distributed. It is important to note that as sample size increases, the *z* scores may become greater than ± 1.96 because standard errors tend to get smaller as sample size increases. As such, *z* score critical values can be increased (e.g., to ± 2.58 or ± 3.29 which assumes an alpha = .01 or .001, respectively) as sample size increases. Skewness and kurtosis values can be calculated in basic statistical software packages (e.g., R, SAS, SPSS, STATA). Both LISREL and M*plus* provide univariate skewness and kurtosis values of the quantitative values used in an analysis. In addition, LISREL provides the *z* scores associated with skewness and kurtosis, which are calculated differently than previously described (see Jöreskog, Olsson, & Wallentin, 2016).

Non-normal data can occur because of the scaling of variables (ordinal rather than interval or ratio), or the limited sampling of participants. When data are deemed to be non-normally distributed, a different estimator (e.g., robust weighed least squares) than the default estimator in SEM software programs (i.e., maximum likelihood) can be used during the analysis to accommodate the non-normality in the endogenous variable(s). Aside from changing the estimator, data transformations are available to handle issues with non-normal data. The more common data transformations are described in Table 2.1.

Table 2.1: Data Transformation Types

y_new = ln(y) or y = log10(y) or y_new = ln(y + 0.5)	Useful with clustered data or cases where the standard deviation increases with the mean
y_new = sqrt(y)	Useful with Poisson counts
y_new = arcsin((y + 0.375)/(n + 0.75))	Useful with binomial proportions [$0.2 < p = y/n < 0.8$]
y_new = 1/y	Useful with gamma-distributed y variable
y_new = logit(y) = ln(y/(1 − y))	Useful with binomial proportions $y = p$
y_new = normit(y)	Quantile of normal distribution for standardized y
y_new = probit(y) = 5 + normit(y)	Most useful to resolve non-normality of data

Note: y is the original non-normal variable; y_new is the transformed variable; probit(y) is same as normit(y) plus 5 to avoid negative values.

One or more of the data transformation methods in Table 2.1 may correct for skewness or leptokurtic values in the data, however, platykurtic data usually requires recoding into categories or obtaining better data. Other potential solutions for skewness and kurtosis are to resample more participants. As previously mentioned, data transformations may not necessarily resolve the distributional problem and the results for transformed data are not easily interpretable. Data transformations can be accomplished in standard statistical packages as well as in LISREL and M*plus* (using the DEFINE command). Both LISREL and M*plus* have the capability of using robust estimators when requested by the user when endogenous (dependent) variables are not normally distributed. Thus, we recommend using a robust estimator when deviations from normality are observed among endogenous variables in order to provide robust estimation of standard errors and goodness of fit measures.

Linearity

Unless otherwise specified, traditional analyses in SEM assume linear relationships among exogenous and endogenous variables. Thus, a standard practice is to visualize the coordinate pairs of data points of two continuous variables by plotting the data in a scatterplot. These bivariate plots depict whether the data are linearly increasing or decreasing. The presence of curvilinear data reduces the magnitude or strength of the *linear* relationship between two quantitative variables, such as that measured by the Pearson correlation coefficient. Simply inspecting bivariate correlations among variables may be misleading because curvilinear relationships will most likely result in the presence of a zero or close to zero correlation. Figure 2.13 shows the importance of visually displaying the bivariate data scatterplot.

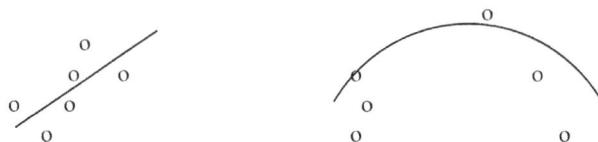

■ **Figure 2.13:** SCATTERPLOTS (LEFT: CORRELATION IS LINEAR. RIGHT: CORRELATION IS NONLINEAR)

The normality of the quantitative variables will also hinder the linear relationship among the variables. If variables are not normally distributed, the scatterplot with the non-normally distributed variables is less likely to demonstrate a linear relationship. Because outliers can also impact the normality of variables, it can impact the linearity of the relationship among variables. If relationships are truly nonlinear and not a function of non-normality or outliers, researchers have the capability to specify and estimate nonlinear relationships in SEM software packages, including LISREL and M*plus* (see e.g., Marsh, Wen, Hau, & Nagengast, 2013).

Missing Data

The statistical analysis of data is affected by missing data in variables. That is, not every participant has an actual value for every variable in the dataset. Missing data can occur for various reasons, including, but not limited to, equipment failure, non-response to survey items, and participant drop out. Rubin (1976) classified 3 types of non-response for missing values: MCAR (missing completely at random), MAR (missing at random), and MNAR (missing not at random). MCAR does not depend on any variable in the data set, meaning that missingness is due simply to chance. MAR usually implies that variable missingness is related to other variables. An example of MAR is when high performing students in an instructor's class do not complete the course evaluation survey items for that instructor's course; but within the group of high performers, the probability of responding is not related to their course evaluation ratings. MNAR occurs when missing data on a variable depends on the variable on which data are missing. For example, if participants in a drug trial stop participating because of adverse side effects due to the drug, the probability of responding is due to effect of the drug. The MCAR mechanism can be tested in a data set using either univariate, independent-samples *t*-tests with complete cases versus missing data cases on other variables in the data set (Dixon, 1988) as the outcome (e.g., performance in a class) or the multivariate extension offered by Little (1988). In contrast to MCAR, MAR and MNAR are untestable missing data mechanisms. Readers are encouraged to consult Enders (2010) for more discussion concerning missing data mechanisms and the testing for MCAR.

It is common practice in standard statistical software packages to have default settings for handling missing values. Caution is needed when using structural equation modeling techniques because SEM software varies on what missing data options are available. Also, an ASCII tab-delimited data file with blanks for missing data can cause data input errors in certain SEM software packages. As far as managing missing data, the researcher has the option of deleting participants who have any missing values, replacing the missing data values, or using robust statistical procedures that accommodate for the presence of missing data. SEM software packages handle missing data differently and have different options for managing missing data. Table 2.2 lists many of the various options for dealing with missing data, some of which are more traditional and some of which are more modern. These options can dramatically affect the number of participants available for analysis, the magnitude and direction of the correlation coefficient, or create problems if means, standard deviations, and correlations are computed based on different sample sizes.

Table 2.2: Options for Dealing with Missing Data

Traditional Missing Data Methods

Listwise deletion	Delete participants with missing data on any variable.
Pairwise deletion	Delete participants with missing data on each pair of variables used.
Mean substitution/mean imputation	Substitute the mean based on non-missing cases for missing values of a variable.
Regression imputation	Substitute a predicted value for the missing value of a variable.
Stochastic regression imputation	Substitute a predicted value for the missing value of a variable with an added residual term.
Hot-deck imputation	Match cases with incomplete data to cases with complete data and randomly draw from those cases to determine a missing value.
Similar response pattern imputation	Match cases with incomplete data to cases with complete data to determine a missing value.
Last observation carried forward	Substitutes the observed value just before dropout for the missing value(s) in repeated measures designs.

Modern Missing Data Methods

Multiple imputation	Multiple imputed data sets (e.g., 20) are generated, model parameters are estimated in each imputed data set, and parameters are pooled across imputed data sets.
Full information maximum likelihood (FIML)	Population parameters are estimated by maximizing the sample loglikelihood, which is based on the available, non-missing data for the participant.

Listwise deletion of cases is not recommended due to the possibility of losing a large number of participants, dramatically reducing the sample size and, in turn, affecting parameter estimates and standard errors. *Pairwise* deletion of cases is also not recommended because it too can reduce sample size and it can result in calculations of variances and covariances that are based on different sample sizes. Single imputation methods fill in a single value for each data point that is missing, such as *mean substitution* (or *mean imputation*), *regression imputation* (or *conditional mean imputation*), *stochastic regression imputation*, *hot-deck imputation*, *similar response pattern imputation*, and *last observation carried forward*. The majority of these methods have been shown to provide biased parameter estimates with MAR data. Although stochastic regression imputation has been shown to provide unbiased parameter estimates with MAR data, some limitations are still associated with the method. As such, modern missing data approaches are preferred methods when managing missing data (Enders, 2010).

Modern methods to deal with missing data include multiple imputation and maximum likelihood (also called full information maximum likelihood). Multiple imputation is an extension of the stochastic regression single imputation approach with Bayesian estimation principles and is comprised of three steps. Basically, multiple imputation starts with substituting a predicted value with an added residual term for a missing data value multiple times, creating multiple imputed data sets that contain different imputed missing data values. It is recommended that at least 20 imputed data sets be created (Enders, 2013), each of which are then used to estimate model parameters. In the final step, the model parameters are pooled across all imputed data sets, arriving at one set of model parameter estimates.

Multiple imputation can be accomplished using the expectation-maximization (EM) algorithm and the Markov Chain Monte Carlo (MCMC) algorithm. As the name implies, there are two steps involved when using the EM algorithm. The expectation (E) phase involves using a set of regression models that predict missing values using observed values similar to that done by stochastic regression imputation. In the maximization (M) phase, the imputed data set is used to estimate model parameters. The E and M phases iteratively continue until estimates (e.g., means and covariance matrix) do not change from one M step to the next. The EM algorithm will not actually create multiple data sets with imputed values, but will result in one final data set with imputed values when completed.

The Markov Chain Monte Carlo (MCMC) algorithm obtains missing data values by randomly sampling from the posterior distributions of the missing data given the observed data. The MCMC algorithm proceeds through several iterations until the distributions are stable. Multiple imputation using the MCMC algorithm will result in multiple imputed data sets. Thus, multiple imputation based on the MCMC algorithm has been advocated as more reliable than multiple imputation based on the EM algorithm (Jöreskog et al., 2016).

The EM and MCMC algorithms are described simply here, but are much more mathematically detailed. Readers are encouraged to consult the following sources for more detailed explanations about multiple imputation using these algorithms: Dempster, Laird, & Rubin (1977); Tanner and Wong (1987); McLachlan and Krishnan (1997); Schafer (1997); and Enders (2010). Multiple imputation is available in LISREL using either the EM or MCMC algorithms. Multiple imputation using the MCMC algorithm is available in M*plus*. Multiple imputation methods are also available outside of using SEM analyses in other statistical software packages (e.g., R, SAS, SPSS, STATA).

In contrast to multiple imputation, maximum likelihood for missing data does not impute missing data values, but estimates parameters based on the complete data. In general, maximum likelihood for missing data (also called full information maximum likelihood: FIML) is an iterative procedure that estimates population parameters by maximizing the sample loglikelihood, which is calculated as the sum of each participant's loglikelihood equation. Each participant's loglikelihood equation is based on the available, non-missing data for the participant. The larger the loglikelihood, the more the participant's observed scores represent data from a multivariate normal distribution with the calculated mean vector and covariance matrix based on their missing data pattern. More discussion of FIML, with several illustrated examples, can be found in Enders (2010, 2013). FIML estimation with missing data is available in both LISREL and M*plus* (as well as in R – lavaan, SAS – PROC CALIS, and STATA – SEM).

Multiple imputation and FIML methods for handling missing data both assume that data are MAR. Given the ease in which FIML can be implemented with missing data in SEM software packages, it is becoming increasingly used in SEM analyses with missing data. Multiple imputation can be used with most models and estimation procedures. Regardless of which of the two modern methods is chosen, we highly recommend comparing any analyses before and after accounting for missing data with multiple imputation or FIML to fully understand the impact missing data values have on the parameter estimates and standard errors. Readers interested in more information about general missing data issues should consult McKnight, McKnight, Sidani and Aurelio (2007) as well as Peng, Harwell, Liou, and Ehman (2007). Davey and Savla (2010) have also published an excellent book with SAS, SPSS, STATA, and M*plus* source programs to handle missing data in SEM, especially in the context of power analysis.

The use of FIML and multiple imputation methods are not appropriate when data are MNAR. Instead, researchers will have to consider different models for MNAR data, such as the pattern mixture model and the selection model. However, models for MNAR data have strict assumptions that are not testable and/or not able to be attained easily. For more information about these models, readers should consult Enders (2010).

SUMMARY

As with other statistical techniques, characteristics of the data that will be modeled have an impact on analyses using structural equation modeling. The measurement scale of the data, restriction of range in the data values, outliers, non-normality, nonlinearity, and missing data can affect the variance–covariance among variables and, thus, can impact the SEM analysis. Researchers should use the built-in menu options in a statistics package or SEM program to examine, graph, and test for any of these problems in the data prior to conducting any SEM analysis. Data screening is a very important first step in structural equation modeling. Table 2.3 lists the data editing issues along with suggestions for correcting any of these issues in the data.

Table 2.3: Data Editing Issues with Suggestions

Issue	Suggestions
Measurement scale	Need to take the measurement scale of the variables into account when computing statistics such as means, standard deviations, and correlations.
Restriction of range	Need to consider range of values obtained for variables, as restricted range of one or more variables can reduce the magnitude of correlations.
Outliers	Need to consider outliers as they can affect statistics such as means, standard deviations, and correlations. They can either be explained, deleted, or accommodated (using robust statistics). Can be detected by visual methods such as box plots, scatterplots, histograms or frequency distributions. Can also be detected by calculating z scores, median absolute deviation (MAD), and Mahalanobis distance (MD).
Non-normality	Need to consider whether the variables are normally distributed, as non-normality can affect resulting SEM statistics. Can be detected visually by examining histograms, stem and leaf plots, and normal Q–Q plots. Can also be detected using skewness and kurtosis statistics with corresponding z scores. Can be dealt with by transformations, additional sampling, or alternative methods of estimation.
Linearity	Need to consider whether variables are linearly related, as nonlinearity can reduce the magnitude of correlations. Can be detected by scatterplots. Can be dealt with by data transformations, deleting outliers, or modeling non-linear relationships.
Missing data	Need to consider missing data on one or more participants for one or more variables as this can affect SEM results. Traditional methods to handle missing data can result in reduced sample sizes and biased parameter estimates. Modern missing data methods are recommended, including multiple imputation or full information maximum likelihood (FIML) with missing data.

EXERCISES

1. Define the following scales of measurement:
   a. Nominal
   b. Ordinal
   c. Interval
   d. Ratio
2. For each of the following variables, identify their scale of measurement:
   a. Number of hours spent studying
   b. Political affiliation
   c. GRE scores
   d. Water measured in gallons
   e. Letter grade achieved in a course
   f. FICO credit score
3. Bland and Altman (2011) conducted a study with diabetic patients examining the relationship between abdominal circumference (measured in centimeters) and body mass index (BMI). Below is a scatterplot between BMI and abdominal circumference using simulated data roughly based on their findings. The correlation between BMI and abdominal circumference with all participants is large and positive ($r = .85$).

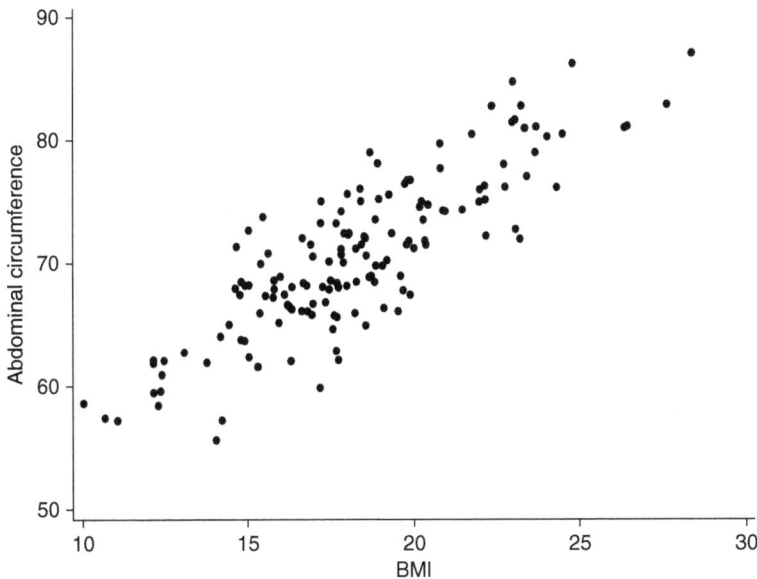

Figure 2.14: SCATTERPLOT BETWEEN BMI AND ABDOMINAL CIRCUMFERENCE

    a.   For participants with BMI values between 10 and 20, does the correlation between BMI and abdominal circumference look smaller (weaker), the same, or larger (stronger) than .85? What would you estimate is the value of the correlation for these participants?

    b.   For participants with BMI values between 20 and 30, does the correlation between BMI and abdominal circumference look smaller (weaker), the same, or larger (stronger) than .85? What would you estimate is the value of the correlation for these participants?

    c.   What is the reason for the difference in the relationship between BMI and abdominal circumference for the entire sample versus the participants with BMI values between 10 and 20 and between 20 and 30?

4.   Using the system file named *fitness.LSF*, run the Data Screening option under the Statistics tab in LISREL. In the output, you should find values of skewness and kurtosis, as well as z scores testing the normality for all variables in the *fitness.dat* file.

    a.   For each of the quantitative variables, identify which are non-normal using suggested skewness and kurtosis cutoff values of ± 2 and ± 7, respectively.

    b.   Which variables are not normally distributed based on the z scores associated with skewness and kurtosis (assuming an alpha = .05)?

    c.   In the *fitness.LSF* Window, run the Univariate option under the Graphs tab. Click on the **Trigl** variable and select "Bar chart." Click the "Plot" button. You should see a histogram. Describe the distribution of the **Trigl** variable.

    d.   Do you notice any atypical data values for the **Trigl** variable? If so, identify to which case number(s) those atypical points belong.

5.   Four of the variables in the *fitness.dat* file have missing values (i.e., Len, Mass, % Fat, and Strength). The missing data points are identified with values of –999999.000. Run the Multiple Imputation option under the Statistics tab. Highlight all of the variables in the Variable List box on the left and Click the "Select>>" button to place them in the Select Variables box on the right. Click the "MCMC" option to remove it from the default EM option. Click on the "Output Options" button. Type in 20 for the number of iterations and Click the "Set seed to" option (type in the value 1234 for the seed). Click "OK." Then, Click the "Run" button.

    a.   Compare the means and standard deviations for the four variables before using multiple imputation (see output from Exercise #4) and after multiple imputation using the MCMC algorithm.

    b.   Go back and run the multiple imputation using the EM algorithm by following these steps: Run the Multiple Imputation option under the Statistics tab. Highlight all of the variables in the Variable List box on the left and Click the "Select>>" button to place them in the Select Variables box on the right. Click the "Run" button. Compare the means and standard deviations for the four variables before using multiple imputation (see output from Exercise #4) and after multiple imputation using the EM algorithm.

6. Explain how each of the following affects statistics:
   a. Restriction of range
   b. Outliers
   c. Non-normality
   d. Non-linearity
   e. Missing data

REFERENCES

Anderson, N. H. (1961). Scales and statistics: Parametric and non-parametric. *Psychological Bulletin*, 58, 305–316.

Bland, J. M., & Altman, D. G. (2011). Correlation in restricted ranges of data. *BMJ (Clinical research ed.)*, 342, d556.

Davey, A., & Savla, J. (2010). *Statistical power analysis with missing data: A structural equation modeling approach.* New York: Routledge, Taylor & Francis Group.

Dempster, A. P., Laird, N. M., & Rubin, D. B. (1977). Maximum likelihood from incomplete data via the EM algorithm. *Journal of the Royal Statistical Society, Series B*, 39, 1–38.

Dixon, W. J. (1988). *BMDP statistical software*. Los Angeles, CA: University of California Press.

Enders, C. K. (2010). *Applied missing data analysis*. New York: Guilford Press.

Enders, C. K. (2013). Analyzing structural equation models with missing data. In G. R. Hancock & R. O. Mueller (Eds.), *Structural equation modeling: A second course* (2nd ed., pp. 493–519). Charlotte, NC: Information Age Publishing.

Hampel, F. R. (1974). The influence curve and its role in robust estimation. *Journal of the American Statistical Association*, 69(346), 383–393.

Ho, K., & Naugher, J.R. (2000). Outliers lie: An illustrative example of identifying outliers and applying robust methods. *Multiple Linear Regression Viewpoints*, 26(2), 2–6.

Jöreskog, K. G., Olsson, U. H., & Wallentin, F. Y. (2016). *Multivariate analysis with LISREL*. Switzerland: Springer.

Jöreskog, K. G., & Sörbom, D. (1996). *PRELIS2: User's reference guide*. Lincolnwood, IL: Scientific Software International.

Judd, C. M., McClelland, G. H., & Ryan, C. S. (2009). *Data analysis: A model comparison approach* (2nd ed.). New York: Routledge.

Little, R. J. A. (1988). A test of missing completely at random for multivariate data with missing values. *Journal of the American Statistical Association*, 83(404), 1198–1202.

Marsh, H. W., Wen, Z., Hau, K.-T., & Nagengast, B. (2013). Structural equation models of latent interaction and quadratic effects. In G. R. Hancock & R. O. Mueller (Eds.), *Structural equation modeling: A second course* (2nd ed., pp. 267–308). Charlotte, NC: IAP Information Age Publishing.

McKnight, P. E., McKnight, K. M., Sidani, S., & Aurelio, J. F. (2007). *Missing data: A gentle introduction.* New York: Guilford.

McLachlan, G. J., & Krishnan, T. (1997). *The EM algorithm and extensions.* New York: Wiley.

Peng, C.-Y. J., Harwell, M., Liou, S.-M., & Ehman, L. H. (2007). Advances in missing data methods and implications for educational research. In S. S. Sawilowsky (Ed.), *Real data analysis.* Charlotte, NC: Information Age.

Rubin, D. B. (1976). Inference and missing data. *Biometrika, 63*(2), 581–592.

Schafer, J. L. (1997). *Analysis of incomplete data.* Boca Raton, FL: Chapman & Hall/ CRC Press.

Stevens, S. S. (1946). On the theory of scales of measurement. *Science, 103,* 677–680.

Tanner, M. A., & Wong, W. H. (1987). The calculation of posterior distributions by data augmentation. *Journal of the American Statistical Association, 82*(398), 528–550.

West, S. G., Finch, J. F., & Curran, P. J. (1995). Structural equation models with nonnormal variables: Problems and remedies. In R. H. Hoyle (Ed.), *Structural equation modeling: Concepts, issues, and applications* (pp. 56–75). Thousand Oaks, CA: Sage Publications.

Yuan, K.-H., & Bentler, P. M. (1998). Robust mean and covariance structure analysis. *British Journal of Mathematical and Statistical Psychology, 51*(1), 63–88.

Yuan, K.-H., & Zhong, X. (2008). Outliers, leverage observations, and influential cases in factor analysis: Using robust procedures to minimize their effect. *Sociological Methodology, 38*(1), 329–368.

Zhong, X., & Yuan, K.-H. (2011). Bias and efficiency in structural equation modeling: Maximum likelihood versus robust methods. *Multivariate Behavioral Research, 56*(2), 229–265.

Chapter 3

CORRELATION AND REGRESSION METHODS

TYPES OF CORRELATION COEFFICIENTS

Sir Francis Galton conceptualized the correlation and regression procedure for examining the covariance between two or more traits, and Karl Pearson (1896) developed the statistical formula for the correlation coefficient and regression based on his suggestion (Tankard, 1984; Crocker & Algina, 1986; Ferguson & Takane, 1989). Shortly thereafter, Charles Spearman (1904) used the correlation procedure to develop a factor analysis technique. The correlation, regression, and factor analysis techniques have, for many decades, formed the basis for generating tests and defining constructs. Today, researchers are expanding their understanding of the roles that correlation, regression, and factor analysis play in theory and construct definition to include confirmatory factor analysis or measurement models, latent variable models, and covariance structure models.

The relationships and contributions of Galton, Pearson, and Spearman to the field of statistics, especially correlation, regression, and factor analysis, are quite interesting (Tankard, 1984). In fact, the basis of association between two variables (i.e., the bivariate correlation) has played a major role in statistics. The Pearson correlation coefficient provides the basis for point estimation (and test of significance), explanation (variance accounted for in a dependent variable by an independent variable), prediction (of a dependent variable from an independent variable through linear regression), reliability estimates (test–retest, equivalence), and validity (factorial, predictive, concurrent). The Pearson correlation coefficient also provides the basis for establishing and testing models among measured and/or latent variables.

DOI: 10.4324/9781003044017-3

Although the Pearson correlation coefficient has had a major impact in the field of statistics, other correlation coefficients have emerged to accommodate the different levels of variable measurement. Stevens (1968) provided the properties of scales of measurement that have become known as nominal, ordinal, interval, and ratio. The distinguishing features of the four levels of measurement were discussed in Chapter 2. The types of correlation coefficients developed for these various levels of measurement are categorized in Table 3.1.

Table 3.1: Types of Correlation Coefficients

Correlation Coefficient	Level of Measurement
Pearson product-moment	Both variables interval
Spearman rank, Kendall's tau	Both variables ordinal
Phi	Both variables dichotomous
Point biserial	One variable interval, one variable dichotomous
Gamma, rank biserial	One variable ordinal, one variable nominal
Biserial	One variable interval, one variable artificial[a]
Polyserial	One variable interval, one variable ordinal with underlying continuity
Tetrachoric	Both variables dichotomous (artificial[a])
Polychoric	Both variables ordinal with underlying continuities

Note: [a] = *artificial* refers to recoding variable values into a dichotomy with underlying continuity.

Many popular computer programs, for example, SAS and SPSS, typically do not compute all of these correlation types. Therefore, you may need to check a popular statistics book or look around for a computer program (R software) that will compute the type of correlation coefficient you need. For example, the phi and biserial coefficients are usually calculated with available macros (e.g., in SPSS or SAS) or using available functions in R. The Pearson correlation, tetrachoric or polychoric correlation, and biserial or polyserial correlation can all be used in SEM analysis with both LISREL and M*plus* software. The correlations that are most commonly taught in a correlation and regression methods course are presented in Table 3.2 with their respective formulas.

Table 3.2: Formulas for the Most Commonly Taught Correlation Coefficients

Correlation Coefficient	Formula
Pearson product-moment	$r_{xy} = \dfrac{\sum_{i=1}^{n}\left(X_i - \bar{X}\right)\left(Y_i - \bar{Y}\right)/(n-1)}{s_x s_y}$
Spearman rank	$r_S = 1 - \left[\dfrac{6\sum_{i=1}^{n} d_i^2}{n\left(n^2 - 1\right)}\right]$

Table 3.2: (*cont.*)

Correlation Coefficient	Formula
Phi	$r_\varphi = \dfrac{ad - bc}{\sqrt{(a+b)(a+c)(b+d)(c+d)}}$
Point biserial	$r_{pb} = \dfrac{(\bar{Y}_1 - \bar{Y}_0)\sqrt{PQ}}{s_Y}$
Biserial	$r_b = \dfrac{(\bar{Y}_1 - \bar{Y}_0)PQ}{h(s_Y)} = r_{pb}\dfrac{\sqrt{PQ}}{h}$

Note: X_i is the score on variable X for participant i; Y_i is the score on variable Y for participant i; n is the total sample size; s_x is the standard deviation of variable X; s_y is the standard deviation of variable Y; d is the difference between each pair of X and Y ranks; a, b, c, and d are frequencies in a 2 x 2 contingency table; $\bar{Y}_1$ is the mean on Y for group 1 and $\bar{Y}_0$ is the mean on Y for group 0; P is the proportion in one group and Q is the proportion in the other group; and h is the height of the normal distribution curve associated with cutting off P and Q on the distribution.

FACTORS AFFECTING CORRELATION COEFFICIENTS

Given the important role that correlation plays in structural equation modeling, we need to understand the factors that affect relationships among multivariable data points. Although the factors we describe will have an impact on all of the correlation coefficients listed in Table 3.1, we use the Pearson correlation coefficient for illustrative purposes. The Pearson correlation coefficient (r_{xy}) assumes a linear relationship between two continuous variables (e.g., X and Y). The Pearson correlation ranges from -1.0 to $+1.0$, with values close to zero representing no linear relationship between X and Y. The larger the magnitude of the correlation coefficient [as it approaches $|1.0|$], the stronger the linear relationship between X and Y. The negative ($-$) and the positive ($+$) sign associated with the correlation coefficient represents the direction of the linear relationship and not the magnitude.

Figure 3.1 illustrates a scatterplot matrix involving three variables (x1, x2, and x3) based on 1,000 observations. The degrees of freedom (*df*) associated with Pearson correlations are $n - 2$. Thus, the *df* for each of the three correlation coefficients is 998, assuming no missing data points.

Scatter Plot Matrix

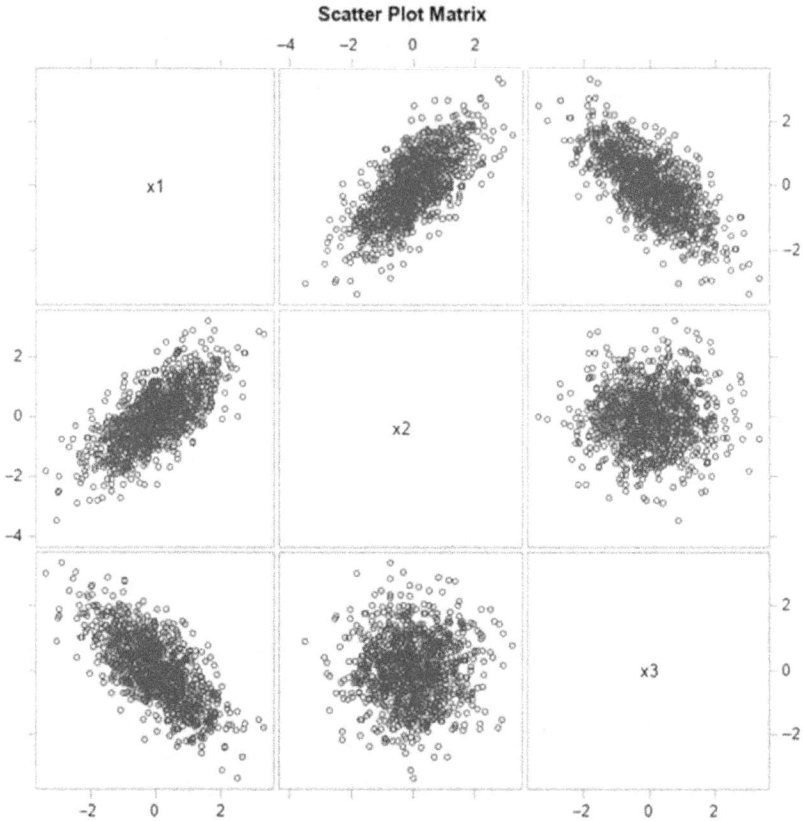

■ **Figure 3.1:** SCATTERPLOT MATRIX BETWEEN THREE VARIABLES

The Pearson correlation between x1 and x2 is $r(998) = .70$, $p < .0001$; the Pearson correlation between x1 and x3 is $r(998) = -.73$, $p < .0001$; and the Pearson correlation between x2 and x3 is $r(998) = 0$, $p = .798$. The magnitude of the relationship between x1 and x2 and between x1 and x3 is approximately equal [around |.70|], which represents a fairly strong linear relationship. This strong linear association is demonstrated in both of their respective scatterplots. Also demonstrated in their respective scatterplots is the difference in direction of the relationships. You will notice that the positive relationship between x1 and x2 shows that as x1 increases, x2 also increases. In contrast, the negative relationship between x1 and x3 shows that as x1 increases, x3 decreases. The magnitude of the correlation between x2 and x3 is around zero, indicating no linear relationship. The points in a scatterplot representing correlations close to zero appear round in shape instead of linear. When two variables are perfectly correlated, the scatterplot results in a perfectly straight line. Figure 3.2 illustrates a perfectly positive ($r = +1.0$; top) and a perfectly negative correlation ($r = -1.0$; bottom) between variables x4 and x5.

$r = +1.0$

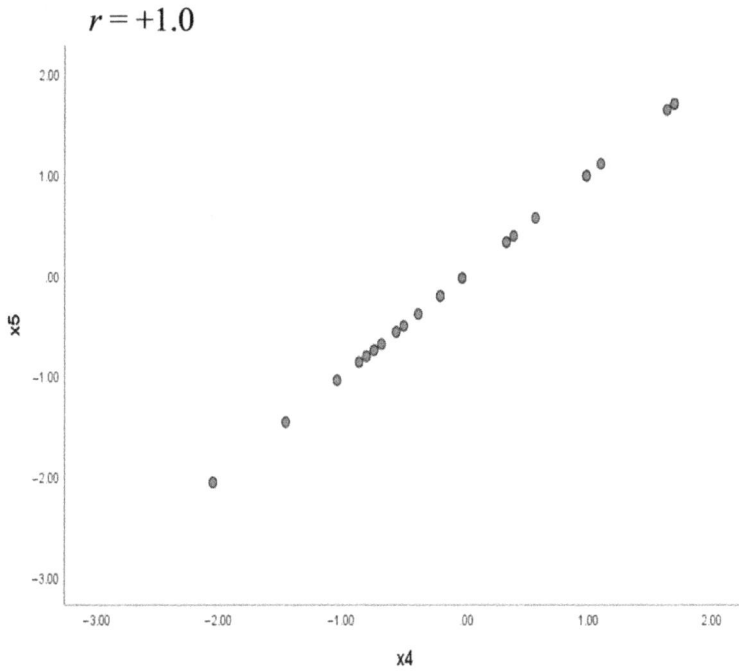

$r = -1.0$

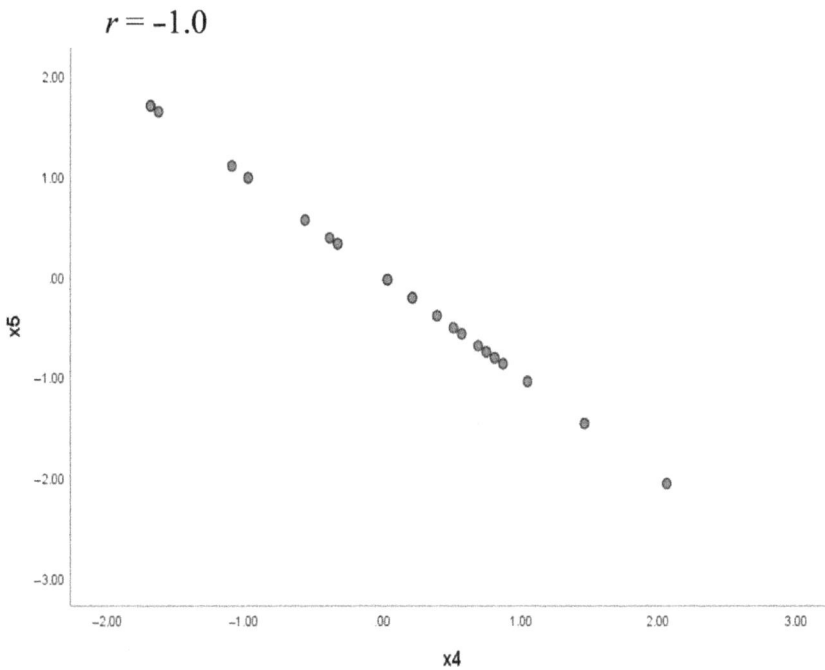

■ **Figure 3.2:** SCATTERPLOTS OF PERFECT LINEAR RELATIONSHIPS

The key factors that can affect the calculation of correlation coefficients are non-linearity, missing data, level of measurement and restriction of range in data values, non-normality, and outliers. Some of these factors are illustrated in Table 3.3, where the complete data set for $n = 10$ pairs of scores indicates a Pearson correlation coefficient between variables X and Y of $r(8) = .782$, $p = .007$. The nonlinear data results in $r(8) = 0.0$, $p = 1.0$; the missing data results in $r(5) = .659$, $p = .10$; the restriction of range in values results in $r(8) = 0.0$, $p = 1.0$; and the sampling effect results in $r(1) = -1.0$, $p < .0001$ (a complete reversal of direction). We see in Table 3.3 the dramatic impact that these factors have on the Pearson correlation coefficient. We describe the impact of these illustrated factors and additional factors to consider that may impact correlation coefficients.

Table 3.3: Heuristic Data Sets

Complete Data ($r = .782$, $p = .007$)		Non-linear Data ($r = 0.0$, $p = 1.0$)		Missing Data ($r = .659$, $p = .108$)	
X	Y	X	Y	X	Y
6	8	1	1	--	8
5	7	2	2	5	7
4	8	3	3	--	8
2	5	4	4	2	5
3	4	5	5	3	4
2	5	5	6	2	5
3	3	4	7	3	3
4	5	3	8	--	5
1	3	2	9	1	3
2	2	1	10	2	2

Range of Data ($r = 0.0$, $p = 1.0$)		Sampling Effect ($r = -1.0$, $p < .00001$)	
X	Y	X	Y
1	3	8	3
2	3	9	2
3	4	10	1
4	4		
1	5		
2	5		
3	6		
4	6		
1	7		
2	7		

Non-linearity

The Pearson correlation coefficient indicates the degree of linear relation between two variables. It is possible that two variables can indicate no correlation if they have a curvilinear relation. Thus, the extent to which the variables deviate from the assumption of a linear relation will affect the size of the correlation coefficient. In Table 3.3, the nonlinear data indicates a correlation of $r(8) = 0.0$, $p = 1.0$, which indicates no relation when the data are actually non-linearly related. It is therefore important to check for linearity of the scores. As discussed in Chapter 2, the common method is to graph the coordinate data points in a scatterplot (see Figure 2.13 for an example of linear and non-linear relationships). If non-linear relationships exist among variables, the non-linearity should be modeled appropriately. For instance, the *eta* coefficient is an index of the non-linear relationship between two variables. Further, multiple regression can be used to estimate linear, quadratic, and/or cubic effects when they are specified in the model.

Missing Data

As seen in Table 3.3, the Pearson correlation coefficient drops to $r(5) = .659$, $p = .108$ for $n = 7$ pairs of scores with listwise deletion of missing data. The Pearson correlation coefficient changes from being statistically significant to not being statistically significant. More importantly, in a correlation matrix with several variables, the various correlation coefficients could be computed on different sample sizes if pairwise deletion is used. As discussed in Chapter 2, the cause of missing data is important to attempt to determine in order to address missing data appropriately during the analysis.

Level of Measurement and Restriction of Range

As discussed in Chapter 2, four types or levels of measurement typically define whether the scale interpretation of a variable is nominal, ordinal, interval, or ratio (Stevens, 1968). Linear regression, path analysis, factor analysis, and structural equation modeling were initially developed under the assumption that the variables involved were measured at the interval or ratio level of measurement. The level of variable measurement and the range of values for the measured variables can have profound effects on your statistical analysis (in particular, on the mean, variance, and correlation), and this is no different in structural equation modeling. The interval or ratio scaled variable values should have a sufficient range of score values to introduce variance. If the range of scores of a variable is restricted, the magnitude of the relationship involving the variable with a restricted range is decreased. In Table 3.3, we see that a restriction of range changed the correlation from $r(8) = .782$ to a value of $r(8) = 0.0$.

Non-normality

If the distributions of the variables are widely divergent from normality, correlations can also be affected. Several data transformations were suggested by Ferguson and Takane (1989) to provide a closer approximation to a normal distribution in the presence of skewed or kurtotic data. Some possible transformations are the square root transformation [sqrt (X)], the logarithmic transformation [log (X)], the reciprocal transformation $(1/X)$, and the arcsine transformation [arcsin (X)]. The probit transformation appears to be most effective in handling univariate skewed data, but it is wise to check which transformation works best.

OUTLIERS

The Pearson correlation coefficient can be drastically affected by a single outlier on either X or Y, or both. For example, the two data sets in Table 3.4 indicate a $Y = 27$ value (Set A) versus a $Y = 2$ value (Set B) for the last subject. In the first set of data, $r = .524$, $p = .37$, whereas in the second set of data, $r = -.994$, $p = .001$. Is the $Y = 27$ data value an outlier based on limited sampling or is it a data entry error? A large body of research has been undertaken to examine how different outliers on X, Y, or both X and Y affect correlational relations and how to better analyze the data using robust statistics (Huber, 1981; Rousseeuw & Leroy, 1987; Staudte & Sheather, 1990; Ho & Naugher, 2000; Anderson & Schumacker, 2003). As described in Chapter 2, the identification of and the diagnostics of outlying scores are important steps to take prior to any statistical analysis in order to avoid potential bias in parameter estimates.

Table 3.4: Outlier Data Sets

Set A (r = .54, p = .37)		Set B (r = −.994, p = .001)	
X	**Y**	**X**	**Y**
1	9	1	9
2	7	2	7
3	5	3	5
4	3	4	3
5	27	5	2

SIMPLE LINEAR REGRESSION

Regression techniques require a basic understanding of sample statistics (mean and variance), standardized variables, and correlation (Houston & Bolding, 1974;

Pedhazur, 1982; Cohen & Cohen, 1983). Simple linear regression involves the prediction of a continuous outcome or dependent variable (Y) by a single predictor variable (X), which can be continuous or categorical. The basic simple linear regression equation can be expressed as follows: $\hat{Y} = a + bX$, where $\hat{Y}$ is the predicted value of the outcome variable; a is the constant or Y-intercept and is equal to $\hat{Y}$ when $X = 0$; b is the slope or unstandardized regression coefficient; and X is the predictor variable. The linear regression equation will result in a regression line that will fit the data the most precisely because the slope $\left[b = \dfrac{\Sigma(X - \bar{X})(Y - \bar{Y})}{\Sigma(X - \bar{X})^2} \right]$ and the intercept $\left[a = \bar{Y} - b\bar{X} \right]$ are calculated to minimize the residuals. The residuals are represented as the difference between observed and predicted Y values: $(Y - \hat{Y})$.

To illustrate these calculations and concepts in simple linear regression, we will use the data from Chatterjee and Yilmaz (1992). The **satisfaction.csv** file contains scores from 24 patients on four variables (V1 = patient's age in years, V2 = severity of illness, V3 = level of anxiety, and V4 = satisfaction level). The descriptive statistics for the data set, including Pearson correlations, are provided in Table 3.5.

Table 3.5: Descriptives and Correlations for Satisfaction Data Set

	1	2	3	4
1. Age	–			
2. Severity	.614**	–		
3. Anxiety	.193	.389	–	
4. Satisfaction	−.765**	−.600**	−.453*	–
M	40.58	51.38	2.28	60.58
SD	9.56	5.22	0.32	16.77

Note: *$p < .05$, **$p < .01$. $n = 24$.

We will use *severity of illness* (X) to predict *satisfaction level* (Y): $\hat{Y}_{Satisfaction} = a + bX_{Severity}$. As you can see from Table 3.5, *severity of illness* and *satisfaction level* have a moderately large, negative, and statistically significant correlation, $r(22) = -.60$, $p < .01$. Below are the calculations for the intercept and slope.

Table 3.6: Calculation of Intercept and Slope for Satisfaction Data Set

ID	Severity (X)	Satis- faction (Y)	$(X-\bar{X})$	$(Y-\bar{Y})$	$(X-\bar{X})(Y-\bar{Y})$	$(X-\bar{X})^2$	$\hat{Y}$	$(Y-\hat{Y})$
1	53	67	1.63	6.42	10.43	2.64	57.45	9.55
2	48	89	-3.38	28.42	-95.91	11.39	67.09	21.91
3	50	77	-1.38	16.42	-22.57	1.89	63.23	13.77
4	62	26	10.63	-34.58	-367.45	112.89	40.11	-14.11
5	48	54	-3.38	-6.58	22.22	11.39	67.09	-13.09
6	50	46	-1.38	-14.58	20.05	1.89	63.23	-17.23
7	54	36	2.63	-24.58	-64.53	6.89	55.53	-19.53
8	43	89	-8.38	28.42	-237.99	70.14	76.72	12.28
9	44	70	-7.38	9.42	-69.45	54.39	74.79	-4.79
10	65	43	13.63	-17.58	-239.57	185.64	34.33	8.67
11	48	66	-3.38	5.42	-18.28	11.39	67.09	-1.09
12	46	57	-5.38	-3.58	19.26	28.89	70.94	-13.94
13	51	48	-0.38	-12.58	4.72	0.14	61.31	-13.31
14	55	47	3.63	-13.58	-49.24	13.14	53.60	-6.60
15	51	51	-0.38	-9.58	3.59	0.14	61.31	-10.31
16	54	57	2.63	-3.58	-9.41	6.89	55.53	1.47
17	49	66	-2.38	5.42	-12.86	5.64	65.16	0.84
18	56	79	4.63	18.42	85.18	21.39	51.67	27.33
19	46	88	-5.38	27.42	-147.36	28.89	70.94	17.06
20	49	60	-2.38	-0.58	1.39	5.64	65.16	-5.16
21	51	49	-0.38	-11.58	4.34	0.14	61.31	-12.31
22	52	77	0.63	16.42	10.26	0.39	59.38	17.62
23	58	52	6.63	-8.58	-56.86	43.89	47.82	4.18
24	50	60	-1.38	-0.58	0.80	1.89	63.23	-3.23

$$\bar{X} = 51.38; \quad \bar{Y} = 60.58; \quad \Sigma(X-\bar{X})(Y-\bar{Y}) = -1209.25; \quad \Sigma(X-\bar{X})^2 = 627.63$$

$$b = \frac{\Sigma(X-\bar{X})(Y-\bar{Y})}{\Sigma(X-\bar{X})^2} = \frac{-1209.25}{627.63} = -1.9267$$

$$a = \bar{Y} - b\bar{X} = 60.58 - (-1.9267)*51.38 = 60.58 + 1.9267*51.38 = 159.57$$

Thus, the linear regression equation is: $\hat{Y}_{Satisfaction} = 159.57 - 1.9267 X_{Severity}$. The scatterplot between *severity of illness* and *satisfaction* with the regression line superimposed is presented in Figure 3.3, demonstrating the negative relationship. The numbers next to the points in the scatterplot are person ID numbers.

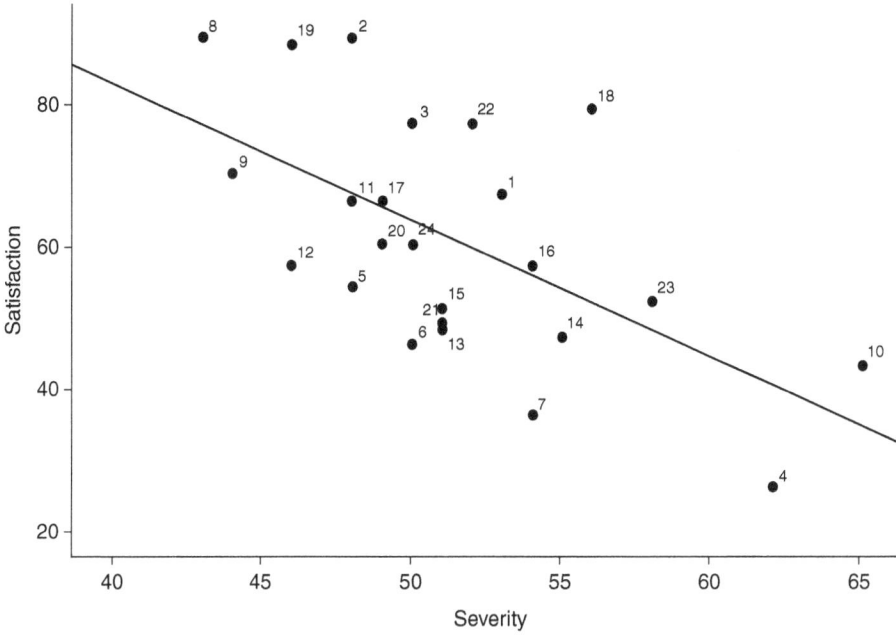

Figure 3.3: SCATTERPLOT FOR SIMPLE LINEAR REGRESSION

The value of $\hat{Y}$ when $X = 0$ is the Y-intercept or the constant in the regression equation $(a = 159.57): \hat{Y}_{Satisfaction} = 159.57 - (1.9267*0)$. The slope (b) indicates that a one-point increase in the *severity of illness* (X) results in an estimated decrease of 1.93 points in *satisfaction* (Y). Using the regression equation, you can predict satisfaction (Y) using a given value of *severity of illness* (X). For instance, if *severity of illness* were equal to 46, the predicted *satisfaction* value for the person with *severity of illness* equal to 46 would be: $\hat{Y}_{Satisfaction} = 159.57 - (1.9267*46) = 70.94$. In Figure 3.4, you can see this point on the regression line where the dotted lines intersect at values of $X = 46$ and $\hat{Y} = 70.94$.

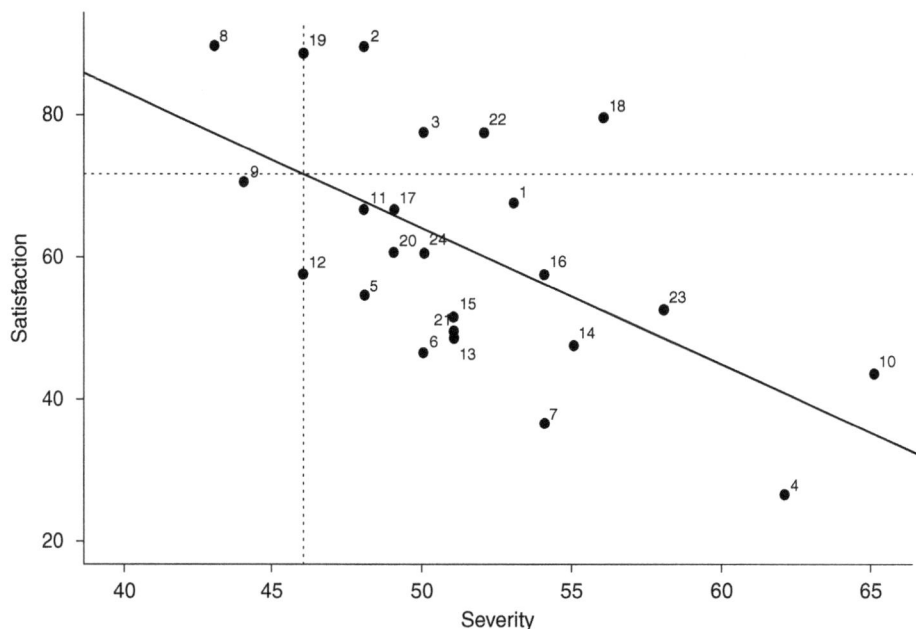

■ **Figure 3.4:**   SCATTERPLOT FOR SIMPLE LINEAR REGRESSION FOR $X = 46$

Two participants, persons 12 and 19, have *severity of illness* values equal to 46 and *satisfaction* scores equal to 57 and 88, respectively. Neither of these points fall on the regression line. As indicated by the correlation coefficient, *severity of illness* is not perfectly correlated with *satisfaction*. Thus, *severity of illness* did not perfectly predict *satisfaction*. This can be seen since the scatterplot between the two variables does not illustrate a straight line. Person 12 has a *satisfaction* score closer to the predicted value of 70.94 than person 19 (see Figure 3.4). The last two columns in Table 3.6 shows the predicted value of *satisfaction* for each observed value of *severity* and the difference between each person's *observed satisfaction* score and their *predicted satisfaction* score, respectively. The difference between the observed satisfaction score and the predicted satisfaction score $\left(Y - \hat{Y}\right)$ is called a residual and represents prediction error. You will notice in Table 3.6 that the residual for person 12 (–13.94) is smaller than the residual for person 19 (17.06). The regression line will minimize the sum of squared residuals $\left(SS_{Res}\right)$: $\Sigma(Y - \hat{Y})^2$, which is why linear regression is referred to as ordinary least squares (OLS) regression.

As with other estimates, the unstandardized regression coefficient can be statistically tested against the value of zero. We first need a standard error estimate for testing the statistical significance of b. This starts with the variance of estimate $\left(s_{Y.X}^2\right)$, which indicates the variability of observed Y scores around predicted Y scores, or the variance of the residuals $\left(Y - \hat{Y}\right)$:

$$s_{Y.X}^2 = \frac{\sum\left(Y - \hat{Y}\right)^2}{n - k - 1} = \frac{SS_{Res}}{n - k - 1},$$

where n is the total sample size and k is the number of predictor variables in the regression model. The standard error associated with b is calculated as follows:

$$s_b = \sqrt{\frac{s_{Y.X}^2}{\sum\left(X - \bar{X}\right)^2}}$$

Below are the calculations involved in the standard error associated with b:

$$\sum\left(Y - \hat{Y}\right)^2 = 4137.96$$

$$S_{Y.X}^2 = \frac{\sum\left(Y - \hat{Y}\right)^2}{n - k - 1} = \frac{SS_{Res}}{n - k - 1} = \frac{4137.96}{24 - 1 - 1} = 188.089$$

$$S_b = \sqrt{\frac{S_{Y.X}^2}{\sum\left(X - \bar{X}\right)^2}} = \sqrt{\frac{188.089}{627.63}} = \sqrt{.2997} = .547$$

The unstandardized regression coefficient is divided by its respective standard error to yield either a z or t statistic: t or $z = \dfrac{-1.9267}{.547} = -3.52$. Depending upon the software used, the standard error and resulting test statistic may differ slightly. M*plus* analyzes variances based on the population formula without $n - 1$ in the denominator whereas LISREL analyzes variances based on the sample formula with $n - 1$ in the denominator. Thus, M*plus* test statistics are z tests whereas LISREL test statistics are either z or t tests, depending on whether you run the regression analyses using SIMPLIS or PRELIS syntax, respectively. The value of –3.52 calculated above corresponds to a t value because sample formulas were used. The absolute value of –3.52 exceeds the critical t value for a two-tailed test (2.074) with 22 *df* (*df* = $n - 2$ = 24 – 2 = 22) at an alpha = .05 (see Table A.2 in Appendix 2). Thus, the regression coefficient is statistically significantly different than zero.

The unstandardized regression slope is expressed in measurement units of the variables involved. Standardized regression coefficients (also called beta weights) may also be calculated. In standard score form (z scores), the simple linear regression equation for predicting the dependent variable Y from a single independent variable X is: $\hat{z}_Y = \beta z_X$, where $\hat{z}_Y$ is the predicted standardized (z) score for Y, β is the standardized regression coefficient or slope, and z_X is the standardized (z) score for X. The basic rationale for using the standard score formula is that variables are converted to the same scale of measurement, the z scale. The formula connecting the Pearson product-moment correlation coefficient, the unstandardized regression coefficient b, and the standardized slope or regression coefficient β in simple linear regression is:

$$\beta = \frac{\sum z_x z_y}{\sum z_x^2} = b \frac{s_x}{s_y} = r_{xy},$$

where s_x and s_y are the sample standard deviations for variables X and Y, respectively. Remember that the correlation between severity of illness and satisfaction level is $r(22) = -.60$, $p < .01$ (see Table 3.5). Using the standard deviations for severity of illness (X) and satisfaction level (Y) in Table 3.5 and the calculated regression coefficient, we see that the formula for the standardized regression coefficient (β) is equal to the Pearson correlation coefficient:

$$\beta = b \frac{s_x}{s_y} = -1.9267 \frac{5.22}{16.77} = r_{xy} = -.60.$$

The interpretation of the standardized regression coefficient is similar to that of the unstandardized regression coefficient, except it is explained in standard deviation units. Thus, as *severity of illness* increases by one standard deviation, *satisfaction level* is estimated to decrease by .60 standard deviations.

When squaring the Pearson correlation coefficient, the resulting value can be interpreted as the proportion of variance in the dependent variable accounted for by the independent variable. This value is called the *coefficient of determination*. In this example, *severity of illness* explains 36% $(-.60^2)$ of the variance in *satisfaction* level. It is important to note that in simple linear regression, if the correlation coefficient is statistically significant, the regression coefficient will also be statistically significant. This is not necessarily the case, however, with multiple linear regression.

MULTIPLE LINEAR REGRESSION

For two independent variables ($X1$ and $X2$), the multiple linear regression equation is:

$$\hat{Y} = a + b_1 X_1 + b_2 X_2,$$

where $\hat{Y}$ is the predicted value of the outcome variable; a is the constant or Y-intercept and is equal to $\hat{Y}$ when both $X1$ and $X2$ are equal to zero; b_1 is the unstandardized partial regression coefficient associated with predictor variable $X1$; and b_2 is the unstandardized partial regression coefficient associated with predictor variable $X2$. The regression coefficients are now called partial regression coefficients because they are calculated with consideration of the relationships among all of the variables in the model as follows:

$$b_1 = \beta_1 \frac{s_Y}{s_{X1}} \quad \text{and} \quad b_2 = \beta_2 \frac{s_Y}{s_{X2}},$$

where β_1 and β_2 are the standardized partial regression coefficients associated with predictor variables $X1$ and $X2$, respectively; s_{X1} and s_{X2} are standard deviations associated with predictor variables $X1$ and $X2$, respectively; and s_Y is the standard deviation of the dependent variable.

As you can see, the standardized partial regression coefficients, β_1 and β_2, must be calculated first:

$$\beta_1 = \frac{r_{Y1} - r_{Y2}r_{12}}{1 - r_{12}^2} \quad \text{and} \quad \beta_2 = \frac{r_{Y2} - r_{Y1}r_{12}}{1 - r_{12}^2},$$

where r_{Y1} is the correlation between $X1$ and Y; r_{Y2} is the correlation between $X2$ and Y; and r_{12} is the correlation between $X1$ and $X2$. The standardized regression equation is:

$$\hat{z}_Y = \beta_1 z_{X1} + \beta_2 z_{X2},$$

where z_{X1} and z_{X2} are standardized (z) scores for predictor variables $X1$ and $X2$, respectively.

The correlation between the observed Y scores and the predicted Y scores ($\hat{Y}$) is called the *multiple correlation coefficient* or *multiple R* and is calculated as:

$$R = \sqrt{\beta_1 r_{Y1} + \beta_2 r_{Y2}}.$$

The squared multiple correlation coefficient or squared multiple R results in R *square* (R^2), which is also called the *coefficient of multiple determination*. R square represents the proportion of variance in the dependent variable accounted for or explained by the set of independent variables (in this case, both $X1$ and $X2$).

R square (R^2) tends to be an overestimate of the variance explained by a set of predictors in the population. One reason for this is because adding predictors to a model will increase the R square value, regardless of the meaningfulness of including the predictor in the model. That is, including a predictor in a model that has a close to zero correlation with the dependent variable will still increase the R square. Thus, an adjusted R square has been formulated to penalize the inclusion of predictors to simply increase R square values and to provide a better estimate of explained variance in the population. The adjusted formula using sample size (n) and the number of predictors (k) is:

$$R^2_{Adj} = R^2 - \frac{k}{n-k-1}\left(1 - R^2\right).$$

From examining this formula, you will notice that adjusted R square will be smaller than R square. You will also notice that as sample size increases, the less different R square will be from the adjusted R square.

BIVARIATE, PART, AND PARTIAL CORRELATIONS IN MULTIPLE REGRESSION

In addition to partial regression coefficients, different types of correlations may be meaningful to calculate when considering how predictor variables relate to the dependent variable in multiple regression scenarios. The different types of correlations presented in Table 3.1 are considered bivariate correlations, or associations between two variables. Cohen and Cohen (1983), in describing correlation research, further presented the correlation between two variables controlling for the influence of a third variable. These correlations are referred to as *part* and *partial* correlations, depending upon how variables are controlled or partialled out. Some of the various ways in which three variables can be depicted are illustrated in Figure 3.5. The diagrams illustrate different situations among variables where: (a) all the variables are uncorrelated (Case 1); (b) only one pair of variables is correlated (Cases 2 and 3); (c) two pairs of variables are correlated (Cases 4 and 5); and (d) all of the variables are correlated (Case 6). It is obvious that with more than three variables, the possibilities become overwhelming. It is therefore important to have a theoretical perspective to suggest how certain variables are correlated in a study. A theoretical perspective is essential in specifying a model and forms the basis for testing a model in SEM.

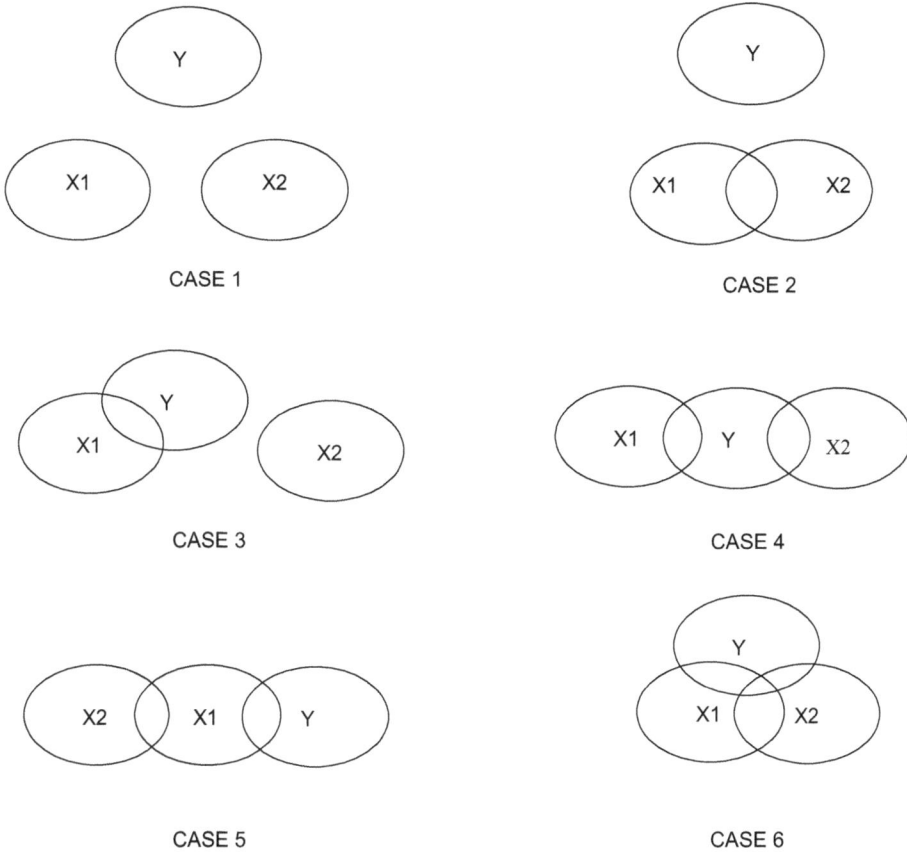

Figure 3.5: POSSIBLE THREE-VARIABLE RELATIONSHIPS

Suppose we have three variables: $X1 = age$, $X2 = reading\ level$, $Y = comprehension$. The correlations among these three variables are provide in Table 3.7. The *partial correlation coefficient* measures the association between two variables while controlling for a third variable. The partial correlation will calculate the association between *age* and *comprehension*, controlling for *reading level*. Controlling for *reading level* in the correlation between *age* and *comprehension* partials out the correlation of *reading level* with *age* and the correlation of *reading level* with *comprehension*. *Part (or semipartial) correlation*, in contrast, is the correlation between *age* and *comprehension* with *reading level* controlled for, where only the correlation between *reading level* and *comprehension* is removed before *age* is correlated with *comprehension*.

Whether a part or partial correlation is used depends on the specific model or research question. Convenient notation helps distinguish these two types of correlation. The partial correlation between *age* and *comprehension* is expressed as

$r_{1Y.2}$ whereas the part correlation between *age* and *comprehension* is expressed as $r_{1(Y.2)}$ or $r_{Y(1.2)}$. Likewise, the partial correlation between *reading level* and *comprehension* is expressed as $r_{2Y.1}$ whereas the part correlation between *reading level* and *comprehension* is expressed as $r_{2(Y.1)}$ or $r_{Y(2.1)}$.

Table 3.7: Correlation Matrix for Age, Reading Level, and Comprehension ($n = 100$)

Variable	Age	Reading Level	Comprehension
Age (X1)	1.00		
Reading level (X2)	.25	1.00	
Comprehension (Y)	.45	.80	1.00

Using the correlations in Table 3.7, we can compute the partial correlation coefficient between *age* and *comprehension*, controlling for *reading level* as:

$$r_{1Y.2} = \frac{r_{1Y} - r_{12}r_{2Y}}{\sqrt{\left(1 - r_{12}^2\right)\left(1 - r_{2Y}^2\right)}} = \frac{.45 - (.25)(.80)}{\sqrt{\left[1 - (.25)^2\right]\left[1 - (.80)^2\right]}} = .43$$

The partial correlation coefficient between *reading level* and *comprehension*, controlling for *age* is calculated as:

$$r_{2Y.1} = \frac{r_{2Y} - r_{12}r_{1Y}}{\sqrt{\left(1 - r_{12}^2\right)\left(1 - r_{1Y}^2\right)}} = \frac{.80 - (.25)(.45)}{\sqrt{\left[1 - (.25)^2\right]\left[1 - (.45)^2\right]}} = .80$$

When squaring the partial correlation, the resulting value represents the proportion of variance in the dependent variable that is explained by the respective predictor that is not explained by the other predictor(s) in the model. Thus, *age* explains about 18% [$(.43)^2$] of the variance in *comprehension* that is not accounted for by *reading level*. *Reading level* explains about 64% [$(.80)^2$] of the variance in *comprehension* that is not accounted for by *age*.

The part (or semipartial) correlation coefficient for *age* [$r_{1(Y.2)}$], or correlation between *age* and *comprehension* where *reading level* is controlled for in *comprehension* only, is computed as:

$$r_{1(Y.2)} = \frac{r_{1Y} - r_{12}r_{2Y}}{\sqrt{1 - r_{2Y}^2}} = \frac{.45 - (.25)(.80)}{\sqrt{1 - .80^2}} = .42;$$

or, in the case of correlating *reading level* with *comprehension* where *age* is controlled for in *comprehension* only is:

$$r_{2(Y.1)} = \frac{r_{2Y} - r_{12}r_{1Y}}{\sqrt{1 - r_{1Y}^2}} = \frac{.80 - (.25)(.45)}{\sqrt{1 - .45^2}} = .77.$$

When squaring the part or semipartial correlation, the resulting value represents the proportion of variance in the dependent variable that is uniquely explained by the respective predictor variable. In other words, it represents the proportion of R square that may be attributed uniquely to the predictor variable. Thus, *age* uniquely explains about 18% [$(.42)^2$] of the variance in *comprehension* whereas *reading level* uniquely explains about 59% [$(.77)^2$] of the variance in *comprehension*.

The partial and part correlation coefficients will generally be smaller in magnitude than the respective Pearson product-moment correlations when a non-zero correlation exists between independent variables. Partial correlations will be greater in magnitude than their respective part correlations, except when independent variables are perfectly uncorrelated with each other ($r_{12} = 0$); then, part correlations are equal to partial correlations.

The correlation, whether zero-order (bivariate), part, or partial can be tested for significance, interpreted as variance accounted for by squaring each coefficient, and diagrammed using Venn or Ballentine figures to conceptualize their relationships. In our example, the squared value of the zero-order relationships among the three variables is represented by the overlapping areas diagrammed in Figure 3.6. However, the squared partial correlation for *age* with *comprehension* controlling for *reading level* would be $(r_{1Y.2})^2 = .18$, or area *a* divided by the combined area of *a* and *e* [$a/(a + e)$] as shown in Figure 3.7. The squared part or squared semipartial correlation for *age* with *comprehension* while controlling for the correlation between *reading level* and *comprehension* would be [$r_{1(Y.2)}]^2 = .18$, or just area *a* as shown in Figure 3.8. More specifically, because $R^2 = a + b + c$, the squared part (semipartial) correlation $= R^2 - (b + c) = a$.

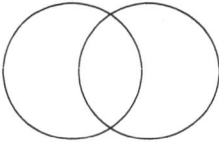

Age and Comprehension

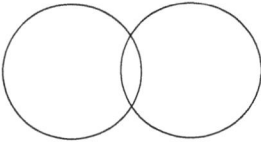

Age and Reading

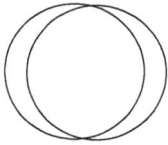

Reading and Comprehension

■ **Figure 3.6:**   BIVARIATE CORRELATIONS AMONG THREE VARIABLES

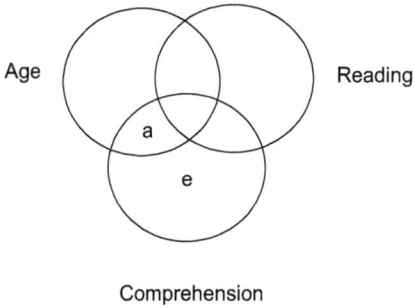

Age   Reading

a

e

Comprehension

■ **Figure 3.7:**   PARTIAL CORRELATION AREA FOR AGE

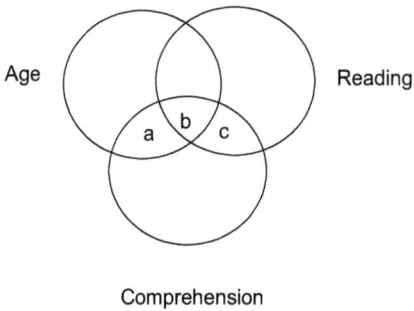

Age   Reading

a b c

Comprehension

■ **Figure 3.8:**   PART OR SEMIPARTIAL CORRELATION AREA FOR AGE

The partial and part correlations measure specific bivariate relations between variables while controlling for the influence of other variables. These examples consider only controlling for one variable when correlating two other variables (partial), or controlling for the impact of one variable on another before correlating with a third variable (part). Other higher-order part correlations and partial correlations are possible (e.g., $r_{1Y.23}$, $r_{1Y(2.3)}$), but are beyond the scope of this book. Readers should refer to the following references for a more detailed discussion of part and partial correlation: Cohen and Cohen (1983); Pedhazur (1997); Hinkle, Wiersma, and Jurs (2003); Lomax and Hahs-Vaughn (2012).

MULTICOLLINEARITY AND SUPPRESSOR VARIABLES

Because predictor variables are typically intercorrelated to some degree, the top Venn diagram in Figure 3.9 represents the circumstance that there is some shared variance between the two moderately correlated predictor variables (represented by the combined area of $a + b$) and there is redundancy in terms of the explanation of variance in the dependent variable by the correlated predictors (represented by area b). Areas c and d represent unique variance in the dependent variable that may be attributed to X1 and X2, respectively. When predictor variables are too highly correlated, neither predictor is able to explain much unique variance in the dependent variable. The bottom Venn diagram in Figure 3.9 represents how it looks when predictor variables are highly correlated. You can see in the diagram that areas a and b become larger whereas areas c and d become smaller relative to the same areas portrayed in the top Venn diagram with moderately correlated predictors.

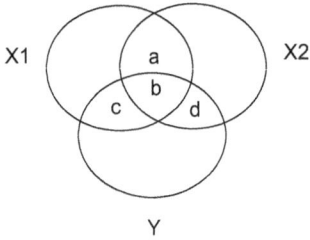

Moderately Correlated Predictors

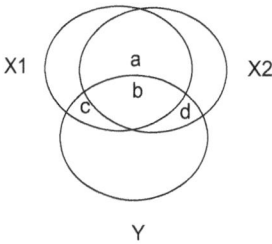

Highly Correlated Predictors

▦ **Figure 3.9:**   REDUNDANT RELATIONSHIPS

Multicollinearity occurs when predictor variables are too highly correlated. This could indicate that the two predictor variables are measuring the same or similar constructs (e.g., career attainment and career achievement). Problems resulting from multicollinearity are large standard errors for the partial regression coefficients associated with predictors that are too highly correlated and difficulty interpreting the partial regression coefficients associated with predictors that are too highly correlated.

Multicollinearity can be detected using *tolerance* or the *variance inflation factor* (VIF). Tolerance is the proportion of a predictor's variance that is not accounted for by other predictor variables in the model. It is computed as $1 - R^2$ when regressing each predictor on the set of remaining predictors. Tolerance values associated with predictor variables that are less than .10 (less than 10% of the variance remaining unexplained by the set of other predictors) suggests severe multicollinearity for the respective predictor. VIF values are calculated as the reciprocal of tolerance $\left(\dfrac{1}{Tolerance} \right)$. Consequently, VIF values associated with predictor variables that are greater than 10 suggest severe multicollinearity for the respective predictor. When predictor variables are flagged for multicollinearity, the regression model will need to be re-specified by either excluding one of the highly correlated predictors or by combining the highly correlated predictors as a composite variable to include in the model.

As mentioned previously, when predictors are correlated, the part correlations for the predictors will be smaller than their respective bivariate correlations with the dependent variable. If the part (or semipartial) correlation coefficient is not smaller than its respective Pearson product-moment correlation, then a *suppressor variable* may be present (Pedhazur, 1997). A suppressor variable correlates near zero with a dependent variable but correlates significantly with other independent variables. This correlation situation serves to control for variance shared with predictor variables and not the dependent variable, resulting in an increase in the strength of the relationship between predictors and the dependent variable. Suppression occurs when the standardized partial regression coefficient for each predictor/independent variable is greater than its bivariate correlation with the dependent variable. Suppression can also result in one of the partial regression coefficients associated with a predictor becoming negative when the predictor is positively correlated with the dependent variable. Another case of suppression occurs when the correlation between predictor variables is negative, but the correlations between each predictor and dependent variable is positive.

Discovering the existence of a suppressor variable in a model can result in difficulty when interpreting the findings. Given the different types of scenarios that may result due to the suppression phenomenon, it is important for researchers to explain why the phenomenon may be happening when discovered in a model. For a more detailed discussion concerning suppression, readers may consult Maassen and Bakker (2001).

CORRELATION VERSUS COVARIANCE

The type of data matrix used for computations in structural equation modeling programs is a covariance matrix. Structural equation modeling software uses the covariance matrix rather than the correlation matrix because Boomsma (1983) found that the analysis of correlation matrices led to imprecise parameter estimates and standard errors of the parameter estimates in a structural equation model. In SEM, incorrect estimation of the standard errors for the parameter estimates could lead to statistically significant parameter estimates and an incorrect interpretation of the model. That is, the parameter divided by the standard error indicates a test statistic (*t* or *z* value), which can be compared to tabled critical *t*-values and *z*-values for statistical significance at different alpha levels, respectively. Browne (1982), Jennrich and Thayer (1973), and Lawley and Maxwell (1971) have suggested corrections for the standard errors when correlations or standardized coefficients are used in SEM. In general, a covariance matrix should be used in structural equation modeling, with some SEM models requiring variable means (e.g., structured means models which test differences in latent variable means).

If a correlation matrix is used as the input data matrix, most of the computer programs by default convert it to a covariance matrix using the provided standard deviations of the variables. A covariance matrix, also called a variance–covariance matrix, is made up of variance terms on the main diagonal and covariance terms on the off-diagonal of the matrix. The number of distinct or non-redundant elements in a covariance matrix (**S**) is $p(p + 1)/2$, where *p* is the number of observed variables. The number of distinct elements in a covariance matrix is important to understand when evaluating model identification and when calculating degrees of freedom for a model, which will be discussed in greater detail subsequently. As an example, the covariance matrix (**S**) for the following three variables, *X*, *Y*, and *Z*, is as follows:

	X	**Y**	**Z**
X	15.80		
Y	10.16	11.02	
Z	12.43	9.23	15.37

S =

It has 3 (3 + 1)/2 = 6 distinct values: 3 variances in the main diagonal and 3 covariance terms in the off-diagonal of the matrix.

A correlation matrix can be thought of as a covariance matrix for standardized data. Specifically, a correlation matrix has values of one in the main diagonal (representing the variance of the standardized data) and correlations in the off-diagonal. The elements in a correlation matrix are computed using the variances of and covariances among the bivariate variables, using the following formula:

$$r = \frac{s_{XY}}{\sqrt{s_X^2 * s_Y^2}}$$

Dividing the covariance between two variables (covariance terms are the off-diagonal values in the matrix) by the square root of the product of the two respective variable variances (variances of variables are on the diagonal of the matrix) yields the following correlations among the three variable pairs:

$r_{xy} = 10.16/(15.80 * 11.02)^{1/2} = .77$
$r_{xz} = 12.43/(15.80 * 15.37)^{1/2} = .80$
$r_{yz} = 9.23/(11.02 * 15.37)^{1/2} = .71$

Dividing the variance of a variable by the square root of the product of the respective variable's variance results in values on the main diagonal using the following formula:

$$r = \frac{s_X^2}{\sqrt{s_X^2 * s_X^2}}$$

Using this formula, you can see that the values on the main diagonal are all values of 1:

$r_{xx} = 15.80/(15.80 * 15.80)^{1/2} = 1.00$
$r_{yy} = 11.02/(11.02 * 11.02)^{1/2} = 1.00$
$r_{zz} = 15.37/(15.37 * 15.37)^{1/2} = 1.00.$

Thus, the correlation matrix (**R**) for these data is as follows:

		X	Y	Z
	X	1.00		
R =	Y	0.77	1.00	
	Z	0.80	0.71	1.00

The researcher has the option to input raw data, a correlation matrix (with standard deviations), or a covariance matrix. Raw data are required as input when missing data are to be accounted for during estimation, when categorical outcomes are used in a model because different estimators should be used to account for categorical outcomes, and when performing bootstrapping methods during data analysis (e.g., to build confidence intervals around indirect effects).

VARIABLE METRICS (STANDARDIZED VERSUS UNSTANDARDIZED)

Researchers have debated the use of unstandardized or standardized variables. The standardized coefficients are thought to be sample specific and not stable across different samples because of changes in the variance of the variables. The unstandardized coefficients permit an examination of change across different samples. The standardized coefficients are useful, however, in determining the relative importance of each variable to other variables for a given sample. Other reasons for using standardized variables are that variables are on the same scale of measurement, are more easily interpreted, and can easily be converted back to the raw scale metric. SEM software provides the option to produce both unstandardized and standardized output. It is recommended that you report both types of output, especially because standard errors are used to compute the statistical significance of the unstandardized parameter estimates. It is important to note that M*plus* does calculate standard errors associated with standardized parameter estimates, which are also tested for statistical significance. It is still recommended, however, that you report both unstandardized and standardized parameter estimates, the unstandardized standard errors, and respective statistical significance results for the unstandardized parameter estimates when using M*plus*.

CORRECTION FOR ATTENUATION

A basic assumption in psychometric theory is that observed data contain measurement error. A test score (observed data) is a function of a true score and measurement error. A Pearson correlation coefficient will have different values depending on whether it was computed with observed scores or the true scores where measurement error has been removed. The Pearson correlation coefficient can be corrected for attenuation or unreliable measurement error in scores, thus yielding a true score correlation; however, the corrected correlation coefficient can become greater than 1.0. Low reliability in the independent and/or dependent variables, coupled with a high correlation between the independent and dependent variable, can result in correlations greater than 1.0. For example, given a correlation of $r = .90$ between the observed scores on X and Y, the Cronbach alpha reliability coefficient of .60 for X scores, and the Cronbach alpha reliability coefficient of .70 for Y scores, the Pearson correlation coefficient, corrected for attenuation (r^*), is greater than 1.0:

$$r^*_{xy} = \frac{r_{xy}}{\sqrt{r_{xx}r_{yy}}} = \frac{.90}{\sqrt{.60(.70)}} = \frac{.90}{.648} = 1.389$$

When this happens, the parameter estimates will not be trustworthy, if estimated. Thus, it is recommended that measurement error of the observed variables used in models be a consideration prior to data collection. Moreover, latent variable modeling can be used when feasible, which will take measurement error into account.

MULTIPLE REGRESSION EXAMPLE

It is important to understand the basic concepts of correlation and multiple regression because they provide a better understanding of hypothesis testing, prediction, and explanation of a dependent variable. A review of multiple regression techniques also helps us to better understand path analysis, and structural equation modeling in general. An example is presented next to further clarify these basic multiple regression computations and interpretations.

A regression model is a theoretical, observed variable path model specified by the researcher. Therefore, the *model specification* involves finding relevant theory and prior research to formulate a theoretical regression model. Model specification directly involves deciding which variables to include or not to include in the theoretical regression model. If the researcher does not select the right variables, then the regression model could be misspecified and lack validity (Tracz, Brown, & Kopriva, 1991). Misspecified models may result in biased parameter estimates or estimates that are systematically different from what they are in the true population model. This bias is a result of *specification error*.

The multiple linear regression example analysis is conducted using the same data from Chatterjee and Yilmaz (1992). The multiple regression model of theoretical interest in our example is to predict the *Satisfaction Level* of patients based on their *Age*, *Severity of Illness*, and *Level of Anxiety* (independent variables). The diagram of the implied regression model is shown in Figure 3.10. The curved arrows indicate covariances or correlations between the observed independent variables. The lines pointing toward *Satisfaction Level* indicate the direct paths for regression weights. The oval circle containing error indicates unexplained variance in *Satisfaction Level* (i.e., $1 - R^2$) not accounted for by the three independent predictor variables, sometimes referred to as the *coefficient of multiple alienation*.

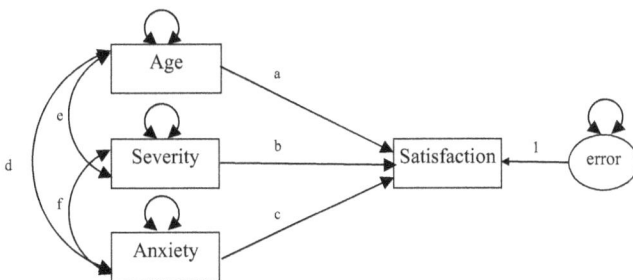

■ **Figure 3.10:** SATISFACTION MULTIPLE REGRESSION MODEL

Once a theoretical regression model is specified, the next concern is *model identification*. Model identification refers to deciding whether a set of unique parameter estimates can be computed for the regression equation. Algebraically, every free parameter in the multiple regression equation can be estimated from the sample covariance matrix. A free parameter is an unknown parameter that you want to estimate in the model. The number of distinct or non-redundant values in the sample covariance matrix equals the number of parameters to be estimated for simple and multiple linear regression models. For example, below is the sample covariance matrix of the four variables for the satisfaction multiple regression example:

$$
\begin{array}{llll}
91.384 & & & \\
30.641 & 27.288 & & \\
0.584 & 0.641 & 0.100 & \\
-122.616 & -52.576 & -2.399 & 281.210
\end{array}
$$

As seen in the above variance–covariance matrix, there are *four* variances in the main diagonal and *six* non-redundant covariances in the off-diagonal of the sample covariance matrix for four observed variables. As shown in Figure 3.10, the regression model includes the variances of the 3 independent variables (indicated by the three circular two-headed arrows associated with each of the independent variables), the 3 covariances (indicated by the three double-headed arrows connecting pairs of the independent variables: *d, e, f*), the 3 partial regression weights for the independent variables (indicated by the one-headed arrows pointing from each independent variable to the dependent variable: *a, b, c*), and 1 error variance (represented by the circular two-headed arrow associated with error). Because the number of unique values in the sample variance–covariance matrix (4 variances + 6 covariances = 10) is equal to the number of parameters to estimate in the regression model (4 variances + 3 covariances + 3 direct effects = 10), this model, and generally all multiple regression models, are considered *just-identified models*. The number of unique values in the sample covariance matrix is also referred to as pieces of information. We will discuss model identification for different types of models in more detail in subsequent chapters.

Given raw data, two different approaches are possible: (a) read in the raw data file; or (b) compute a correlation or covariance matrix for input into the software. For this example, we chose to compute and input a covariance matrix into the software program. SEM software does not output all of the same information and related diagnostic results for multiple regression that you may be used to viewing in SAS, SPSS, STATA, etc. For example, SEM software does not provide an overall *F* test for the multiple regression model.

In LISREL, we can write a SIMPLIS program to compute the regression weights in the regression model. The SIMPLIS program includes a TITLE command; an OBSERVED VARIABLE command to specify variable names and the order of variables; a SAMPLE SIZE command; and a COVARIANCE MATRIX command. The EQUATION command specifies the regression equation with the dependent variable on the left-hand side of the equation. The NUMBER OF DECIMALS and PATH DIAGRAM commands are optional. The END OF PROBLEM command ends the program. In LISREL, covariates or predictor variables in the regression model will be correlated by default.

The SIMPLIS program commands can be saved in a file (***satisfaction.spl***) and then run in LISREL by clicking the "Run LISREL" button. The basic program setup is:

```
Regression Analysis Example (without intercept term)
Observed variables: Age Severity Anxiety Satisfac
Sample size: 24
Covariance matrix:
91.384
30.641 27.288
0.584   0.641 0.100
-122.616 -52.576 -2.399 281.210
Equation: Satisfac = Age Severity Anxiety
Number of decimals = 3
Path Diagram
End of Problem
```

The abbreviated regression output without an *intercept* term in the regression equation is:

```
Satisfac =  - 1.153*Age  - 0.267*Severity  - 15.546*Anxiety,  Errorvar.= 88.515,  R² = 0.685
Standerr     (0.273)        (0.531)           (7.058)                       (27.316)
Z-values    -4.231         -0.503            -2.203                          3.240
P-values     0.000          0.615             0.028                          0.001

                        Goodness-of-Fit Statistics

Degrees of Freedom for (C1)-(C2)                    0
Maximum Likelihood Ratio Chi-Square (C1)          0.0 (P = 1.0000)
Browne's (1984) ADF Chi-Square (C2_NT)            0.0 (P = 1.0000)

              The Model is Saturated, the Fit is Perfect !
```

The partial regression weights are listed in front of each independent variable (*Age*, *Severity*, *Anxiety*). Thus, as a person's *Age* increases by one year, their *Satisfaction* score is estimated to decrease by approximately 1.15 points, while controlling for

Severity and *Anxiety*. Below each regression weight is their standard error in parenthesis. For example, the regression weight associated with *Age* has a standard error of .273; with the *z*-value indicated below that, and a *p*-value listed below the *z*-value. The *z*-value is computed as the parameter estimate divided by the standard error ($z = -1.153/.273$). If testing each regression weight at the critical $z = 1.96$, $\alpha = .05$ level of significance, then *Age* and *Anxiety* are statistically significant, but *Severity* is not ($z = -.503$). The R^2 for *Satisfaction* is .685, meaning that approximately 69% of the variability in *Satisfaction* scores is explained by *Age*, *Severity*, and *Anxiety*. This example is further explained in Jöreskog and Sörbom (1993, pp. 1–6).

Model testing involves determining the fit of the theoretical regression model. As seen in the abbreviated LISREL output, goodness-of-fit statistics are provided. As previously discussed, this model is just-identified, also called *saturated*, and has zero degrees of freedom. Because this model is saturated, it will reproduce the data in the covariance matrix and fit the data perfectly. Specifically, there are as many pieces of information (variances and non-redundant covariances in the variance–covariance matrix) as there are things to estimate in the model.

In M*plus*, we can write an input program to compute the regression weights in the regression model. The M*plus* program includes a TITLE command; a DATA command to specify the location and data file name; a VARIABLE command to specify variable names; a MODEL command to specify the regression model; and an OUTPUT command to request different types of output. The variance–covariance matrix was saved as ***satisfaction.dat*** and should look like the matrix shown in Figure 3.11.

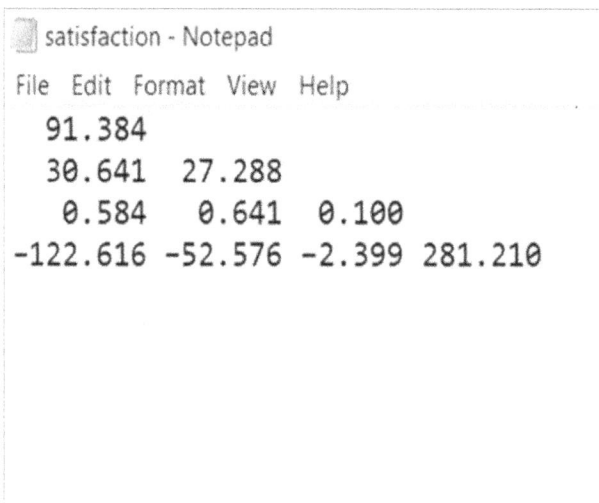

```
satisfaction - Notepad

File  Edit  Format  View  Help
   91.384
   30.641   27.288
    0.584    0.641   0.100
 -122.616  -52.576  -2.399  281.210
```

▦ **Figure 3.11:**   COVARIANCE MATRIX IN ***SATISFACTION.DAT***

Subcommands are needed in the DATA command in order to read in the correct type of summary data. **Type is Covariance** indicates that the summary data consists of the lower triangle of the covariance matrix (see Figure 3.11). **Nobservations** indicates the number of observations or sample size ($n = 24$). In the VARIABLE command, the user must specify the order in which to read the variables as saved in the data file using the **Names are** subcommand. In the MODEL command, it is specified that *Satisfaction* is regressed ON *Age*, *Severity*, and *Anxiety*. **Stdyx** in the OUTPUT command requests the fully standardized output to be printed, including R^2 (only unstandardized output is printed out by default in M*plus*). Just as in LISREL, covariates or predictor variables in the regression model will be correlated by default in M*plus*.

In M*plus*, three types of standardizations are available. When specifying STDYX in the OUTPUT command, the standardization uses the variances of background variables and outcome variables, as well as variances of continuous latent variables (when applicable). When specifying STDY in the OUTPUT command, the standardization uses the variances of outcome variables, as well as variances of continuous latent variables (when applicable). When specifying STD in the OUTPUT command, the standardization uses the variances of continuous latent variables (when applicable). STDYX standardization is recommended with continuous covariates whereas STDY standardization is recommended with binary covariates.

Semicolons are needed after lines of code in M*plus*. The code in the M*plus* input file should look as follows:

```
TITLE:     Regression Analysis Example (without intercept term)
DATA:      FILE IS C:\Chapter 3\Satisfaction.Dat;
           Type is Covariance;
           Nobservations=24;

VARIABLE:  Names are Age Severity Anxiety Satisfac;

MODEL:     Satisfac on Age Severity Anxiety;

OUTPUT:    Stdyx;
```

The input file was saved as ***satisfaction.inp*** and run in M*plus* using the "RUN" button. The unstandardized M*plus* output without an *intercept* term in the regression equation is:

```
MODEL RESULTS
                                                    Two-Tailed
                     Estimate      S.E.    Est./S.E.  P-Value
SATISFAC ON
   AGE                -1.153      0.255     -4.523     0.000
   SEVERITY           -0.267      0.497     -0.538     0.591
   ANXIETY           -15.545      6.602     -2.355     0.019

Residual Variances
   SATISFAC           84.825     24.487      3.464     0.001
```

The values in the column labeled "Estimate" are unstandardized parameter estimates; the values in the column labeled "S.E." are standard errors associated with the unstandardized parameter estimates; the values in the column labeled "Est./S.E." are z values associated with the unstandardized parameter estimate; and the values in the column labeled "Two-Tailed P-Value" are the p-values associated with testing the unstandardized parameter estimate against the value of zero. You will notice that the M*plus* parameter estimates are the same as those from LISREL, with the exception of the residual variance associated with *Satisfaction*. Error variances associated with endogenous variables are labeled as residual variances in M*plus*. The standard errors are slightly different across the two programs, which result in slightly different z values and *p*-values. M*plus* assumes that the data are from the population and uses N instead of $n-1$ when computing variance estimates. Accordingly, there will be subtle differences between M*plus* and other SEM software that use $n-1$ in their computations.

The standardized M*plus* output for the regression equation is:

```
STDYX Standardization
                                                    Two-Tailed
                     Estimate      S.E.    Est./S.E.  P-Value
SATISFAC ON
   AGE                -0.657      0.134     -4.898     0.000
   SEVERITY           -0.083      0.158     -0.526     0.599
   ANXIETY            -0.293      0.130     -2.262     0.024

Residual Variances
   SATISFAC            0.315      0.109      2.897     0.004

R-SQUARE
   Observed                                          Two-Tailed
   Variable          Estimate      S.E.    Est./S.E.  P-Value

   SATISFAC            0.685      0.109      6.306     0.000
```

Thus, as *Age* increases by one standard deviation, *Satisfaction* is estimated to decrease by .657 standard deviations, controlling for *Severity* and *Anxiety*. The R^2 value in M*plus* is the same as that in LISREL ($R^2 = .685$).

Because the R^2 value is not 1.0 (perfect explanation or prediction), additional variables could be added if additional research indicated that another variable was relevant to a patient's satisfaction level. For example, the number of psychological assessment visits could be added as an additional predictor variable. The unexplained error variance (88.515 and 84.825 in LISREL and M*plus*, respectively) was statistically significant, indicating a statistically significant amount of unexplained variance in *Satisfaction* scores. The standardized residual variance estimate in M*plus* is .315, indicating that there is approximately 32% [$1 - R^2 = 1 - .685 = .315$] of the variance in *Satisfaction* scores unexplained. Obviously, more variables can be added in the model modification process, but a theoretical basis should be established by the researcher for the additional variables.

SEM software can output both standardized and unstandardized parameter estimates. Both should be reported. In order to request standardized output in the SIMPLIS program, you would simply modify the code by including the following statement: **Options: SC**. The new SIMPLIS program looks as follows:

```
Regression Analysis Example (without intercept term)
Observed variables: Age Severity Anxiety Satisfac
Sample size: 24
Covariance matrix:
 91.384
 30.641 27.288
 0.584   0.641 0.100
 -122.616 -52.576 -2.399 281.210
Equation: Satisfac = Age Severity Anxiety
Number of decimals = 3
Options: SC
Path Diagram
End of Problem
```

Below is the standardized solution from LISREL:

```
Standardized Solution

        GAMMA
                    Age      Severity    Anxiety
                 --------    --------   --------
Satisfac           -0.657      -0.083     -0.293
```

You can see that the standardized estimates (under the Gamma output) match those from M*plus*.

Figure 3.12 is the diagram generated by LISREL after running the SIMPLIS program. The estimates are the ten unstandardized estimates for the multiple regression model, including the 4 variances, the 3 direct effects, and the 3 covariances. Using the View menu options, standardized estimates as well as the *t*-values associated with the unstandardized parameter estimates can be displayed in the diagram.

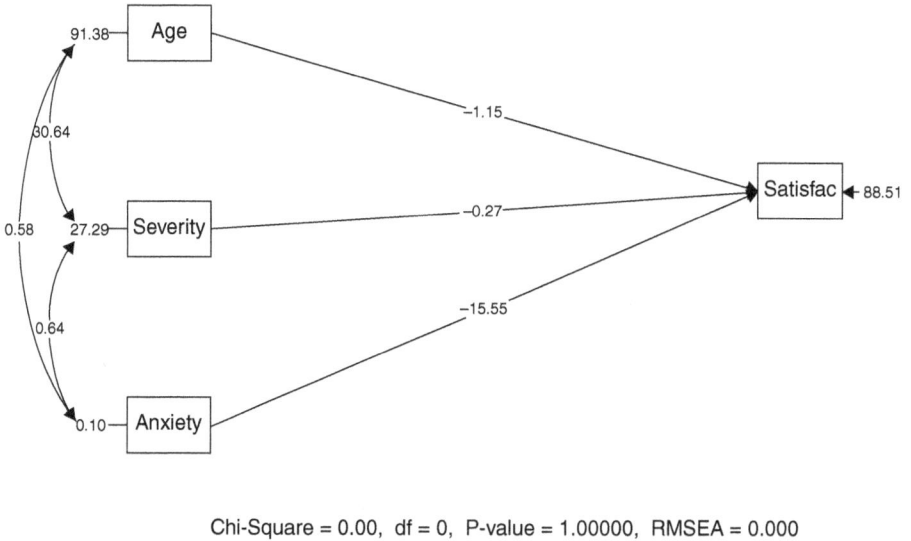

Chi-Square = 0.00, df = 0, P-value = 1.00000, RMSEA = 0.000

▦ **Figure 3.12:** LISREL DIAGRAM FOR SATISFACTION MULTIPLE REGRESSION

In the next chapter, we will extend these multiple regression models to models with more than one observed dependent variable and/or observed intervening variables.

SUMMARY

We have described some of the basic correlation and regression concepts underlying structural equation modeling. We introduced the various types of bivariate correlation coefficients that may be calculated depending upon the variables involved in the bivariate relationship. The factors that affect the Pearson correlation coefficient, including nonlinearity, missing data, level of measurement, restriction of range in the scores, non-normality, and outliers, were briefly illustrated. The foundations of simple and multiple linear regression were presented. Part and partial correlations were defined and illustrated using an example. The different types of relationships among predictor variables in multiple regression models were briefly introduced, such as multicollinearity and suppression.

Regression models are considered saturated or just-identified models because each of the variables is connected with the remaining variables in the model, either by a covariance and/or a direct effect. Thus, multiple regression models fit the data perfectly. A multiple regression analysis was conducted in LISREL and in M*plus*. We introduced the basic elements to include in SIMPLIS and M*plus* code, which will be extended when using models presented in subsequent chapters.

CHAPTER FOOTNOTE

Regression Model with Intercept Term

In both LISREL and M*plus*, when using summary data (e.g., covariance matrix) without also providing the means of the variables, the analyses will not include a mean structure. In the case of multiple regression, the constant or Y-intercept will not be estimated without knowledge of the observed variable means. To estimate the constant or Y-intercept, you will need to include the observed variable means along with the summary data. Including the mean structure in SEM models will not generally impact model estimates or model fit. We discuss the mean structure in more detail in subsequent chapters.

In SIMPLIS, the following code will include means in the summary information and will estimate a constant or Y-intercept (called CONST in LISREL) in the multiple regression model:

```
Regression Analysis Example (with intercept term)
Observed variables: Age Severity Anxiety Satisfac
Sample size: 24
Means: 40.583 51.375 2.283 60.583
Covariance matrix:
 91.384
 30.641 27.288
 0.584   0.641 0.100
 -122.616 -52.576 -2.399 281.210
Equation: Satisfac = Const Age Severity Anxiety
Number of decimals = 3
Options: SC
Path Diagram
End of Problem
```

You will notice that Const was included in the equation and will now be estimated.

In M*plus*, the mean structure will be estimated with raw data or when including means with summary data. The data file to be used with M*plus* will look as follows:

```
40.583 51.375 2.283 60.583
  91.384
  30.641   27.288
   0.584    0.641   0.100
-122.616 -52.576 -2.399 281.210
```

The means will be on the first row of the data file with the covariance matrix immediately below the row of means. These data are saved in the ***satisfaction_ means.dat*** file. The M*plus* code with the mean structure included is as follows:

```
TITLE:     Regression Analysis Example (with intercept term)

DATA:      FILE IS C:\Chapter 3\Satisfaction_means.Dat;
           Type is Covariance Means;
           Nobservations=24;

VARIABLE:  Names are Age Severity Anxiety Satisfac;

MODEL:     Satisfac on Age Severity Anxiety;

OUTPUT:    stdyx;
```

You will notice that "Type is Covariance Means" is specified in the DATA command. M*plus* reads the data files from the bottom up. That is, the covariance is immediately below the row of means.

EXERCISES

1. Given the Pearson correlation coefficients $r_{12} = .6$, $r_{13} = .7$, and $r_{23} = .4$, compute the partial and part correlations: $r_{12.3}$ and $r_{1(2.3)}$, respectively.
2. Compare the variance explained in the bivariate, partial, and part correlations in Exercise 1.
3. Given the following covariance matrix, compute the following Pearson correlation coefficients: r_{XY}, r_{XZ}, and r_{YZ}:

	X	Y	Z
X	15.80		
Y	10.16	11.02	
Z	12.43	9.23	15.37

4. The covariance matrix for the hypothesized regression model in Figure 3.13 is provided in Table 3.8. The theoretical regression model specifies that the dependent variable, gross national product (GNP), is predicted by labor, capital, and time (three independent variables).
   a. How many distinct or non-redundant elements are in the sample covariance matrix?
   b. How many parameters need to be estimated in the model?
   c. Write the code in LISREL or M*plus* to estimate parameters for the multiple regression model.

Table 3.8: Covariance Matrix ($n = 23$)

Covariance Matrix				
GNP	4256.530			
Labor	449.016	52.984		
Capital	1535.097	139.449	1114.447	
Time	537.482	53.291	170.024	73.747

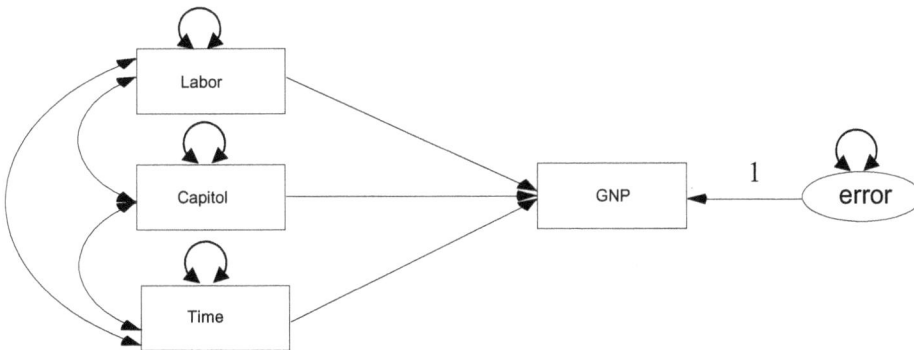

■ **Figure 3.13:** GNP MULTIPLE REGRESSION MODEL

REFERENCES

Anderson, C., & Schumacker, R. E. (2003). A comparison of five robust regression methods with ordinary least squares regression: Relative efficiency, bias, and test of the null hypothesis. *Understanding Statistics, 2*, 77–101.

Boomsma, A. (1983). *On the robustness of LISREL against small sample size and nonnormality.* Amsterdam: Sociometric Research Foundation.

Browne, M. W. (1982). *Covariance structures.* In D. M. Hawkins (Ed.), *Topics in applied multivariate analysis* (pp. 72–141). Cambridge: Cambridge University Press.

Chatterjee, S., & Yilmaz, M. (1992). A review of regression diagnostics for behavioral research. *Applied Psychological Measurement, 16*, 209–227.

Cohen, J., & Cohen, P. (1983). *Applied multiple regression/correlation analysis for the behavioral sciences* (2nd ed.). Hillsdale, NJ: Lawrence Erlbaum Associates.

Crocker, L., & Algina, J. (1986). *Introduction to classical and modern test theory.* New York: Holt, Rinehart & Winston.

Ferguson, G. A., & Takane, Y. (1989). *Statistical analysis in psychology and education* (6th ed.). New York: McGraw-Hill.

Hinkle, D. E., Wiersma, W., & Jurs, S. G. (2003). *Applied statistics for the behavioral sciences* (5th ed.). Boston: Houghton Mifflin.

Ho, K., & Naugher, J. R. (2000). Outliers lie: An illustrative example of identifying outliers and applying robust methods. *Multiple Linear Regression Viewpoints, 26*(2), 2–6.

Houston, S. R., & Bolding, J. T., Jr. (1974). Part, partial, and multiple correlation in commonality analysis of multiple regression models. *Multiple Linear Regression Viewpoints, 5*, 36–40.

Huber, P. J. (1981). *Robust statistics.* New York: Wiley.

Jennrich, R. I., & Thayer, D. T. (1973). A note on Lawley's formula for standard errors in maximum likelihood factor analysis. *Psychometrika, 38*, 571–580.

Jöreskog, K. G., & Sörbom, D. (1993). *LISREL8: Structural equation modeling with the SIMPLIS command language.* Chicago, IL: Scientific Software International.

Lawley, D. N., & Maxwell, A. E. (1971). *Factor analysis as a statistical method.* London: Butterworth.

Lomax, R. G. & Hahs-Vaughn, D.L. (2012). *An introduction to statistical concepts* (3rd ed.). New York: Routledge Press (Taylor & Francis Group).

Maassen, G. H., & Bakker, A. B. (2001). Suppressor variables in path models: Definitions and interpretations. *Sociological Methods & Research, 30*(2), 241–270.

Pearson, K. (1896). Mathematical contributions to the theory of evolution. Part 3. Regression, heredity and panmixia. *Philosophical Transactions, A, 187*, 253–318.

Pedhazur, E. J. (1982). *Multiple regression in behavioral research: Explanation and prediction* (2nd ed.). New York: Holt, Rinehart, & Winston.

Pedhazur, E. J. (1997). *Multiple regression in behavioral research: Explanation and prediction* (3rd ed.). Fort Worth: Harcourt Brace.

Rousseeuw, P. J., & Leroy, A. M. (1987). *Robust regression and outlier detection.* New York: Wiley.

Spearman, C. (1904). The proof and measurement of association between two things. *American Journal of Psychology*, 15, 72–101.

Staudte, R. G., & Sheather, S. J. (1990). *Robust estimation and testing.* New York: Wiley.

Stevens, S. S. (1968). Measurement, statistics, and the schempiric view. *Science*, 161, 849–856.

Tankard, J. W., Jr. (1984). *The statistical pioneers.* Cambridge, MA: Schenkman.

Tracz, S. M., Brown, R., & Kopriva, R. (1991). Considerations, issues, and comparisons in variable selection and interpretation in multiple regression. *Multiple Linear Regression Viewpoints*, 18, 55–66.

Chapter 4

PATH MODELS

PATH MODEL DIAGRAMS

Path models adhere to certain common drawing conventions that are utilized in SEM models. These drawing conventions were introduced in Chapter 1 (see Figure 1.1). Specifically, observed variables are enclosed by rectangles, and errors of prediction are indicated by oval shapes. Lines drawn from one observed variable to another observed variable denote *direct effects* or the direct influence of one variable on another. A two-headed arrow between two independent observed variables indicates *covariance* or correlation. The covariance among independent observed variables also suggests that such variable relations are influenced by other variables exogeneous to or external to the path model. Because those influences are not studied in the path model, it is reasonable to expect that some unmeasured variables may influence the independent variables. Each dependent variable in a path model requires an error term, which is denoted by an oval around the error term that points toward the dependent variable. The error term, which is latent or unobserved, indicates the unexplained error variance of each dependent variable.

Model specification is necessary in examining multiple variable relations in path models, just as in the case of multiple regression. Many different relations among a set of variables can be hypothesized with many different parameters being estimated. In a simple three-variable model, for example, many possible path models can be postulated on the basis of different hypothesized relations among the three variables. For example, in Figure 4.1, we see three different path models where X1 influences X2. In the first model, X1 influences X2, which in turn influences Y. Here, X2 serves as a mediator between X1 and Y. In the second model, an additional path is drawn from X1 to Y, such that X1 has both a direct and an indirect effect upon Y. The direct effect is represented by the one-headed arrow from X1 to Y (no variables intervene between X1 and Y in this connection), whereas the indirect effect is represented by X1 directly influencing X2 which

DOI: 10.4324/9781003044017-4

subsequently directly influences Y (X2 intervenes between X1 and Y in this connection). In the third model, X1 influences both X2 and Y; however, X2 and Y are not related directly or indirectly. If we were to switch X1 and X2 around, this would generate three more plausible path models. Other path models are also possible. For example, in Figure 4.2, X1 does *not* influence X2. In the first model, X1 and X2 both influence Y, but are uncorrelated. In the second model, X1 and X2 both influence Y and they are correlated, representing a multiple regression model with two predictors.

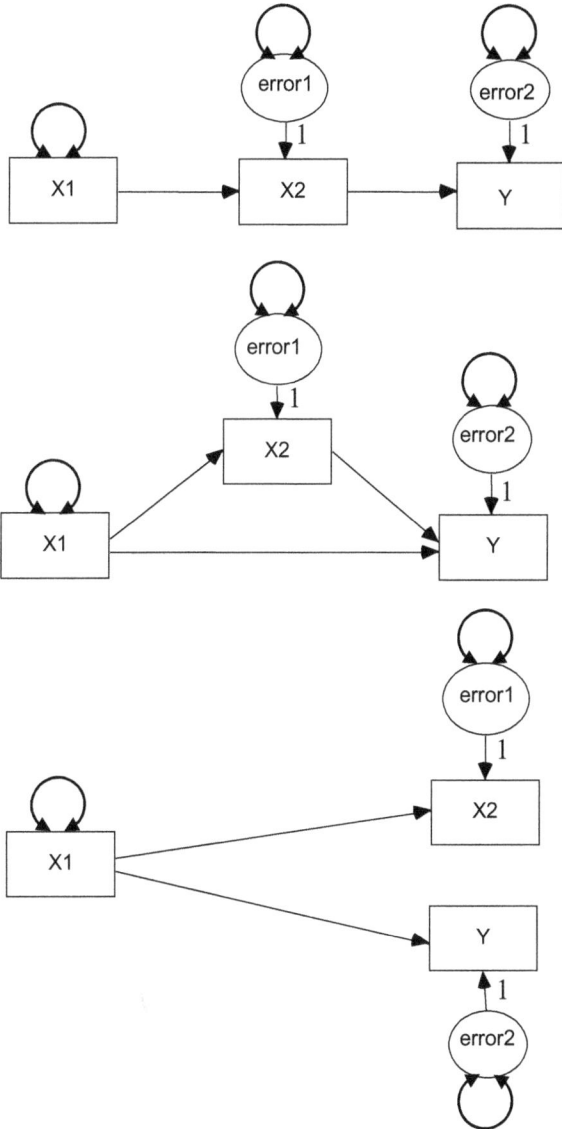

Figure 4.1: POSSIBLE THREE-VARIABLE MODELS (WHERE X1 INFLUENCES X2)

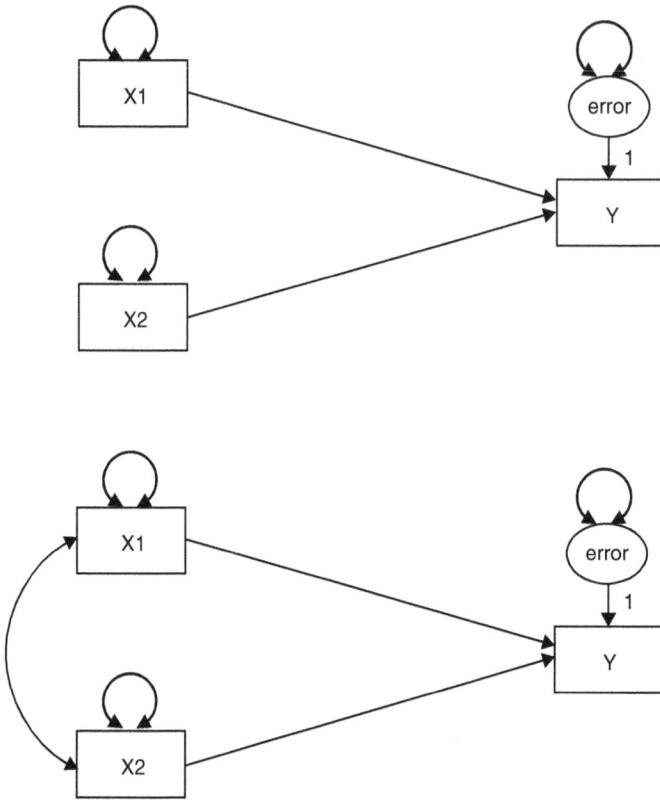

■ **Figure 4.2:** POSSIBLE THREE-VARIABLE MODELS (WHERE X1 DOES NOT INFLUENCE X2)

How can one determine which model is correct? *Model specification* involves using theory, previous research, and temporal considerations to justify variable relations in a hypothesized path model. Path analysis does not provide a way to specify the model, but rather estimates the effects of the variables once the model has been specified by the researcher on the basis of theoretical considerations. For this reason, model specification is a critical part of SEM modeling.

DECOMPOSITION OF THE CORRELATION MATRIX

Traditional path analysis progressed along a correlational track. That is, coefficients in path models were originally derived from the values of a Pearson product-moment correlation coefficient and/or a standardized partial regression coefficient (Wright, 1918, 1934; Wolfle, 1977). For example, in the standardized path model in Figure 4.3, the path coefficients (p) are depicted by arrows from X1 to Y and X2 to Y, respectively, as:

$$\beta 1 = p1Y$$
$$\beta 2 = p2Y$$

and the curved arrow between X1 and X2 is denoted as:

$$r_{12} = p_{12}.$$

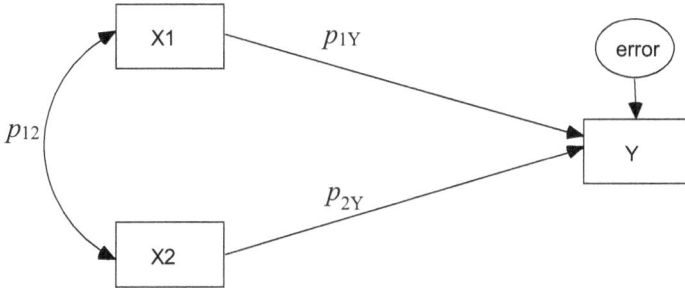

■ **Figure 4.3:** MULTIPLE REGRESSION PATH MODEL WITH PATH COEFFICIENTS

The direct effects, when specified in standard score form, become standardized partial regression coefficients. In multiple regression, a dependent variable is regressed in a single analysis on all of the independent variables. In path analysis, one or more regression equations are analyzed depending on the variable relations specified in the path model. Path coefficients are therefore computed only on the basis of the particular set of independent variables that lead to the dependent variable under consideration. In the path model in Figure 4.3, two standardized partial regression coefficients (path coefficients) are computed, p_{1Y} and p_{2Y}. The curved arrow represents the correlation between the two independent variables, p_{12}.

In path models, estimation of parameters permits a decomposition of the correlation matrix using *path tracing*. Wright (1934) proposed path tracing rules to follow in order to correctly compute the correlation between two variables based on the relationship between the variables hypothesized in the path model:

1. Once you move forward on an arrow in the model, you cannot then move backward on an arrow.
2. You are not allowed to pass through the same variable more than once.
3. Only one two-headed arrow is allowed in a compound path.

A compound path consists of single path coefficients (p_{12}) or products of path coefficients ($p_{12}p_{1Y}$) used to connect two variables. When determining the relationships among two variables in a model, they can be comprised of *direct*, *indirect*, and *spurious* effects. Summing the direct and indirect effects results in what is called a *total effect;* and summing direct, indirect, and spurious effects results in what is called the *total correlation.* Direct effects consist of a single one-headed arrow connecting two variables $(X \rightarrow Y)$; indirect effects consist of two or more one-headed arrows going in the same direction with one or more intervening variables, respectively $(X \rightarrow Y \rightarrow Z)$; and spurious effects consist of two-headed arrows $(X1 \leftrightarrow X2)$ or other legal paths in a connection. Spurious associations consist of relationships between two variables with no *causal* relation. For instance, the connection between X2 and Y in the third model in Figure 4.1 is a spurious relationship. X2 and Y are spuriously related because X1 directly influences both variables. The connection is also consistent with the path tracing rules because you can go backward on an arrow and then go forward on an arrow when connecting X2 and Y $(X2 \leftarrow X1 \rightarrow Y)$. If the arrows were reversed, the resulting connection would violate the path tracing rule by going forward and then backward (e.g., $X2 \rightarrow X1 \leftarrow Y$).

The original correlation matrix can be completely reproduced if all of the parameters in a path model are specified, resulting in a just-identified model. For example, take the path model in Figure 4.3 where X1 and X2 predict Y and are intercorrelated. We can decompose all pairwise relationships based on this hypothesized model. For instance, given the model in Figure 4.3, how are the variables X1 and Y connected? There is a direct effect from X1 to Y (p_{1Y}). There is also a connection from X1 to Y because of the correlation between X1 and X2. Specifically, X1 is spuriously connected to Y through its correlation with X2 (p_{12}) *and* the direct influence of X2 to Y (p_{2Y}). Thus, the total correlation for X1 and Y (r_{1Y}) is $p_{1Y} + p_{12}p_{2Y}$. Given the model in Figure 4.3, how are the variables X2 and Y connected? There is a direct effect from X2 to Y (p_{2Y}). There is also a connection from X2 to Y because of the correlation between X1 and X2. That is, X2 is spuriously connected to Y through its correlation with X1 (p_{12}) and the direct influence of X1 to Y (p_{1Y}). Thus, the total correlation for X2 and Y (r_{2Y}) is $p_{2Y} + p_{12}p_{1Y}$. Lastly, how are X1 and X2 connected given the model in Figure 4.3? They are connected spuriously because they are intercorrelated. Thus, the total correlation for X1 and X2 (r_{12}) is p_{12}.

Suppose that the observed correlations are as follows: $r_{12} = .224$, $r_{1Y} = .507$, and $r_{2Y} = .480$. We can use Table 4.1 below to organize the decompositions for this path model.

Table 4.1: Decomposition Table for Multiple Regression Path Model

Connection	Direct Effect	Indirect Effect	Total Effect	Spurious Association	Total Correlation	Observed Correlation
X1, Y (r_{1Y})	p_{1Y}	–	p_{1Y}	$p_{12}p_{2Y}$	$p_{1Y} + p_{12}p_{2Y}$	.507
X2, Y (r_{2Y})	p_{2Y}	–	p_{2Y}	$p_{12}p_{1Y}$	$p_{2Y} + p_{12}p_{1Y}$	.480
X1, X2 (r_{12})	–	–	–	p_{12}	p_{12}	.224

If you recall from Chapter 3, multiple regression models (as with the model in Figure 4.3) are just-identified. Because this model is just-identified, we can actually solve for the path coefficients and reproduce the observed correlations. To do this, we will first set the *total correlation* equal to the *observed correlation* as:

$$r_{1Y} = p_{1Y} + p_{12}\,p_{2Y} = .507$$

$$r_{2Y} = p_{2Y} + p_{12}\,p_{1Y} = .480$$

$$r_{12} = p_{12} = .224$$

The r values are the actual observed correlations and the p values are the path coefficients (standardized estimates).

When discussing identification of multiple regression models in Chapter 3, we included variance information in terms of parameters requiring estimation. With standardized data, the variances of the variables are equal to 1 and not considered in terms of estimation, making path tracing simpler. Nonetheless, model identification can be determined similarly with standardized models as with unstandardized models. Specifically, we have three correlations and we have three path coefficients to solve for (p_{1Y}, p_{2Y}, and p_{12}). The number of correlations equals the number of path coefficients to solve, signifying that this is a just-identified model.

Now we need to solve for the path coefficients. The correlation between X1 and X2 is already solved for us ($r_{12} = p_{12} = .224$)! The remaining path coefficients require some algebra:

First, we can substitute p_{12} and then isolate p_{1Y} as follows:

$$r_{1Y} = p_{1Y} + (.224)p_{2Y} = .507 \quad \rightarrow \quad p_{1Y} = .507 - (.224)(p_{2Y})$$

Next, we can substitute p_{1Y} and simplify:

$$r_{2Y} = p_{2Y} + (.224)p_{1Y} = .480$$

$$r_{2Y} = p_{2Y} + (.224)[.507 - (.224)(p_{2Y})] = .480 \rightarrow p_{2Y} + .113 - (.05)(p_{2Y}) = .480$$

$$.367 = .95(p_{2Y}) \rightarrow p_{2Y} = .386$$

Finally, we can substitute p_{2Y} and simplify:

$$p_{1Y} = .507 - (.224)(p_{2Y})$$
$$p_{1Y} = .507 - (.224)(.386) \rightarrow p_{1Y} = .421$$

Now that we have solved for the path coefficients, it is important to discuss what reproducing the correlation matrix means in the case of just-identified models. We have our sample correlation matrix (**R**):

$$
\begin{array}{ccc}
\text{X1} & \text{X2} & \text{Y}
\end{array}
$$
$$
\begin{bmatrix}
1 & .224 & .507 \\
.224 & 1 & .480 \\
.507 & .480 & 1
\end{bmatrix},
$$

and we have what is called our model-implied correlation matrix based on the hypothesized connections in the model:

$$
\begin{array}{ccc}
\text{X1} & \text{X2} & \text{Y}
\end{array}
$$
$$
\begin{bmatrix}
1 & p_{12} & p_{1Y} + p_{12}p_{2Y} \\
p_{12} & 1 & p_{2Y} + p_{12}p_{1Y} \\
p_{1Y} + p_{12}p_{2Y} & p_{2Y} + p_{12}p_{1Y} & 1
\end{bmatrix}.
$$

Plugging the solved path coefficient values into the model-implied correlation matrix will reproduce the correlation values observed in the sample:

$$
\begin{array}{ccc}
\text{X1} & \text{X2} & \text{Y}
\end{array}
\qquad
\begin{array}{ccc}
\text{X1} & \text{X2} & \text{Y}
\end{array}
$$
$$
\begin{bmatrix}
1 & .224 & .421 + (.224)(.386) \\
.224 & 1 & .386 + (.224)(.421) \\
.421 + (.224)(.386) & .386 + (.224)(.421) & 1
\end{bmatrix}
=
\begin{bmatrix}
1 & .224 & .507 \\
.224 & 1 & .480 \\
.507 & .480 & 1
\end{bmatrix}.
$$

Thus, estimates for just-identified models will be specific values that result in perfectly reproducing the observed correlation matrix.

You could also compute R^2 associated with the standardized model using path tracing rules by attempting to decompose how the dependent variable, Y, varies with itself or how the model explains variance in Y. Using the model in Figure 4.3, you can go backward and forward on p_{1Y} when connecting Y with itself $\left(p_{1Y}^2\right)$. You can also go backward and forward on p_{2Y} when connecting Y with itself $\left(p_{2Y}^2\right)$. Finally, you can travel away from Y by going backward on p_{1Y}, through p_{12}, and forward on p_{2Y} and then you will need to get back to where you started on Y by traveling backward on p_{2Y}, through p_{12}, and forward on p_{1Y} $\left[2\left(p_{1Y}p_{12}p_{2Y}\right)\right]$. Plugging in the solved path coefficient values results in the following R^2 value:

$$\mathrm{R}^2 = p_{1Y}^2 + p_{2Y}^2 + 2\left(p_{1Y}p_{12}p_{2Y}\right) = .421^2 + .386^2 + 2(.421*.224*.386) = .399.$$

Hence, approximately 40% of the variance in Y is explained by X1 and X2 combined, whereas approximately 60% of the variance in Y is unexplained $(1 - R^2 = 1 - .40 = .60)$.

When a variable relation is left out of a path model; for example, leaving p_{12} out in the model in Figure 4.3, then the correlations would not be completely reproduced. Specifically, there would be three correlations with only two path coefficients (p_{1Y} and p_{2Y}) requiring estimation, resulting in what is called an *over-identified* model with 1 degree of freedom (*df*; 3 correlations minus 2 parameters to estimate). A researcher typically does not specify all possible variable relations in a path model based on their theoretical framework. In this case, there will be a difference between the sample correlation matrix and the implied path model correlation matrix. SEM tests the difference between these two matrices using a chi-square test, χ^2 (discussed more in Chapter 5). As such, the goal is to test a hypothesized path model that does not contain all variable relations, yet closely approximates the variable relations given by the sample covariance matrix. Consequently, if we specified all variable relations in a path model, the chi-square test would equal zero ($\chi^2 = 0$). Generally, researchers are not interested in testing just-identified models that specify all variable relations. The correlation decomposition is a useful way to understand the model specification process in path analysis. For further details on the correlation decomposition approach, we recommend reading Duncan (1975).

DECOMPOSITION FOR UNSTANDARDIZED PATH MODELS

Because the distributional properties of variables are important considerations in SEM, we will continue our discussion of models in the textbook with the consideration of unstandardized models that do incorporate information about variable variances. Recall that SEM software programs use the sample covariance matrix during the estimation of parameters as the default. Using the unstandardized version of the model in Figure 4.3, the model in Figure 4.4 now includes circular, double-headed arrows associated with exogenous observed variables (e.g., X1 and X2) and exogenous unobserved variables (i.e., errors) to emphasize that variances are estimated for all exogenous variables in SEM models. Thus, decomposition of the unstandardized path model will need to incorporate variances. Using the model in Figure 4.4, we can determine the elements of the model-implied covariance matrix as we did before with the standardized model, but now with the inclusion of the variances and covariances associated with the exogenous variables in the respective connection.

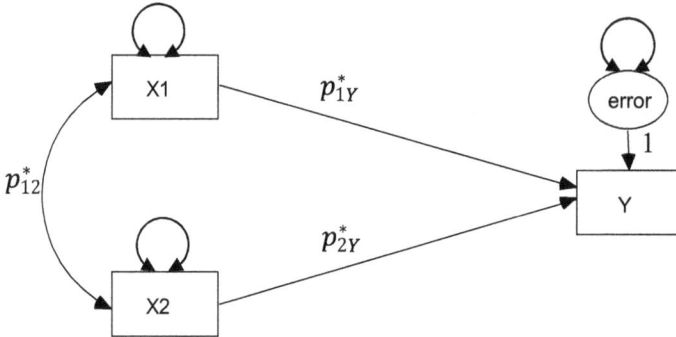

Figure 4.4: MULTIPLE REGRESSION PATH MODEL WITH UNSTANDARDIZED PATH COEFFICIENTS

Since we are transitioning from standardized to unstandardized path models, we will use asterisks to represent unstandardized path coefficients (e.g., p_{12}^*). Given the model in Figure 4.4, why do the variables X1 and Y covary? There is the direct effect from X1 to Y $\left(p_{1Y}^*\right)$ and the spurious connection through the covariance between X1 and X2 and the direct effect of X2 to Y $\left(p_{12}^* p_{2Y}^*\right)$. But now we have to include either a covariance or a variance associated with the exogenous variable in the connection for each compound path. For the direct effect from X1 to Y, the variance associated with X1 is included because it is exogenous in that connection $\left(p_{1Y}^* \sigma_1^2\right)$. The spurious connection already includes the covariance between X1 and X2. Thus, the reason that X1 and Y covary is $p_{1Y}^* \sigma_1^2 + p_{12}^* p_{2Y}^*$. Why do X2 and Y covary? There is the direct effect from X2 to Y $\left(p_{2Y}^*\right)$ and the spurious connection through the covariance between X1 and X2 and the direct effect of X1 to Y $\left(p_{12}^* p_{1Y}^*\right)$. For the direct effect from X2 to Y, the variance associated with X2 is included because it is exogenous in that connection $\left(p_{2Y}^* \sigma_2^2\right)$. Again, the spurious connection already contains the covariance between X1 and X2. Hence, the reason that X2 and Y covary based on the model is $p_{2Y}^* \sigma_2^2 + p_{12}^* p_{1Y}^*$. Finally, X1 and X2 covary because of their covariance $\left(p_{12}^*\right)$.

We now have to determine the elements for the model-implied covariance matrix on the main diagonal, which are the variances of X1, X2, and Y given the model. This will be similar to the method used when calculating R^2 for Y in the standardized model. For each of the variables, we need to figure out what is explaining their variance given the model, or how they vary with themselves. For X1, there are no arrows pointing to it. You could try to move forward on the path directed to Y and then backward, but that violates the path tracing rule (you cannot go backward on a path once you start forward). You could also try to move through the covariance between X1 and X2, forward on the direct effect from X2 to Y, and backward on the path connecting X1 and Y. Again, this would violate the path tracing rule (you cannot go backward once you have started forward). Accordingly, the only thing explaining variability in X1 in the model is the variance associated with X1 (σ_1^2; hence, its circular two-headed arrow). Similarly, because there are no causal influences on X2, the only reason it varies is because it has variance $\left(\sigma_2^2\right)$. There are causal influences on Y. Just as we did before when computing R^2 for Y in the standardized model, you can go backward and forward on p_{1Y}^* when connecting Y with itself $\left(p_{1Y}^{*2}\right)$. You can also go backward and forward on p_{2Y}^* when connecting Y with itself $\left(p_{2Y}^{*2}\right)$. Finally, you can travel away from Y by going backward on p_{1Y}^* through p_{12}^*, and forward on p_{2Y}^*; and then you will need to get back to where you started with Y by traveling backward on p_{2Y}^*, through p_{12}^*, and forward on $p_{1Y}^* \left[2\left(p_{1Y}^* p_{12}^* p_{2Y}^*\right)\right]$. Finally, there is an error associated with Y. Assuming a value of 1 is on the path (this will be explained in more detail later), we can go backward and then forward on the path from error, with the variance of the error included in this compound path because it is exogenous in that connection $(1 * \sigma_{error}^2)$. Without all variances included, this would look as follows: $p_{1Y}^{*2} + p_{2Y}^{*2} + 2\left(p_{1Y}^* p_{12}^* p_{2Y}^*\right) + \sigma_{error}^2$. We need to include a variance or covariance with each compound path. For the compound path p_{1Y}^{*2}, X1 is the exogenous variable in that connection; hence, the compound path will include the variance of X1 $\left(p_{1Y}^{*2}\sigma_1^2\right)$. Similarly, for the compound path p_{2Y}^{*2}, the variance of X2 will be included $\left(p_{2Y}^{*2}\sigma_2^2\right)$.

Now we have the elements for our model-implied or reproduced covariance matrix $(\hat{\Sigma})$ based on the hypothesized connections in the model:

$$
\begin{array}{ccc}
\text{X1} & \text{X2} & \text{Y} \\
\end{array}
$$

$$
\begin{bmatrix}
\sigma_1^2 & p_{12}^* & p_{1Y}^*\sigma_1^2 + p_{12}^* p_{2Y}^* \\
p_{12}^* & \sigma_2^2 & p_{2Y}^*\sigma_2^2 + p_{12}^* p_{1Y}^* \\
p_{1Y}^*\sigma_1^2 + p_{12}^* p_{2Y}^* & p_{2Y}^*\sigma_2^2 + p_{12}^* p_{1Y}^* & p_{1Y}^{*2}\sigma_1^2 + p_{2Y}^{*2}\sigma_2^2 + 2\left(p_{1Y}^* p_{12}^* p_{2Y}^*\right) + \sigma_{error}^2
\end{bmatrix}
$$

The parameters that need to be solved for in the model include: σ_1^2, σ_2^2, σ_{error}^2, p_{1Y}^*, p_{2Y}^*, and p_{12}^*. Suppose we had the following observed sample covariance matrix (**S**):

$$
\begin{array}{ccc}
X1 & X2 & Y
\end{array}
$$

$$
\begin{bmatrix}
20 & 5 & 12.42 \\
5 & 25 & 13.15 \\
12.42 & 13.15 & 30
\end{bmatrix}.
$$

We have six distinct or non-redundant elements in the sample covariance matrix (3 variances and 3 covariances in the lower triangle of the matrix). The values in the model-implied covariance matrix ($\hat{\Sigma}$) will be solved, resulting in perfectly reproducing the observed sample covariance matrix (**S**) because the model is just-identified. The difference between the sample covariance matrix and the model-implied covariance matrix is referred to as the residual covariance matrix, which will have values of zero for each element with just-identified models ($\mathbf{S} - \hat{\Sigma}$).

Again, researchers are not interested in testing just-identified models that specify all variable relations. Instead, researchers would prefer to test a hypothesized path model that does not contain all variable relations. We present an example of an over-identified path model next.

PATH MODEL EXAMPLE

The path model for this example was taken from McDonald and Clelland (1984) who collected data on the sentiments toward unions of Southern non-union textile laborers ($n = 173$). This example was analyzed in the LISREL 8 manual (Jöreskog & Sörbom, 1993, pp. 12–15, example 3) and explained by Bollen (1989, pp. 82–83). The path model consists of five observed variables. The independent/exogenous variables are the number of years worked in the textile mill (actually log of years, denoted simply as *years*) and worker age (*age*). The dependent/endogenous variables are deference to managers (*deference*), support for labor activism (*support*), and sentiment toward unions (*sentiment*). The diagram of the theoretical union sentiment path model is shown in Figure 4.5.

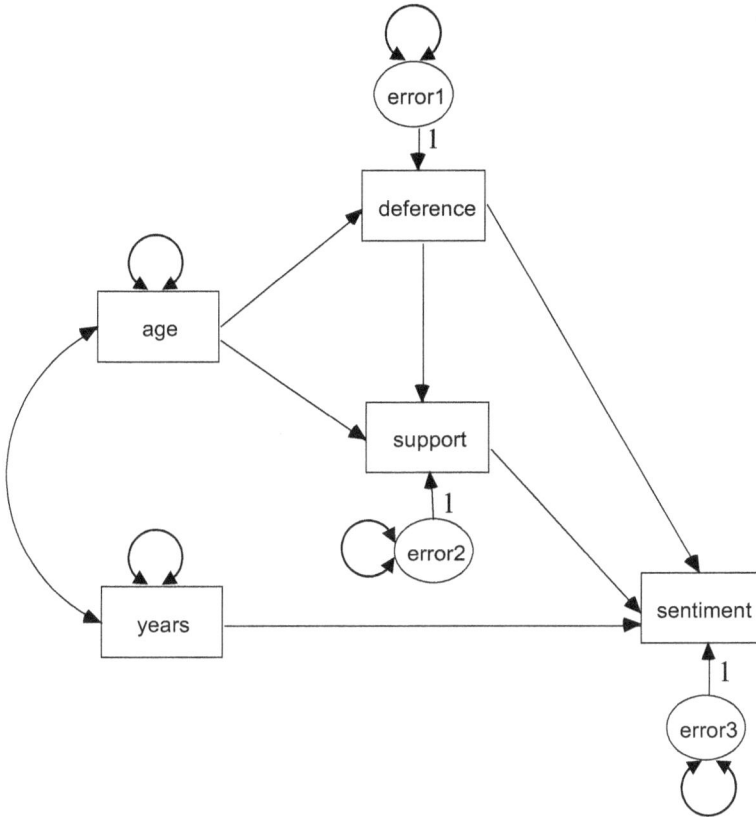

■ **Figure 4.5:** UNION SENTIMENT PATH MODEL

The path model in Figure 4.5 indicates that *age* and *years* covary. *Age* has a direct effect on *deference* and *support*. *Deference* has a direct effect on *support*. *Years*, *deference*, and *support* have a direct effect on *sentiment*. Since *deference*, *support*, and *sentiment* are being predicted, they each have an error term. There are three equations in the model, one for each of the three dependent/endogenous variables (*deference*, *support*, and *sentiment*). Using the variable names, the equations may be written as:

deference = age + error1

support = age + deference + error2

sentiment = years + support + deference + error3.

The hypothesized path model does not contain all possible variable relations, only the ones based on the researchers' theoretical perspective. Many possible path models could be specified for this set of observed variables, but theory guided the *model specification* for testing this hypothesized path model.

The SIMPLIS code for the union sentiment model is provided below:

```
Union Sentiment of Textile Workers
Observed Variables:
Deference Support Sentiment Years Age
Covariance Matrix
14.610
-5.250 11.017
-8.057 11.087 31.971
-0.482 0.677 1.559 1.021
-18.857 17.861 28.250 7.139 215.662
Sample Size = 173
Relationships
Support=Deference Age
Sentiment=Deference Support Years
Deference=Age
Print Residuals
OPTIONS SC
Path Diagram
End of Problem
```

The M*plus* code for the union sentiment model is provided below:

```
TITLE: Union Sentiments Path Model

DATA: FILE IS C:\union.txt;
      Type is Covariance;
      Nobservations=173;

VARIABLE: Names are Deferenc Support Sentimen Years Age;

MODEL: Deferenc on Age;
       Support on Age Deferenc;
       Sentimen on Years Deferenc Support;
       Years with Age;

OUTPUT: Stdyx Residual;
```

The data set used is a covariance matrix, which is included in the SIMPLIS code and is saved as ***union.txt*** for the M*plus* program and is available on the book website. It must be noted that variables names are limited to eight characters in the M*plus* input files to avoid error messages related to the eight-character limit for variable names. Standardized estimates were requested using the **SC** option in the OPTIONS command in LISREL and the **STDYX** statement in the OUTPUT command in M*plus*: Recall from Chapter 3 that M*plus* offers different types of standardized output. The **STDYX** standardization is recommended with continuous covariates whereas the **STDY** standardization is recommended with binary/dichotomous covariates. Because *age* and *years* are continuous covariates, the **STDYX** standardization was used in the M*plus* code. Further, you will notice that the MODEL command in the M*plus* code includes the following statement: Years with Age. **With** statements allow covariances among exogenous variables in M*plus*. This statement will allow the estimated covariance (and correlation) between *years* and *age* to be printed in the output.

Model identification is another concern in path models. The *model identification* problem is whether we can obtain estimates for the path coefficients in the model. Basically, given the sample data (sample covariance matrix) and the theoretical path model, can a unique set of parameter estimates be found? In the union sentiment path model, we have chosen to not estimate some relations between variables, which correspond to parameters. These parameters are fixed. An example of a fixed parameter is when no path or direct relation is specified. For instance, there is no relationship between *age* and *sentiment*, consistent with fixing the relationship between the two variables equal to zero. Free parameters are those to be estimated in the model. An example of a free parameter is the path for *age* predicting *deference*.

In model identification, we consider the *t-rule* or the *counting rule*. The *t-rule* specifies that the number of free parameters to be estimated must be less than or equal to the number of distinct or non-redundant values in the sample covariance matrix. In the union sentiment path model, we have the following free parameter values:

6 path coefficients (direct effects)
3 error variances (for 3 dependent/endogenous variables)
1 covariance among the independent/exogenous variables (between age and years)
2 independent/exogenous variable variances (for age and years)

There are 12 free parameters in the path model. The number of distinct values in the sample covariance matrix is computed by:

$$[p (p + 1)]/2 = [5 (5 + 1)]/2 = 15,$$

where p is the number of observed variables. The number of distinct values in the sample variance–covariance matrix, 15, is greater than the number of free parameters, 12, in the path model. The *t*-rule is satisfied.

The *degrees of freedom* (*df*) for a path model is the difference between the number of distinct values and the number of free parameters. The *df* = 15 − 12 = 3 for this path model. The path model is an *over-identified* model because there are more distinct values in the sample covariance matrix than free parameters to be estimated in the path model. The degrees of freedom are equal to zero for just-identified models and equal to or greater than 1 for over-identified models. Negative degrees of freedom are associated with *under-identified* models, requiring more parameters to estimate in the model than distinct values available in the sample covariance matrix. In the case of an under-identified model, the parameter estimates cannot be estimated. In SEM, we prefer over-identified models rather than saturated or just-identified models, because it allows for model testing. Many of the SEM software programs provide condition codes and error messages when there are problems with model identification. For instance, after running the union sentiment path model in M*plus*, if you do not see the following message in the output, there will be an error message instead:

```
THE MODEL ESTIMATION TERMINATED NORMALLY.
```

We recommend determining whether a hypothesized model is identified prior to data collection. If feasible, we also suggest that the hypothesized model have several degrees of freedom associated with it in case of model modifications that may subsequently be imposed after model testing.

Model estimation involves choosing the estimation method to estimate the parameters in the path model. SEM software programs default to the maximum likelihood (ML) estimation method, but this is not always a wise choice, especially when data are not normally distributed or endogenous variables are ordered categorical variables. The choice of an estimation method can affect whether a parameter estimate is attenuated or statistically significant; that is, the value of the parameter and its associated standard error estimate may be biased. Different estimation methods are available given the level of measurement of each variable, non-normality, and type of model.

In the union sentiment path model, the default maximum likelihood estimation method was used to obtain the parameter estimates. The path model parameters are shown in Table 4.2, which were all statistically significant at $p < .05$. *Age* had a significant direct effect on both *deference* and *support*; *deference* had a significant direct effect on *sentiment*; *years* had a significant direct effect on *sentiment*; and *support* had a significant direct effect on *sentiment*. We would interpret the path coefficients similar to regression coefficients, so the positive and negative coefficients are meaningful. For example, as *age* increased by one year, *deference* was estimated to decreased by .087 points, controlling for all other variables. Thus, younger workers tended to have less deference for union management than older workers. *Support* and number of *years* working in the textile mill had positive path coefficients in predicting *sentiment* toward unions. This implies that more support for labor activism and longer years of working at the textile mill are related to increased sentiment for unions. *Deference* had a negative path coefficient in predicting both *support* and *sentiment*. Using the standardized estimates, as *deference* increased by 1 standard deviation, *support* and *sentiment* were estimated to decrease by .328 and .147 standard deviations, respectively, holding all else constant. This implies that more deference for management is related to decreased support for labor activism and sentiment for unions.

Table 4.2: Union Sentiment Path Model Estimates

Paths	Standardized Parameter Estimates	Unstandardized Parameter Estimates	Standard Error	z	p
Age → deference	−.336	−.087	.019	−4.66	<.001
Age → support	.256	.058	.016	3.59	<.001
Deference → support	−.328	−.285	.062	−4.59	<.001
Years → sentiment	.154	.861	.341	2.52	<.05
Deference → sentiment	−.147	−.218	.097	−2.23	<.05
Support → sentiment	.499	.850	.112	7.55	<.001
Years ↔ age	.481	7.14	1.259	5.67	<.001
Error variances					
Deference		12.96			
Support		8.49			
Sentiment		19.45			

The three equations, with reported unstandardized path coefficients, standard errors, z-values, and p-values, from LISREL output are as follows:

```
Structural Equations

Deference =  - 0.0874*Age,      Errorvar.= 12.961, R² = 0.113
Standerr     (0.0187)                    (1.402)
Z-values    -4.664                       9.247
P-values     0.000                       0.000

Support =   - 0.285*Deference + 0.0579*Age,    Errorvar.= 8.488, R² = 0.230
Standerr     (0.0619)           (0.0161)                 (0.918)
Z-values    -4.598              3.597                    9.247
P-values     0.000              0.000                    0.000

Sentiment = - 0.218*Deference + 0.850*Support + 0.861*Years,  Errorvar.= 19.454, R² = 0.39
Standerr     (0.0974)           (0.112)         (0.341)                 (2.104)
Z-values    -2.235              7.555           2.526                   9.247
P-values     0.025              0.000           0.012                   0.000
```

While the direct effects are shown in the "Structural Equations" output above, the covariances among independent variables are shown in the "Covariance Matrix" output in LISREL:

```
        Covariance Matrix of Independent Variables
                     Years              Age

                 ------------       ------------
     Years           1.021
                    (0.110)
                     9.247

       Age           7.139            215.662
                    (1.259)          (23.323)
                     5.669             9.247
```

The unstandardized estimates, with standard errors, *z*-tests, and *p*-values, from M*plus* output are as follows:

```
MODEL RESULTS
                                                      Two-Tailed
                 Estimate    S.E.    Est./S.E.         P-Value
DEFERENC ON
  AGE             -0.087     0.019    -4.691            0.000

SUPPORT ON
  AGE              0.058     0.016     3.618            0.000
  DEFERENC        -0.285     0.062    -4.625            0.000

SENTIMEN ON
  YEARS            0.861     0.339     2.540            0.011
  DEFERENC        -0.218     0.097    -2.248            0.025
  SUPPORT          0.850     0.112     7.599            0.000

YEARS WITH
  AGE              7.098     1.245     5.702            0.000

Variances
  YEARS            1.015     0.109     9.301            0.000
  AGE            214.416    23.054     9.301            0.000

Residual Variances
  DEFERENC        12.886     1.386     9.301            0.000
  SUPPORT          8.439     0.907     9.301            0.000
  SENTIMEN        19.342     2.080     9.301            0.000
```

Again, you may notice subtle differences between some of the standard error and variance estimates in LISREL and M*plus* because M*plus* uses *n* instead of *n* – 1 in their calculations.

The standardized estimates from LISREL are as follows:

```
Standardized Solution

        BETA

              Deference    Support   Sentiment
              ---------  ---------  ---------
   Deference      - -        - -        - -
     Support   -0.328        - -        - -
   Sentiment   -0.147      0.499        - -

        GAMMA

                 Years        Age
              ---------  ---------
   Deference      - -     -0.336
     Support      - -      0.256
   Sentiment    0.154        - -

Correlation Matrix of Y and X

              Deference    Support   Sentiment      Years        Age
              ---------  ---------  ---------  ---------  ---------
   Deference    1.000
     Support   -0.414      1.000
   Sentiment   -0.379      0.587      1.000
       Years   -0.162      0.176      0.266      1.000
         Age   -0.336      0.366      0.307      0.481      1.000

PSI
    Note: This matrix is diagonal.

              Deference    Support   Sentiment
              ---------  ---------  ---------
                  0.887      0.770      0.610
```

In the standardized output in LISREL, the "BETA" matrix presents standardized direct effects among endogenous variables in the model and the "GAMMA" matrix presents the standardized direct effects from independent/exogenous to dependent/endogenous variables in the model. The correlation between *age* and *years* is presented in the "Correlation Matrix" and the standardized residual variances associated with endogenous variables are shown in the "PSI" matrix. You will notice that the standardized residual variance for an endogenous variable is equal to one minus their respective R square value $(1 - R^2)$, which is presented in the unstandardized "Structural Equations." The standardized estimates from M*plus* are provided below:

```
STANDARDIZED MODEL RESULTS

STDYX Standardization
```

	Estimate	S.E.	Est./S.E.	Two-Tailed P-Value
DEFERENC ON				
AGE	-0.336	0.067	-4.981	0.000
SUPPORT ON				
AGE	0.256	0.069	3.714	0.000
DEFERENC	-0.328	0.068	-4.843	0.000
SENTIMEN ON				
YEARS	0.154	0.060	2.555	0.011
DEFERENC	-0.147	0.065	-2.261	0.024
SUPPORT	0.499	0.059	8.445	0.000
YEARS WITH				
AGE	0.481	0.058	8.234	0.000
Variances				
YEARS	1.000	0.000	999.000	999.000
AGE	1.000	0.000	999.000	999.000
Residual Variances				
DEFERENC	0.887	0.045	19.576	0.000
SUPPORT	0.770	0.056	13.727	0.000
SENTIMEN	0.610	0.058	10.598	0.000

R-SQUARE Observed Variable	Estimate	S.E.	Est./S.E.	Two-Tailed P-Value
DEFERENC	0.113	0.045	2.490	0.013
SUPPORT	0.230	0.056	4.089	0.000
SENTIMEN	0.390	0.058	6.779	0.000

As previously mentioned, M*plus* computes standard errors (with resulting z-tests) associated with the standardized estimates, whereas LISREL only computes standard errors associated with unstandardized estimates (as do most SEM software packages).

The **Print Residuals** command in the SIMPLIS program and the **Residual** option in the M*plus* OUTPUT command requests that the model-implied covariance matrix ($\widehat{\boldsymbol{\Sigma}}$) and the residual covariance matrices be printed in the output. Recall that the residual covariance matrix is the difference between the elements of the observed sample covariance (**S**) and the model-implied covariance matrix ($\widehat{\boldsymbol{\Sigma}}$). With just-identified models, you may recall that the elements in the residual covariance matrix will be comprised of zeros because the sample covariance matrix will be equal to the model-implied covariance matrix. With over-identified models, the elements in the model-implied matrix will typically not be equal to the elements in the observed covariance matrix. Thus, the values in the residual covariance matrix will not all be equal to zero. Values closer to zero in the residual matrix are desired for over-identified models because that suggests that the hypothesized connections imposed in the path model are effectively explaining the variances of and the covariances among the observed variables. Hence, the residual covariance matrix is a helpful tool that may be used to evaluate how well relationships among variables are being explained by the theoretical path model.

The values of the elements for the different matrices, including the observed covariance matrix, the model-implied covariance matrix, and residual covariance matrices (unstandardized and standardized) are reported in Tables 4.3 and 4.4 from LISREL and M*plus* output, respectively.

Table 4.3: Original, Model Implied, Residual, Standardized Residual, and Normalized Residual Covariance Matrices for Union Sentiment Path Model based on LISREL Output

Original Covariance Matrix

Variable	Deference	Support	Sentiment	Years	Age
Deference	14.610				
Support	−5.250	11.017			
Sentiment	−8.057	11.087	31.971		
Years	−0.482	0.677	1.559	1.021	
Age	−18.857	17.861	28.250	7.139	215.662

Model Implied Covariance Matrix

Variable	Deference	Support	Sentiment	Years	Age
Deference	14.610				
Support	−5.250	11.017			
Sentiment	−8.179	11.013	31.899		
Years	−0.624	0.591	1.517	1.021	
Age	−18.857	17.861	25.427	7.139	215.662

Residual Covariance Matrix

Variable	Deference	Support	Sentiment	Years	Age
Deference	−				
Support	−0.000	−0.000			
Sentiment	0.122	0.074	0.072		
Years	0.142	0.086	0.042	−	
Age	−	0.000	2.823	−	−

Table 4.3: (*cont.*)

Standardized Residual Covariance Matrix

Variable	Deference	Support	Sentiment	Years	Age
Deference	–				
Support	–	–			
Sentiment	0.585	0.411	0.208		
Years	0.585	0.411	0.208	–	
Age	–	–	0.703	–	–

Normalized Residual Covariance Matrix

Variable	Deference	Support	Sentiment	Years	Age
Deference	–				
Support	–0.000	–0.000			
Sentiment	0.070	0.045	0.021		
Years	0.477	0.330	0.093	–	
Age	–	0.000	0.427	–	–

Table 4.4: Original, Model Implied, Residual, Standardized Residual, and Normalized Residual Covariance Matrices for Union Sentiment Path Model based on M*plus* Output

Original Covariance Matrix

Variable	Deference	Support	Sentiment	Years	Age
Deference	14.610				
Support	–5.250	11.017			
Sentiment	–8.057	11.087	31.971		
Years	–0.482	0.677	1.559	1.021	
Age	–18.857	17.861	28.250	7.139	215.662

Model Implied Covariance Matrix

Variable	Deference	Support	Sentiment	Years	Age
Deference	14.526				
Support	–5.220	10.953			
Sentiment	–8.132	10.950	31.715		
Years	–0.621	0.588	1.508	1.015	
Age	–18.748	17.758	25.280	7.098	214.416

Residual Covariance Matrix

Variable	Deference	Support	Sentiment	Years	Age
Deference	0.000				
Support	0.000	0.000			
Sentiment	0.122	0.073	0.072		
Years	0.141	0.085	0.042	0.000	
Age	0.000	0.000	2.806	0.000	0.000

Table 4.4: (cont.)

Standardized Residual Covariance Matrix

Variable	Deference	Support	Sentiment	Years	Age
Deference	999.000				
Support	999.000	999.000			
Sentiment	0.627	0.344	0.173		
Years	0.591	0.409	0.206	999.000	
Age	999.000	999.000	0.685	999.000	999.000

Normalized Residual Covariance Matrix

Variable	Deference	Support	Sentiment	Years	Age
Deference	0.000				
Support	0.000	0.000			
Sentiment	0.070	0.045	0.021		
Years	0.481	0.330	0.093	0.000	
Age	0.000	0.000	0.423	0.000	0.000

The elements in the residual covariance matrix are computed using the unstandardized values, which makes it difficult to interpret whether non-zero values are substantially different from a value of zero. As a result, LISREL (version 11) and M*plus* both provide standardized and normalized residual covariance matrices in the output to help researchers better interpret the non-zero values in the residual matrix. The standardized and normalized residuals may be interpreted similarly to z scores, with values greater than 1.96 or 2.58 indicating that the respective residual element is statistically significantly different from zero using an alpha of .05 or .01, respectively. This would suggest that a particular variable relation is not well accounted for in the path model.

You will notice some differences between the standardized values computed by LISREL and M*plus*, but the values are fairly comparable. The standardized residual values in LISREL may appear to be the same for more than one relationship (e.g., .585 for the relationship between *years* and *deference* and between *sentiment* and *deference*). The results are printed out to three decimal places in the output using the **ND = 3** option. Printing the results out to more decimal places (e.g., **ND = 6**) does show differences in the standardized residual values in the output.

The normalized residuals in M*plus* are more conservative than their standardized residual counterparts (see M*plus*, 2007 for detailed calculations of standardized and normalized residuals in M*plus*). Hence, the normalized residual values will be smaller in magnitude than their respective standardized residual values. Dashes in the LISREL standardized covariance matrix indicates a value equal to zero.

When a standardized residual cannot be computed in M*plus* because a negative variance estimate is computed, you will see values of "999.000" printed in the residual covariance matrix. One important consideration when examining the standardized or normalized residuals is that you will likely see values greater than 1.96 as the number of variables increases because the number of tests also increases. Moreover, the denominator for the standardized and normalized residuals will tend to become smaller as sample size increases. We recommend assessing the magnitude of the standardized or normalized residuals relative to other values in the matrix for detection of potential misspecification and increasing z score cutoffs with more complex models (e.g., from 1.96 to 2.58). Further, it is suggested that researchers use these tests as one method to potentially modify the model in addition to other indices (to be discussed later) and, most importantly, theory.

When examining the standardized and normalized residuals, the largest value is associated with the relationship between *years* and *deference*. As seen in Figure 4.5, there is no direct or indirect relationship between *years* and *deference*. There is only a spurious association through the correlation with *age* and the direct effect of *age* to *deference* connecting these two variables. Hence, if the model did not fit well and were to be modified, this could suggest a potential connection to include in the model if supported by theory. Nonetheless, the printed standardized and normalized residual values in Tables 4.2 and 4.3 are small and do not exceed 1.96, suggesting no distinct misspecification among relationships in the model.

INDIRECT EFFECTS

Indirect effects, also referred to as mediating effects, are also present in the union sentiment path model, but were not formally tested. There are five different indirect effects that can be tested given the hypothesized union sentiment path model. Specifically, 1) *deference* is a mediator variable between *age* and *sentiment*; 2) *deference* is a mediator variable between *age* and *support*; 3) *support* is a mediating variable between *age* and *sentiment*; 4) both *deference* and *support* are serving as mediators in that order (referred to as serial mediators) between *age* and *sentiment*; and 5) *support* is the mediator between *deference* and *sentiment*.

Indirect effect estimates are commonly obtained by multiplying the respective path coefficients. For instance, if the path from *age* to *deference* was labeled as a_1 and the path from *deference* to *sentiment* was labeled as b_1, the indirect effect from *age* to *sentiment* mediated by *deference* would be represented as $a_1 b_1$. The value of the unstandardized indirect effect from *age* to *sentiment* mediated by *deference* would be computed as: $(-.087)*(-.218) = .019$. Traditionally, normal theory tests were used to test indirect effects wherein the unstandardized indirect effect was divided by its respective standard error estimate. One of the most widely used normal theory tests of the indirect effect estimate uses Sobel's (1982) first-order standard error solution (derived with the multivariate delta method) in the denominator, which is calculated as a z score as:

$$z = \frac{a_1 b_1}{\sqrt{a_1^2 \sigma_{b1}^2 + b_1^2 \sigma_{a1}^2}},$$

where a_1^2 and b_1^2 are squared values of the unstandardized a_1 and b_1 parameters and σ_{a1}^2 and σ_{b1}^2 are squared values of the standard errors associated with the unstandardized a_1 and b_1 parameter estimates, respectively. The calculated z value is then compared to +/–1.96 to test for significance of the indirect effect at an alpha = 0.05. Other normal theory tests of the product of the coefficients have been proposed that calculate the standard error differently than the Sobel method, including those proposed by Aroian (1944) and Goodman (1960). Both LISREL and M*plus* use the delta method when computing the standard errors associated with indirect effects (see MacKinnon, 2008 for more details).

To request the normal theory tests (i.e., product of the coefficient divided by its standard error estimate) of the indirect effects in LISREL, the following option after the OPTIONS command is needed: **EF**:

```
Union Sentiment of Textile Workers
Observed Variables:
Deference Support Sentiment Years Age
Covariance Matrix
14.610
-5.250 11.017
-8.057 11.087 31.971
-0.482 0.677 1.559 1.021
-18.857 17.861 28.250 7.139 215.662
Sample Size = 173
Relationships
Support=Deference Age
Sentiment=Deference Support Years
Deference=Age
Print Residuals
OPTIONS SC EF AD=OFF
Path Diagram
End of Problem
```

To request the normal theory tests of the indirect effects in M*plus*, the MODEL INDIRECT command must be used as follows:

```
TITLE: Union Sentiments Path Model

DATA: FILE IS C:\union.txt;
      Type is Covariance;
      Nobservations=173;

VARIABLE: Names are Deferenc Support Sentimen Years Age;

MODEL:  Deferenc on Age;
        Support on Age Deferenc;
        Sentimen on Years Deferenc Support;
        Years with Age;

Model Indirect: Sentimen IND Age;
                Sentimen IND Deferenc;
                Support  IND Age;

OUTPUT: Stdyx Residual;
```

You will notice in the **MODEL INDIRECT** command, the M*plus* code starts with the final variable involved in the indirect effect backward to the beginning variable involved in the indirect effect, separating the final variable from the first variable in the mediating path with the term **IND** (**VIA** can also be used). That is, *sentiment* is indirectly impacted by *age* and indirectly impacted by *deference*; and *support* is indirectly impacted by *age*. More specifically, Sentimen IND Age will capture the three indirect effects from *age* to *sentiment* (through *support*; through *deference;* and through *deference* followed by *support*). Sentimen IND Deferenc will capture the indirect effect from *deference* to *sentiment* through *support*. Finally, Support IND Age will capture the indirect effect from *age* to *support* through *deference*.

The output for the unstandardized indirect effects in LISREL and M*plus* is illustrated in Tables 4.5 and 4.6, respectively. In LISREL, total effects (direct effect + indirect effect) as well as indirect effects are presented. For instance, the total effect from *deference* to *sentiment* (found in the "Total Effects of Y on Y" section) includes the direct effect (–.218) plus the indirect effect (–.242), equaling –.460. The three indirect effects from *age* to *sentiment* (through *support;* through *deference;* and through *deference* followed by *support*) are tested simultaneously and not individually (found in the "Indirect Effects of X on Y" section). The sum of the three unstandardized indirect effects is .089 and is statistically significantly different from zero ($z = 4.886$). The unstandardized indirect effect of *age* to *support* through *deference* is computed by taking the product of the unstandardized direct effect of *age* to *deference* and the unstandardized direct effect of *deference* to *support* [(−.087)*(−.285) = .025], which is statistically significant (*z* = 3.284). The indirect effect of *deference* to *sentiment* through *support* is illustrated in the "Indirect Effects of Y on Y" table because both variables are endogenous. The unstandardized indirect effect is computed by taking the product of the unstandardized direct effect of *deference* to *support* and the unstandardized direct effect of *support* to *sentiment* [(−.285)*(.850) = −.242], which is also statistically significant (*z* = −3.939). Although the normal theory tests are not provided for the specific indirect effects in LISREL, they could be calculated by hand or by using the Sobel test calculator on Kristopher Preacher's webpage (www.quantpsy.org/sobel/sobel.htm).

Table 4.5: LISREL Output for Normal Theory Tests of Unstandardized Indirect Effects

Total Effects of X on Y

	Years	Age
Deference	− −	−0.087
		(0.019)
		−4.678
Support	− −	0.083
		(0.016)
		5.165
Sentiment	0.861	0.089
	(0.340)	(0.018)
	2.533	4.886

Indirect Effects of X on Y

	Years	Age
Deference	− −	− −
Support	− −	0.025
		(0.008)
		3.284
Sentiment	− −	0.089
		(0.018)
		4.886

Total Effects of Y on Y

	Deference	Support	Sentiment
Deference	− −	− −	− −
Support	−0.285	− −	− −
	(0.062)		
	−4.612		

Table 4.5: (*cont.*)

Sentiment	−0.460	0.850	− −
	(0.104)	(0.112)	
	−4.434	7.577	

Indirect Effects of Y on Y

	Deference	Support	Sentiment
Deference	− −	−	− −
Support	− −	− −	− −
Sentiment	−0.242	−	− −
	(0.061)		
	−3.939		

Table 4.6: M*plus* Output for Normal Theory Tests of Unstandardized Indirect Effects

TOTAL, TOTAL INDIRECT, SPECIFIC INDIRECT, AND DIRECT EFFECTS

	Estimate	S.E.	Est./S.E.	Two-Tailed P-Value
Effects from AGE to SENTIMEN				
Total	0.089	0.018	4.901	0.000
Total indirect	0.089	0.018	4.901	0.000
Specific indirect 1 SENTIMEN DEFERENC AGE	0.019	0.009	2.028	0.043
Specific indirect 2 SENTIMEN SUPPORT AGE	0.049	0.015	3.267	0.001
Specific indirect 3 SENTIMEN SUPPORT DEFERENC AGE	0.021	0.007	3.022	0.003
Effects from DEFERENC to SENTIMEN				
Total	−0.460	0.103	−4.447	0.000
Total indirect	−0.242	0.061	−3.951	0.000
Specific indirect 1 SENTIMEN SUPPORT DEFERENC	−0.242	0.061	−3.951	0.000

Table 4.6: (*cont.*)

Direct				
SENTIMEN				
DEFERENC	-0.218	0.097	-2.248	0.025
Effects from AGE to SUPPORT				
Total	0.083	0.016	5.180	0.000
Total indirect	0.025	0.008	3.294	0.001
Specific indirect 1				
SUPPORT				
DEFERENC				
AGE	0.025	0.008	3.294	0.001
Direct				
SUPPORT				
AGE	0.058	0.016	3.618	0.000

The output from M*plus* in Table 4.6 tests specific as well as total indirect effects. It also presents direct effects (if they are modeled between the two respective variables) as well as the total effect (i.e., direct effect + indirect effect(s)). In Table 4.6, the total unstandardized indirect effect of *age* to *sentiment* is computed by summing the three specific unstandardized indirect effects (.019 + .049 + .021 = .089), which are also individually tested. The total indirect effect from *age* to *sentiment* as well as all three specific indirect effects from *age* to *sentiment* are statistically significant (assuming alpha = .05). The two additional indirect effects (one from *deference* to *sentiment* and one from *age* to *support*) are also statistically significant. The unstandardized indirect effect estimates and corresponding standard errors are similar to those in LISREL.

LISREL and M*plus* both compute standardized indirect effect estimates as well. The standardized estimates are based on the products of the standardized path coefficients involved in the specific indirect effect. The output for the standardized indirect effects in LISREL and M*plus* is illustrated in Tables 4.7 and 4.8, respectively. Again, you will notice that M*plus* calculates standard errors associated with the standardized indirect effects.

Table 4.7: LISREL Output for Standardized Indirect Effects

Standardized Total and Indirect Effects

Standardized Total Effects of X on Y

	Years	Age
Deference	- -	-0.336
Support	- -	0.366
Sentiment	0.154	0.232

Table 4.7: (*cont.*)

Standardized Indirect Effects of X on Y

	Years	Age
Deference	– –	– –
Support	– –	0.110
Sentiment	– –	0.232

Standardized Total Effects of Y on Y

	Deference	Support	Sentiment
Deference	– –	– –	– –
Support	−0.328	– –	– –
Sentiment	−0.311	0.499	– –

Standardized Indirect Effects of Y on Y

	Deference	Support	Sentiment
Deference	– –	– –	– –
Support	– –	– –	– –
Sentiment	−0.164	– –	– –

Table 4.8: M*plus* Output for Standardized Indirect Effects

STANDARDIZED TOTAL, TOTAL INDIRECT, SPECIFIC INDIRECT, AND DIRECT EFFECTS

STDYX Standardization

	Estimate	S.E.	Est./S.E.	Two-Tailed P-Value
Effects from AGE to SENTIMEN				
Total	0.232	0.044	5.303	0.000
Total indirect	0.232	0.044	5.303	0.000
Specific indirect 1 SENTIMEN DEFERENC AGE	0.050	0.024	2.048	0.041
Specific indirect 2 SENTIMEN SUPPORT AGE	0.128	0.038	3.364	0.001
Specific indirect 3 SENTIMEN SUPPORT DEFERENC AGE	0.055	0.017	3.154	0.002

Table 4.8: (cont.)

Effects from DEFERENC to SENTIMEN				
Total	-0.311	0.067	-4.651	0.000
Total indirect	-0.164	0.039	-4.163	0.000
Specific indirect 1				
SENTIMEN				
SUPPORT				
DEFERENC	-0.164	0.039	-4.163	0.000
Direct				
SENTIMEN				
DEFERENC	-0.147	0.065	-2.261	0.024
Effects from AGE to SUPPORT				
Total	0.366	0.066	5.567	0.000
Total indirect	0.110	0.032	3.438	0.001
Specific indirect 1				
SUPPORT				
DEFERENC				
AGE	0.110	0.032	3.438	0.001
Direct				
SUPPORT				
AGE	0.256	0.069	3.714	0.000

An inherent problem associated with the product of the path coefficients estimate for the indirect effect (a_1b_1) is that it is generally not normally distributed (Craig, 1936; Springer, 1979). A consequence of this problem is that the normal theory statistical significance tests and the confidence intervals associated with indirect effects are questionable. Because of this problem, bootstrapping techniques have been suggested when calculating indirect effects (Preacher & Hayes, 2008). Two commonly implemented bootstrapping methods that test the significance of an indirect effect first randomly draw data (with replacement) from the existing data set until n is equal to the number of participants in the original data set ($n = 173$ in the union sentiment example). This is repeated a set number of times so that there are, for instance, 5,000 data sets that were randomly drawn with replacement. Each of these data sets will contain, for example, data for 173 participants which could contain duplicate rows of data since random draws are done with replacement. The a_1 and b_1 parameters are then estimated in each of the 5,000 resampled data sets with 173 participants. The 5,000 resampled a_1b_1 products are then sorted in order from lowest to highest. A 95% confidence interval (CI) for the percentile bootstrap method is built by obtaining the 2.5th and the 97.5th percentiles of the sorted a_1b_1 product distribution that represent the lower and upper limits, respectively. The percentile bootstrapped CIs can become asymmetric given that they are based on a sampling distribution of the a_1b_1 product. Consequently, bias-corrected bootstrapped confidence intervals have been recommended. Bias-corrected bootstrapped CIs are calculated using an adjustment to the original percentile bootstrapped CIs. Readers are encouraged to see Efron (1987) for the technical details concerning the bias-corrected bootstrapped CIs for mediation. Once the lower and upper limits of the CIs are obtained, intervals that do not contain a value of zero signify a significant indirect effect. Another approach that has shown promise is called the joint significance test, which signifies mediation if both the a_1 and b_1 coefficients are statistically significant (Cohen & Cohen, 1983; Allison, 1995; Leth-Steensen & Gallitto, 2016).

Simulation research concerning the testing of an indirect effect has generally demonstrated that the bias-corrected bootstrapped CIs for mediation is associated with superior statistical power, followed by the percentile bootstrapped CI and joint significance test methods (Fritz & MacKinnon, 2007; Taylor, MacKinnon, & Tein, 2008; Hayes & Scharkow, 2013; Valente, Gonzalez, Miočević, & MacKinnon, 2016). The bias-corrected bootstrapped CI method has also been associated with inflated Type I errors in certain scenarios (Fritz & MacKinnon, 2007; Fritz, Taylor, & MacKinnon, 2012; Hayes & Scharkow, 2013). Nonetheless, the bias-corrected bootstrapped CI method is more commonly recommended.

Both LISREL and M*plus* allow for bootstrapping. It is important to note that you will need the raw data when conducting bootstrapping techniques in both LISREL and M*plus*. In LISREL, a two-step process must be followed when performing bootstrapping. First, PRELIS is used to obtain bootstrapped estimates of the covariance matrix to be analyzed. Second, the moment matrices generated by PRELIS are then used in LISREL to obtain the bootstrapped parameter estimates of interest. Nonetheless, LISREL does not provide bootstrapped confidence intervals as output for testing indirect effects. Instead, users can build percentile bootstrapped confidence intervals based on the bootstrapped estimates (see Lau & Cheung, 2012 for an example).

M*plus* computes both the percentile and bias-corrected bootstrapped confidence intervals associated with parameter estimates, including indirect effects. Bootstrapping in M*plus* can be done in one step in the same program when analyzing the path model. Bootstrapping cannot be conducted using summary data (i.e., the covariance matrix). Thus, to illustrate the use of bootstrapping of indirect effects in M*plus*, we generated data in M*plus* using the parameter estimates from the union sentiment model. The M*plus* code to run the union sentiment model with bias-corrected bootstrapped confidence intervals is as follows:

```
TITLE: Union Sentiment Path Model with Bias-
Corrected Bootstrapped Confidence Intervals for
Indirect Effects
```

```
DATA: FILE IS C:\Users\unionraw.dat;

VARIABLE: Names are Deferenc Support Sentimen Years Age;

ANALYSIS: Bootstrap = 5000;

MODEL:  Deferenc on Age;
        Support on Age Deferenc;
        Sentimen on Years Deferenc Support;
        Years with Age;

Model Indirect: Sentimen IND Age;
                Sentimen IND Deferenc;
                Support IND Age;

OUTPUT: Stdyx Residual Sampstat Cinterval(BCBOOTSTRAP);
```

The raw data in ***unionraw.dat*** are tab delimited without variable names on the top row. You will notice that you do not need to specify the type of data you are using as input with raw data because the default is **Individual** data, meaning each person's data is contained on a single row. You also do not need to specify the sample size with raw data because M*plus* will count the number of observations. Always make sure that the sample size is accurate in the "SUMMARY OF ANALYSIS" section of the M*plus* output. The **Sampstat** option (can be abbreviated as Samp) in the OUTPUT command will provide univariate descriptive statistics (e.g., mean, variance) in the output for each of the variables in the data file. The ANALYSIS command with **Bootstrap = 5000** specifies that the number of bootstrapped draws (with replacement) should be equal to 5,000. Thus, the program will take 5,000 bootstrapped draws with replacement, where the sample size in each of the data sets drawn is equal to the original number of observations. A minimum of 5,000 bootstrapped draws has been recommended (Preacher & Hayes, 2008). The **Cinterval** option in the OUTPUT command requests confidence intervals to be constructed around parameter estimates and the **(BCBOOTSTRAP)** option with **Cinterval** requests that bias-corrected bootstrapped confidence intervals be constructed around parameter estimates. If you want percentile bootstrapped confidence intervals instead, the following option in the OUTPUT command should be used: **Cinterval(BOOTSTRAP)**. It is important to note that confidence intervals will not only be constructed around the indirect effects, but they will be constructed around *all* parameter estimates when this option is used in the OUTPUT command.

The output from M*plus* with bias-corrected bootstrapped confidence intervals for the unstandardized indirect effects is presented in Table 4.9. In Table 4.9, you can see the unstandardized indirect effect estimate under the "Estimate" column heading. The estimate will not change, regardless of using bootstrapping procedures. You may, however, notice some changes in the standard errors as compared to the normal theory output because they are calculated using bias-corrected bootstrapping. You will notice in Table 4.9 that M*plus* computes 90% (Lower 5%, Upper 5%), 95% (Lower 2.5%, Upper 2.5%), and 99% (Lower .5%, Upper .5%) confidence limits around the parameter estimate. The lower and upper 95% confidence limits associated with the five indirect effects (and total indirect effect) do not encompass a value of zero, indicating that the five indirect effects are statistically significant (assuming alpha = .05).

Table 4.9: M*plus* Output for Bias-corrected Bootstrapped Confidence Intervals for Unstandardized Indirect Effects

CONFIDENCE INTERVALS OF TOTAL, TOTAL INDIRECT, SPECIFIC INDIRECT, AND DIRECT EFFECTS

	Lower .5%	Lower 2.5%	Lower 5%	Estimate	Upper 5%	Upper 2.5%	Upper .5%
Effects from AGE to SENTIMEN							
Total	0.048	0.058	0.064	0.089	0.125	0.132	0.143
Total indirect	0.048	0.058	0.064	0.089	0.125	0.132	0.143
Specific indirect 1 SENTIMEN DEFERENC AGE	−0.002	0.003	0.005	0.019	0.034	0.038	0.046
Specific indirect 2 SENTIMEN SUPPORT AGE	0.020	0.028	0.032	0.049	0.082	0.088	0.098
Specific indirect 3 SENTIMEN SUPPORT DEFERENC AGE	0.007	0.010	0.011	0.021	0.035	0.038	0.044
Effects from DEFERENC to SENTIMEN							
Total	−0.749	−0.678	−0.642	−0.460	−0.296	−0.263	−0.207
Total indirect	−0.449	−0.400	−0.374	−0.242	−0.166	−0.151	−0.122
Specific indirect 1 SENTIMEN SUPPORT DEFERENC	−0.449	−0.400	−0.374	−0.242	−0.166	−0.151	−0.122
Direct SENTIMEN DEFERENC	−0.464	−0.410	−0.374	−0.218	−0.054	−0.022	0.038
Effects from AGE to SUPPORT							
Total	0.050	0.060	0.065	0.083	0.117	0.123	0.133
Total indirect	0.009	0.012	0.014	0.025	0.039	0.042	0.047
Specific indirect 1 SUPPORT DEFERENC AGE	0.009	0.012	0.014	0.025	0.039	0.042	0.047
Direct SUPPORT AGE	0.023	0.034	0.040	0.058	0.091	0.097	0.106

When presenting the results of path models, the results may be illustrated as in Table 4.10 as well as visually in a diagram. For instance, the union sentiment path model with standardized estimates, with asterisks denoting statistical significance, is illustrated in Figure 4.6. When interpreting results, some readers would like to see the standard errors associated with the parameter estimates. The standard errors can be incorporated into figures in parentheses with their respective parameter estimates or presented in a table. You will notice that the standardized error variance estimates (located in the PSI matrix for standardized results in LISREL) are presented in Figure 4.6. An alternative to presenting the standardized error variances is to report the R^2 (1 – standardized error/residual variance) associated with each endogenous variable. The path value from error to its respective endogenous variable could also be presented, which is calculated by taking the square root of the error variance estimate (i.e., $\sqrt{1 - R^2}$). If reporting the unstandardized results in a figure, the unstandardized direct effects, covariances, and error variances would simply replace the corresponding standardized results in Figure 4.6. Table 4.10 illustrates the results associated with testing the indirect effects in the union sentiment path model with bias-corrected bootstrapped confidence intervals, which could be presented similarly in a research paper.

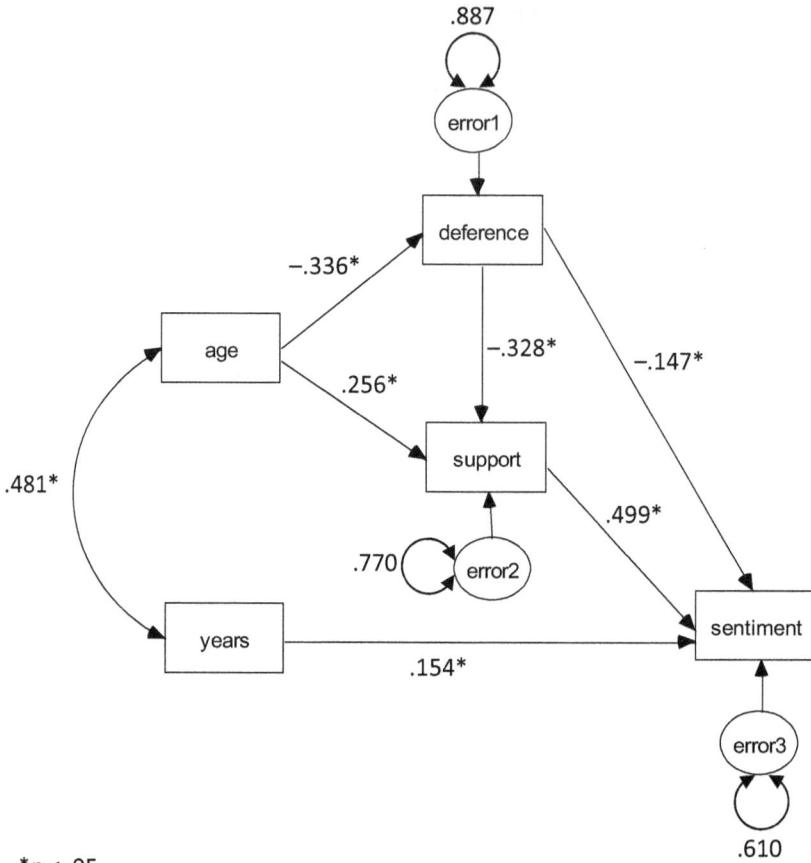

*p < .05

■ **Figure 4.6:** UNION SENTIMENT MODEL WITH STANDARDIZED ESTIMATES

Table 4.10: Summary of Unstandardized Indirect Effects (with Corresponding 95% BC-Bootstrapped Confidence Intervals), Standard Errors, and Standardized Indirect Effects for the Union Sentiment Path Model

Association	Unstandardized Indirect Effect (95% BC-B CI)	S.E.	Standardized Indirect Effect
Age → Deference → Support	.025* (.012, .042)	.008	.110
Age, Sentiment	TI = .089* (.058, .132)	.018	.232
Age → Deference → Sentiment	.019* (.003, .038)	.009	.050
Age → Support → Sentiment	.049* (.028, .088)	.015	.128

Table 4.10: (*cont.*)

Association	Unstandardized Indirect Effect (95% BC-B CI)	S.E.	Standardized Indirect Effect
Age → Deference → Support → Sentiment	.021* (.010, .038)	.007	.055
Deference → Support → Sentiment	−.242* (−.400, −.151)	.061	−.164

Note: BC-B CI = Bias-Corrected Bootstrapped Confidence Interval; S.E. = Standard Error; TI = Total Indirect. * $p < .05$.

When testing for mediation, the temporal ordering of variables involved in the mediating relationships is imperative for stronger claims associated with the findings concerning indirect effects. In the following mediating relationship: $X \rightarrow M \rightarrow Y$, it is recommended that data for variable X be collected prior to when data are collected for variable M, which should occur prior to collecting data for variable Y. If temporal ordering of the variables is not feasible and data are gathered cross-sectionally, implications of the findings involving the indirect effects will be limited and should be stated as such. This is not to say that important findings cannot be found with cross-sectional data concerning indirect effects because they could inform subsequent research studies in which temporal ordering of the variables can be achieved.

CAUSATION ASSUMPTIONS AND LIMITATIONS

Path analysis was originally presented as a method for testing causal outcomes. Today, however, it is understood as a method to test theoretical models that depict relations among variables, not necessarily as a *causal modeling* technique. A causal model would have to meet several criteria, which is generally not available using bivariate correlations among variables. A causal model would need to include the following conditions:

1. Temporal ordering of variables.
2. Covariation or correlation is present among variables.
3. Other causes are controlled for, either statistically or based on design.
4. X is manipulated, which causes a change in Y.

Obviously, a model that is tested over time (as in longitudinal research) and manipulates certain variables to assess the change in other variables (as in experimental research) more closely approaches the idea of causation. In the social and behavioral sciences, causation does have the potential to be modeled. Pearl (2009) argues for causation in the behavioral sciences. His rationale is that causation is a process that can be expressed in mathematical equations for computer analysis, which fits into the testing of theoretical path models.

Philosophical differences exist between assuming causal versus inferential relationships among variables, and the resolution of these issues requires a sound theoretical perspective. Bullock, Harlow, and Mulaik (1994) provided an in-depth discussion of causation issues related to structural equation modeling research. Bollen and Pearl (2013) addressed eight common *myths* related to using SEM and causal inference, such as "no causation without manipulation," "SEMs are less applicable to experiments with randomized treatments," and "SEMs do not test any major part of the theory against the data." Bollen and Pearl (2013) "conclude that the current capabilities of SEMs to formalize and implement causal inference tasks are indispensable; its potential to do more is even greater" (p. 301). Specifically, Bollen and Pearl (2013) explained that:

> researchers do not derive causal relations from an SEM. Rather, the SEM represents and relies upon the causal assumptions of the researcher. These assumptions derive from the research design, prior studies, scientific knowledge, logical arguments, temporal priorities, and other evidence that the researcher can marshal in support of them. The credibility of the SEM depends on the credibility of the causal assumptions in each application.
>
> *p. 309*

More discussion of testing these causal assumptions is beyond the scope of this chapter. In addition to reading Bollen and Pearl's (2013) chapter, readers are encouraged to learn more about SEM and causal inference (see e.g., Halpern & Pearl, 2005a, 2005b; Mulaik, 2009; Pearl, 2009, 2012; Pearl, Glymour, & Jewell, 2016; Pearl & Mackenzie, 2018).

SUMMARY

Path models permit the specification of theoretically meaningful relations among observed variables, including direct, indirect, as well as resulting spurious associations. Path analysis developed along a correlational track where the relationships in a model are easier to decompose into correlations. Nevertheless, SEM analyses uses the covariance matrix during estimation in order to account for the distributional properties of the variables in a model. When decomposing the relationships among variables in unstandardized models, variance information is simply (and importantly) incorporated. Further, decomposing the relationships in an unstandardized model can help to provide some insight into how parameters are estimated in SEM.

The union sentiment path model example provided a better understanding of model specification, model identification, and interpretation of results. We recommend that researchers make sure that models are identified prior to collecting data in order to avoid complications during the SEM analysis phase for a research project. When fitting over-identified models, the residual covariance matrices can be used to help detect potential misspecification in a model, in addition to theoretical justification. Indirect effects in path models may be tested using normal theory tests or using bootstrapping techniques. We recommend using bootstrapping techniques when testing indirect effects for more accurate results. It is also important to consider temporal ordering of the variables involved in mediating relationships in order to more strongly support the implications related to the findings.

Path models were initially called causal models, which created much confusion regarding inference versus causation. It is important to test causal assumptions associated with any modeling framework in which a researcher is working. We agree with Bollen and Pearl (2013) that SEM can be invaluable when testing causal inferences under the right conditions.

EXERCISES

1. Using the unstandardized union sentiment model [re-expressed in Figure 4.7 with path coefficients (a*–g*) and variable labels (X1, X2, Y1, Y2, and Y3) for simplicity], decompose all covariances among the five variables and their variances as we did for the unstandardized path model in Figure 4.4. These are the decomposition equations that would be inserted into the model-implied covariance matrix. There should be ten covariances and five variances given the model as follows:

$\sigma_{X1X2}, \sigma_{X1Y1}, \sigma_{X1Y2}, \sigma_{X1Y3}, \sigma_{X2Y1}, \sigma_{X2Y2}, \sigma_{X2Y3}, \sigma_{Y1Y2}, \sigma_{Y1Y3}, \sigma_{Y2Y3},$
$\sigma^2_{X1}, \sigma^2_{X2}, \sigma^2_{Y1}, \sigma^2_{Y2}, \sigma^2_{Y3}$

Be sure to include all direct effects, all indirect effects, and all spurious relationships in the covariance decompositions. Be sure to use all legal paths from a variable away and back to itself for the variance decompositions. Make sure that the correct variance or covariance is included in each compound path. Don't forget the path tracing rules!

Once you have completed the decompositions, set the decomposition equations equal to the values in the observed sample covariance matrix (these can be found in Table 4.3, Table 4.4, in the SIMPLIS code for the example, or in the *union.txt* file).

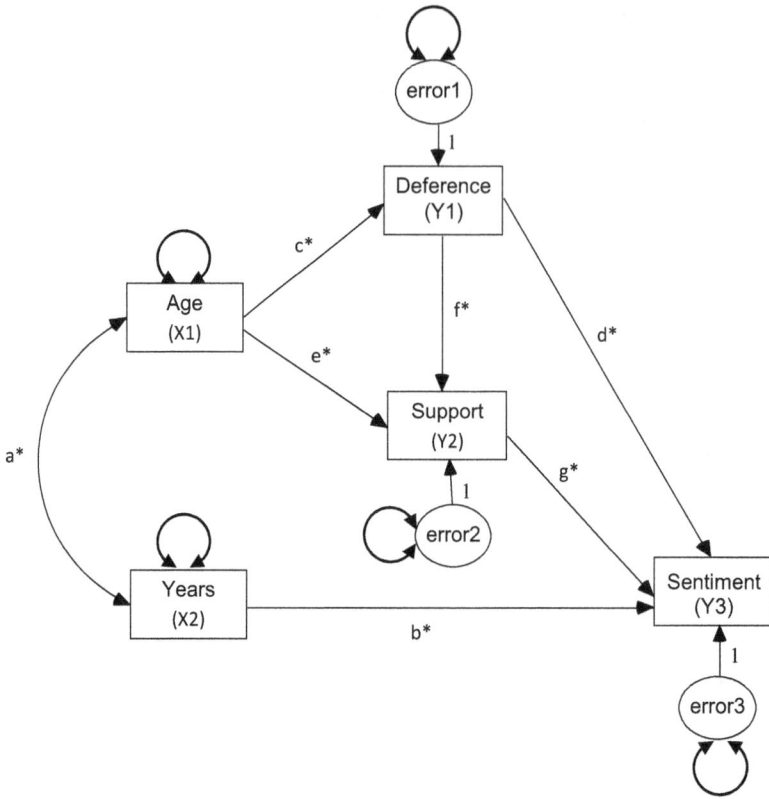

■ **Figure 4.7:** RE-EXPRESSED UNSTANDARDIZED UNION SENTIMENT MODEL

2. a. Using the model in Figure 4.8, which variables (observed and latent) are independent/exogenous variables and dependent/endogenous variables?

   b. Using the path coefficients as labeled in the model (e.g., a*–f*), write the equations associated with the hypothesized path model. Make sure you include the errors in these equations.

   c. How many distinct (non-redundant) elements are in the observed sample covariance matrix?

   d. How many parameters need to be estimated (how many free parameters are in the model)?

   e. What are the degrees of freedom (*df*) associated with the hypothesized model?

   f. Is the model under-identified, just-identified, or over-identified? Explain your answer.

   g. The covariance matrix for this model is provided below:

   ```
   25.500
   20.500 38.100
   22.480 24.200 42.750
   16.275 13.600 13.500 17.000
   ```

Write a LISREL or M*plus* program to analyze the data (using maximum likelihood estimation) and run the program. You will also need to model all the indirect effects.

h. Draw the achievement path model with the standardized parameter estimates, similar to the diagram in Figure 4.6.

i. Using the output associated with testing the indirect effects, which indirect effects are statistically significant using the normal theory tests?

j. What would be a better method of testing the indirect effects than the normal theory tests?

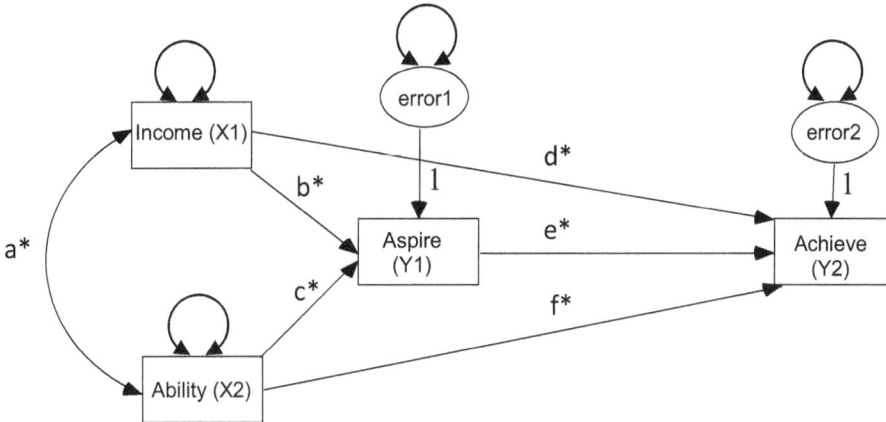

■ **Figure 4.8:** ACHIEVEMENT PATH MODEL

REFERENCES

Allison, P. D. (1995). The impact of random predictors on comparison of coefficients between models: Comment on Clogg, Petkova, and Haritou. *American Journal of Sociology*, 100(5), 1294–1305.

Aroian, L. A. (1944). The probability function of the product of two normally distributed variables. *Annals of Mathematical Statistics,* 18(2), 265–271.

Bollen, K. A. (1989). *Structural equations with latent variables.* New York: Wiley.

Bollen, K. A., & Pearl, J. (2013). Eight myths about causality and structural equation models. In S. L. Morgan (Ed.), *Handbook of causal analysis for social research* (pp. 301–328). Dordrecht, Netherlands: Springer.

Bullock, H. E., Harlow, L. L., & Mulaik, S. A. (1994). Causation issues in structural equation modeling research. *Structural Equation Modeling*, 1(3), 253–267.

Cohen, J., & Cohen, P. (1983). *Applied multiple regression/correlation analysis for the behavioral sciences.* Hillsdale, NJ: Erlbaum.

Craig, C. C. (1936). On the frequency function of xy. *Annals of Mathematical Statistics*, 7(1), 1–15.

Duncan, O. D. (1975). *Introduction to structural equation models.* New York: Academic.

Efron, B. (1987). Better bootstrap confidence intervals. *Journal of the American Statistical Association,* 82(397), 171–185.

Fritz, M. S., & MacKinnon, D. P. (2007). Required sample size to detect the mediated effect. *Psychological Science,* 18(3), 233–239.

Fritz, M. S., Taylor, A. B., & MacKinnon, D. P. (2012). Explanation of two anomalous results in statistical mediation analysis. *Multivariate Behavioral Research,* 47(1), 61–87.

Goodman, L. A. (1960). On the exact variance of products. *Journal of the American Statistical Association,* 55(292), 708–713.

Halpern, J. Y., & Pearl, J. (2005a). Causes and explanations: A structural-model approach. Part I: Causes. *The British Journal for the Philosophy of Science,* 56(4), 843–887.

Halpern, J. Y., & Pearl, J. (2005b). Causes and explanations: A structural-model approach. Part II: Explanations. *The British Journal for the Philosophy of Science,* 56(4), 889–911.

Hayes, A. F., & Scharkow, M. (2013). The relative trustworthiness of inferential tests of the indirect effect of statistical mediation analysis: Does method really matter? *Psychological Science,* 24(10), 1918–1927.

Jöreskog, K. G., & Sörbom, D. (1993). *LISREL 8: Structural equation modeling with the SIMPLIS command language.* Chicago, IL: Scientific Software International.

Lau, R. S., & Cheung, G. W. (2012). Estimating and comparing specific mediation effects in complex latent variable models. *Organizational Research Methods,* 15(1), 3–16.

Leth-Steensen, C., & Gallitto, E. (2016). Testing mediation in structural equation modeling: The effectiveness of the test of joint significance. *Educational and Psychological Measurement,* 76(2), 339–351.

MacKinnon, D. P. (2008). *Introduction to statistical mediation analysis.* New York: Erlbaum.

McDonald, J. A., & Clelland, D. A. (1984). Textile workers and union sentiment. *Social Forces,* 63, 502–521.

M*plus* (2007). *Standardized residuals in Mplus.* Retrieved from M*plus* website: www.statmodel.com/download/StandardizedResiduals.pdf.

Mulaik, S. A. (2009). *Linear causal modeling with structural equations.* New York: CRC Press.

Pearl, J. (2009). *Causality: Models, reasoning, and inference* (2nd ed.). Cambridge University Press: London.

Pearl, J. (2012). The causal foundations of structural equation modeling. In R. H. Hoyle (Ed.), *Handbook of structural equation modeling* (pp. 68–91). New York: Guilford Press.

Pearl, J., Glymour, M., & Jewell, N. P. (2016). *Causal inference in statistics: A primer.* New York: Wiley.

Pearl, J., & Mackenzie, D. (2018). *The book of why: The new science of cause and effect.* New York: Basic Books.

Pedhazur, E. J. (1982). *Multiple regression in behavioral research: Explanation and prediction* (2nd ed.). New York: Holt, Rinehart & Winston.

Preacher, K. J., & Hayes, A. F. (2008). Asymptotic and resampling strategies for assessing and comparing indirect effects in multiple mediator models. *Behavior Research Methods*, 40(3), 879–891.

Sobel, M. E. (1982). Asymptotic confidence intervals for indirect effects in structural equation models. In S. Leinhart (Ed.), *Sociological methodology* (pp. 290–312). San Francisco, CA: Jossey-Bass.

Specht, D. A. (1975). On the evaluation of causal models. *Social Science Research*, 4, 113–133.

Springer, M. D. (1979). *The algebra of random variables*. New York: Wiley.

Taylor, A. B., MacKinnon, D. P., & Tein, J-Y. (2008). Tests of three-path mediated effect. *Organizational Research Methods*, 11(2), 241–269.

Valente, M. J., Gonzalez, O., Miočević, M., & MacKinnon, D. P. (2016). A note on testing mediated effects in structural equation models: Reconciling past and current research on the performance of the test of joint significance. *Educational and Psychological Measurement*, 76(6), 889–911.

Wolfle, L. M. (1977). An introduction to path analysis. *Multiple Linear Regression Viewpoints*, 8, 36–61.

Wolfle, L. M. (1979). Unmeasured variables in path analysis. *Multiple Linear Regression Viewpoints*, 9, 20–56.

Wright, S. (1918). On the nature of size factors. *Genetics*, 3(4), 367–374.

Wright, S. (1921). Correlation and causation. *Journal of Agricultural Research*, 20, 557–585.

Wright, S. (1934). The method of path coefficients. *Annals of Mathematical Statistics*, 5, 161–215.

Wright, S. (1960). Path coefficients and path regression: Alternative or complementary concepts? *Biometrics*, 16, 189–202.

Chapter 5

SEM BASICS

SEM MODELING STEPS

The steps a researcher takes in conducting SEM involves *model specification, model identification, model estimation, model testing,* and *model modification/ re-specification.* The SEM modeling steps generally occur as outlined in the following five basic steps.

Model Specification

Models in SEM are specified based on prior research and theory. This comprises the review of literature which substantiates selection of observed variables to be modeled in observed variable path models, or as indicators of latent variables in measurement or latent variable models. The review of the literature also substantiates the theory behind testing the relations among the observed and/or latent variables in a structural equation model.

Model Identification

A model is identified if the degrees of freedom (*df*) are equal to or greater than zero and latent or unobserved variables have been assigned a scale of measurement. A *df* = 0 indicates a saturated model; thus, all parameters are being estimated, which is also called a just-identified model. An under-identified model would have negative degrees of freedom because more parameters are to be estimated than distinct values in the covariance matrix. We are generally interested in an over-identified model, which specifies fewer paths or variable relations than distinct elements in the sample covariance matrix, hoping that the model-implied (reproduced) covariance matrix is close to the sample covariance matrix. For more details concerning additional rules of model identification, see Bollen (1989) as well as Kenny and Milan (2012).

DOI: 10.4324/9781003044017-5

Assigning latent variables (including errors and disturbances) a scale is necessary for identification. Because latent variables are unobserved, they do not have a scale of measurement. Recall from Chapter 4 that the variances of the errors are included in the decomposition equations in the model-implied covariance matrix and must be estimated. Because the errors do not have an observed metric, the estimation of all of the model parameters would be prohibited unless we set the scale of measurement for the errors. You may recall that in the figures presented in Chapter 4, values of 1 are inserted in the direct paths from the errors to their respective endogenous variable. Thus, the errors are scaled by setting the direct paths to their corresponding variable to values of 1, essentially assigning the error variance the variance of its respective endogenous variable. This also means that summing the explained and unexplained variance in an endogenous variable will equal its total variance. The scaling of errors and disturbances is done by default in SEM software and does not need to be specified by the user. Model identification will continue to be discussed when presenting specific models in subsequent chapters.

Model Estimation

As previously mentioned, SEM tests whether a sample covariance matrix is similar to the covariance matrix implied by a theoretical model. Model parameters are estimated through an iterative process. The software program will iterate through successive cycles, attempting to calculate improved estimates. Specifically, the elements of the model-implied covariance matrix which are computed using the parameter estimates from each iteration will become more similar to the elements of the observed covariance matrix. A hypothesized theoretical model can have parameters estimated using several different estimation methods. The most widely used estimation procedure in SEM, and typically the default in SEM software (e.g., AMOS, EQS, LISREL, and M*plus*), is the maximum likelihood (ML) estimation procedure. A key issue in estimating model parameters is that the appropriate estimation procedure is selected to ensure accurate model test statistics and standard errors of the parameter estimates. More information about the types of estimation procedures available and when to use them will be provided later in this chapter.

Model Testing

Model fit in SEM is a complex and considerable topic of research. The assessment of model fit in SEM continues to be a matter of extensive debate and new model fit methods in SEM continue to be introduced. Generally, a model is tested for fit based on various criteria, including a test of overall model fit (using the chi-square statistic), model fit indices, and appropriate parameter estimates. Model testing can also be evaluated by comparing a set of theoretically plausible models. Assessing model fit and model comparisons will be discussed in more detail subsequently in this chapter.

Model Modification/Re-specification

When testing the fit of a theoretical model in SEM, it may not represent the relationships in the data adequately. At this point, a researcher may attempt to modify the model in order to improve fit to the data. Although statistical tools are available to help researchers when modifying a model, such as residual values in the residual covariance matrix and modification indices, theory should provide the most guidance when making changes to a model. When models are modified and retested, this is a re-specification of the originally hypothesized model. As such, considerations about model modifications and re-specifications are an important point of discussion in research papers. We will provide more discussion about model modification and re-specification in this chapter and in subsequent chapters using examples.

MODEL ESTIMATION

In structural equation modeling, the unknown parameters in a model are estimated by minimizing a discrepancy or loss function between the sample covariance matrix (S) and the model-implied covariance matrix ($\widehat{\Sigma}$). The discrepancy between the sample covariance matrix and the model-implied covariance matrix may be captured in the residual covariance matrix:

$$\left(S - \widehat{\Sigma}\right),$$

with values closer to zero suggesting better fit of the model to the data (Bollen, 1989).

The estimation process of model parameters in SEM is iterative, with the software program going through successive cycles in order to compute improved estimates. Hence, the elements in the model-implied covariance matrix, which are computed simultaneously using the parameter estimates from each iteration, will become more similar to the corresponding elements in the observed covariance matrix. Although different estimators weight the discrepancies between elements in the sample covariance matrix and the corresponding elements in the model-implied covariance matrix differently, the discrepancy or loss function used by each estimator is essentially the sum of squared differences between corresponding elements in the sample covariance matrix (**S**) and the model-implied covariance matrix ($\hat{\Sigma}$), resulting in a single value once the solution has converged.

Recall that if the model is just-identified, the discrepancy value will be equal to zero because the elements in the sample covariance matrix will be the same as the elements in the model-implied covariance matrix, $\mathbf{S} = \hat{\Sigma}$ (Bollen, 1989). For over-identified models, the sample covariance matrix will not be the same as the model-implied covariance matrix, but the estimation process will continue through cycles until the change in the discrepancy value from one iteration cycle to the next is less than the specific default value used by the SEM software program (e.g., < .000001 in LISREL; < .00005 in M*plus*) or until the maximum number of default iterations has been reached (e.g., number of parameters*3 in LISREL; 1,000 in M*plus*).

The most widely used estimation procedure in SEM, and typically the default in SEM software (e.g., AMOS, EQS, LISREL, and M*plus*), is the maximum likelihood (ML) estimation procedure. Ferron and Hess (2007) provide a detailed illustration using ML estimation. ML estimation assumes that the observed variables in the model are multivariate normal. Under conditions with small sample sizes, ML estimators may become biased; however, they are asymptotically unbiased. That is, as sample size becomes larger, the expected values associated with the ML estimates represent the population values. ML estimators are also consistent, meaning that as sample size nears infinity, the probability that the estimate is close to the population value gets higher. Another important characteristic of ML estimators is asymptotic efficiency. Specifically, the ML estimator has the lowest asymptotic variance compared to other consistent estimators. Moreover, the ML estimator is scale invariant, meaning that the value of the ML discrepancy or loss function will be the same with any change in the scale among the observed variables. For more details concerning ML estimation, readers are referred to Bollen (1989) and Eliason (1993).

Unweighted least squares (ULS) estimation does not depend on multivariate normality. However, ULS estimation is not widely used because its estimates are not efficient, nor are they scale invariant or scale free. Generalized least squares (GLS) estimation assumes multivariate normality (for a review of GLS, see Bollen, 1989). Under conditions of multivariate normality, ML and GLS produce estimates that are asymptotically equivalent. Hence, as sample size increases, the GLS estimates are approximately equivalent to the ML estimates. Nonetheless, ML estimation has been shown to perform better than GLS estimation under conditions with model misspecification (Olsson, Foss, Troye, & Howell, 2000).

When violating the multivariate normality assumption among variables, ML parameter estimates are largely robust and consistent (Dolan, 1994; DiStefano, 2002; Beauducel & Herzberg, 2006). Nevertheless, variables having skewed and kurtotic distributions can at times produce an incorrect asymptotic covariance matrix containing parameter estimates (Bollen, 1989). In addition, increased skewness (e.g., greater than 2) and/or kurtosis (e.g., greater than 7) levels essentially invalidates the asymptotic efficiency of the ML-based estimated parameters, resulting in incorrect model test statistics (e.g., the chi-square test). Hence, non-normally distributed variables may impact the test of overall model fit in addition to the consistency and efficiency of parameter estimates.

Other estimation procedures that provide asymptotically efficient estimates have been proposed that do not assume multivariate normality. One of these procedures is the weighted least squares (WLS) estimator (Browne, 1984), also called asymptotically distribution free (ADF) estimation (see Browne, 1984 and Muthén & Kaplan, 1985 for more information about WLS estimation). WLS was intended as an efficient estimator for any type of distribution among the variables, including ordinal variables (Browne, 1984). WLS estimation has produced more consistent and efficient estimates than ML with non-normal, ordinal variables (Muthén & Kaplan, 1985). Nonetheless, the performance of WLS under certain conditions in other research studies has been questionable. For example, tests of model fit with WLS have been found to falsely reject the correct factor model too often, even with normal variable distributions and in small sample size conditions (Hu, Bentler, & Kano, 1992). WLS has been shown to progressively overestimate the expected value associated with the model fit test statistic as model misspecification and non-normality become worse (Curran, West, & Finch, 1996; Olsson et al., 2000). Although WLS has demonstrated better efficiency with non-normal variable distributions relative to ML (Muthén & Kaplan, 1985; Chou, Bentler, & Satorra, 1991), the efficiency of WLS is negatively impacted in conditions with increasing non-normality, small samples, and large model size (Muthén & Kaplan, 1992). WLS estimation requires extremely large sample sizes (e.g., 2,500 to 5,000) for accurate tests of model fit and parameter estimates, which may not be feasible for most researchers (Hu et al., 1992; Loehlin, 2004; Finney & DiStefano, 2006). Moreover, WLS estimation is more computationally demanding than other estimators because it needs to take the inverse of a full weight matrix during the estimation process, which becomes greater in size with increases in the number of variables (Loehlin, 2004).

Robust WLS approaches were developed to fix the problems associated with full WLS estimation (see Muthén, du Toit, & Spisic, 1997 and Jöreskog & Sörbom, 1996 for more information concerning robust WLS). In general, the robust WLS approaches use a diagonal weight matrix instead of the full weight matrix during estimation. Robust WLS has performed better than full WLS in terms of the test of overall model fit (i.e., the chi-square test) and accuracy of parameter estimates under conditions of non-normal data and small sample sizes (Bandalos, 2014; Flora & Curran, 2004; Forero & Maydeu-Olivares, 2009; Rhemtulla, Brosseau-Laird, & Savalei, 2012).

Because non-normality can negatively impact the test of overall model fit and standard errors, the Satorra and Bentler (1994) scaling correction was proposed as an adjustment that may be used with ML and robust WLS estimators. Satorra and Bentler's scaling correction adjusts the overall model test statistic (i.e., the chi-square test) to provide a test statistic that more accurately approximates the chi-square distribution. Their scaling correction also adjusts the standard errors to increase robustness under non-normality conditions when using ML or robust WLS estimation. The Satorra and Bentler scaled chi-square test has been found to perform better than ML or WLS chi-square tests under conditions with increasing non-normality (Chou et al., 1991; Hu et al., 1992) and under conditions with model misspecification (Curran et al., 1996). The Satorra-Bentler adjusted standard errors have also been found to be more robust than those estimated using ML and WLS under conditions with increasing non-normality (Chou et al., 1991).

Under conditions of normality and minor conditions of non-normality with continuous endogenous/dependent variables, ML estimation is generally recommended. If you are working with continuous endogenous variables that are extremely non-normal (e.g., skew greater than 2 and/or kurtosis greater than 7), the Satorra and Bentler (1994) scaling correction is recommended with ML estimation. This estimation method with the scaling correction is available in LISREL as Robust ML and is **MLM** (ML Mean-adjusted) in M*plus*. When estimating models that include ordered categorical variables as endogenous or outcome variables (e.g., Likert scale responses), particularly with less than 4 response option categories, a robust WLS approach is advised with the Satorra-Bentler scaling correction. In LISREL, **DWLS** (Diagonally Weighted Least Squares) is the robust WLS estimator. In M*plus*, the robust WLS estimator is **WLSMV** (WLS Mean- and Variance-adjusted) and is the default estimator with ordinal outcomes. M*plus* can also implement ML estimation with ordinal outcomes upon request, but model fit information is not available. When ML estimation is used with ordinal data, logistic regression is performed. In contrast, probit regression is performed when WLSMV estimation is used with ordinal data.

To request Robust ML estimation in LISREL, you can write the following command in the SIMPLIS program with the default ML estimator: `Robust Estimation`. In M*plus*, you would write the following: `ANALYSIS: Estimator = MLM`. It is important to note that the raw data are needed when using robust estimators. Examples using the estimators for categorical outcomes (i.e., DWLS and WLSMV) will be illustrated with models in Chapter 6.

MODEL TESTING

Structural equation modeling (SEM) tests a hypothesized theoretical model, which has its basis in testing theory. Models in SEM are tested for overall fit to the data, they are tested in terms of the reasonableness of parameters estimated in the model, and they can be tested by comparing a set of theoretically feasible models through model comparisons. We provide a summary of tests and indices that may be used when assessing global model fit, a brief discussion of the plausibility of parameter estimates, and methods for conducting model comparisons. Thorough discussions of model fit have been written by Hu and Bentler (1999) and West, Taylor, and Wu (2012).

Model Fit

Model fit determines the degree to which the model fits the observed data. Model-fit criteria that are commonly used include a chi-square (χ^2) test statistic of model fit accompanied by some combination of model fit indices, which are described below in more detail. We also provide some recommendations when assessing and reporting model fit information.

Chi-square (χ^2)

The fundamental hypothesis in SEM is that the population covariance matrix (Σ) is equal to the model-implied covariance matrix, $\Sigma(\theta)$: $\Sigma = \Sigma(\theta)$. These population matrices are estimated using their respective matrices in the sample, S and $\hat{\Sigma}$, respectively. Once an estimation procedure converges on a solution or completes the estimation process when it meets the convergence criterion, the weighted sum of squared differences between corresponding elements in the sample covariance matrix and the model-implied covariance matrix will result in a single fit or discrepancy function value (F; referred to as F_{ML} or F_{GLS} with ML and GLS estimators, respectively). This fit function value is multiplied by ($N-1$), providing an approximately χ^2 distributed statistic (Bollen, 1989):

$$\chi^2_{model} = F(N-1).$$

Chi-square $\left(\chi^2_{model}\right)$ is considered the only statistical test of significance for testing the theoretical model. The χ^2_{model} has degrees of freedom (df_{model}) equal to the number of distinct or non-redundant elements in the sample covariance matrix minus the number of free parameters. The theoretical model is rejected if:

$$\chi^2_{model} > critical_\alpha,$$

where $critical_\alpha$ is the chi-square critical value associated with df_{model} and α, which is the significance level of the test (e.g., $\alpha = .05$). The χ^2_{model} is testing the null hypothesis that the population covariance matrix is equal to the model-implied covariance matrix, $H_0: \Sigma = \Sigma(\theta)$. Thus, the alternative hypothesis is that the two matrices are not equal, $H_1: \Sigma \neq \Sigma(\theta)$. Because the null hypothesis means that our theoretical model explains the relationships among the variables in our data set well, we do *not* want to reject the null hypothesis. Hence, we want the χ^2_{model} to be non-significant, which would provide support for our theoretical model.

Another chi-square test available in SEM is the chi-square test of the *null, baseline,* or *independence* model, χ^2_{null}. The null or independence model is a model wherein all of the observed variables are treated as exogenous or independent and are not related with any of the other variables. Also, only the variances of the exogenous variables are estimated in the model. For instance, the null model for the union sentiment path model originally presented in Figure 4.5 would look instead like the model in Figure 5.1. Thus, the model-implied covariance matrix would consist of variances for each variable on the main diagonal and values of zero in the off-diagonal elements.

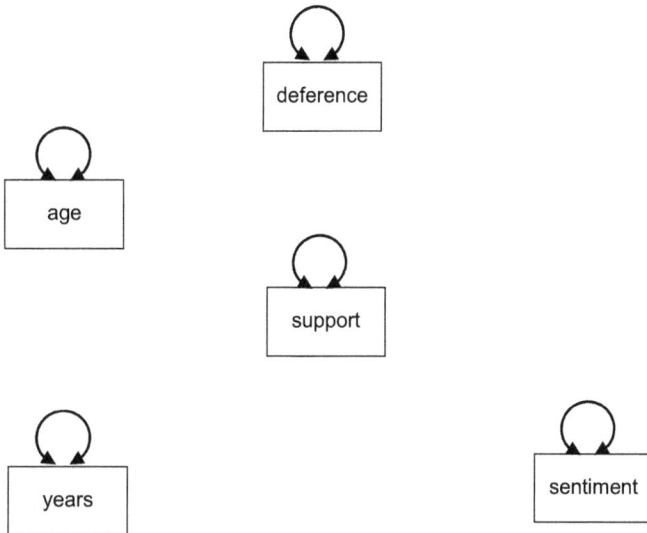

■ **Figure 5.1:** NULL OR BASELINE UNION SENTIMENT PATH MODEL

The degrees of freedom for the null union sentiment model (df_{null}) are calculated as:

$$df_{null} = \frac{p(p-1)}{2} = \frac{5(5-1)}{2} = 10,$$

where p is the number of observed variables. The null hypothesis is the same for the test of the null or baseline model as it is for the theoretical model [i.e., H_0: $\Sigma = \Sigma(\theta)$]. Hence, we *want* to reject the null hypothesis when testing the fit of the null or independence model. Specifically, if our sample covariance matrix is similar to the model-implied covariance matrix for a null model for which no relationships are hypothesized among the variables, this would imply that our variables are not related in the sample data (i.e., they do not covary). Rejecting the null hypothesis for the null or baseline model would provide support for the existence of interrelationships among the variables in our data set. If not, modeling connections among variables that do not actually covary is debatable.

The χ^2_{model} value ranges from zero for a saturated model with all paths included to a maximum value for the independence model with no paths included. Your theoretical model chi-square value will be somewhere between these two extreme values. This can be visualized as follows:

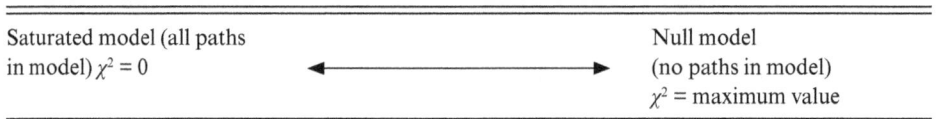

Saturated model (all paths in model) $\chi^2 = 0$	← ————————————————— →	Null model (no paths in model) χ^2 = maximum value

The χ^2_{model} is sensitive to sample size because as sample size increases (generally above 200), the χ^2_{model} test statistic is more likely to detect small differences between the observed and model-implied covariance matrices as being significant. The chi-square statistic is therefore affected by sample size, as noted by its calculation: $\chi^2_{model} = F(N-1)$, where F is a fit function for an estimator (e.g., F_{ML} for the maximum likelihood fit function). The χ^2_{model} statistic is also sensitive to departures from multivariate normality of the observed variables.

Model Fit Indices

The sensitivity of the χ^2_{model} statistic to sample size has led to the proposal of alternative fit indices that may be used to evaluate model fit in addition to the χ^2_{model} test (Bentler & Bonett, 1980; Hu & Bentler, 1999). Although different taxonomies of the model fit indices exist, Hu and Bentler (1998, 1999) organized model fit indices as *absolute* or *incremental* fit indices. Table 5.1 provides the formulas, features, and suggested cutoff criteria for commonly used SEM model fit indices.

Table 5.1: Model-fit Criteria Formulas, Features, and Acceptable Fit Interpretation

Model-fit Criterion	Features	Interpretation
Test of Overall Model Fit		
Theoretical Model Chi-square $$\chi^2_{model} = F(N-1)$$	Is equal to or greater than zero (≥ 0); sensitive to sample size	Non-significant χ^2_{model} reflects good model fit
Incremental Fit Indices		
Comparative Fit Index (CFI) $$1 - \frac{\max(\chi^2_{model} - df_{model}, 0)}{\max(\chi^2_{null} - df_{null}, \chi^2_{model} - df_{model}, 0)}$$	Range is 0 – 1	Value greater than .90 reflects good model fit[a]
Incremental Fit Index $$\frac{\chi^2_{null} - \chi^2_{model}}{\chi^2_{null} - df_{model}}$$	Theoretical range is 0 – 1; can exceed 1; sensitive to small sample size	Value greater than .90 reflects good model fit[a]
Non-normed Fit Index (NNFI) / Tucker-Lewis Index (TLI) $$\frac{\left(\chi^2_{null}/df_{null}\right) - \left(\chi^2_{model}/df_{model}\right)}{\left(\chi^2_{null}/df_{null}\right) - 1}$$	Theoretical range is 0 – 1; can fall outside of theoretical range	Value greater than .90 reflects good model fit[a]
Normed Fit Index (NFI) $$\frac{\chi^2_{null} - \chi^2_{model}}{\chi^2_{null}}$$	Range is 0 – 1; sensitive to sample size	Value greater than .90 reflects good model fit[a]
Absolute Fit Indices		
Goodness-of-Fit Index (GFI) $$1 - \left[\frac{tr(\widehat{\Sigma}^{-1}\mathbf{S} - \mathbf{I})^2}{tr(\widehat{\Sigma}^{-1}\mathbf{S})^2}\right]$$ where tr = trace function; $\widehat{\Sigma}$ = estimated covariance matrix; $\mathbf{S}$ = sample covariance matrix; $\mathbf{I}$ = identity matrix.	Theoretical range is 0 – 1; can become negative; sensitive to sample size	Value greater than .90 reflects good model fit[a]
Adjusted GFI (AGFI) $$1 - \left[p(p+1)/2df_{model}\right](1 - GFI)$$	Theoretical range is 0 – 1; can become negative; sensitive to sample size	Value greater than .90 reflects good model fit[a]
McDonald's Fit Index (MFI) $$\exp\left[\frac{-.5\left(\chi^2_{model} - df_{model}\right)}{N-1}\right]$$	Theoretical range is 0 – 1; can exceed 1; sensitive to sample size	Value greater than .90 reflects good model fit[a]
Root Mean-Square Error of Approximation (RMSEA) $$\sqrt{\frac{\max\left(\chi^2_{model} - df_{model}, 0\right)}{df_{model}(N-1)}}$$	Is equal to or greater than zero (≥ 0); sensitive to small sample size	.05 or less = close fit; .05 – .08 = adequate fit; non-significant p-value reflects good model fit
Standardized Root Mean-Square Residual (SRMR) $$\sqrt{\frac{\sum_{i=1}^{p(p+1)/2}(s\_res^2_i)^2}{p(p+1)/2}}$$ where s_res = standardized residual.	Is equal to or greater than zero (≥ 0); sensitive to sample size	0 = perfect fit; .05 or less = good fit; .05 – .10 = acceptable fit

Note: [a]indicates that values of .95 or greater have been recommended based on Hu and Bentler's (1999) study.

Incremental fit indices measure the improvement in a model's fit to the data by comparing a specific theoretical model to a null model. Some incremental fit indices that are commonly used in SEM include the Comparative Fit Index (CFI; Bentler, 1990), Bollen's (1986) Incremental Fit Index (IFI), the Normed Fit Index (NFI; Bentler & Bonnett, 1980), and the Non-normed Fit Index (NNFI), which is also the Tucker-Lewis Index (TLI; Tucker & Lewis, 1973). Conventionally, models with incremental fit values of .90 and above have been deemed as fitting the data acceptably.

Absolute model fit indices measure how well a model reproduces the observed data. Some absolute model fit indices that are commonly used include the Goodness-of-Fit Index (GFI) and its corresponding adjusted version, the Adjusted Goodness-of-Fit Index (AGFI; Jöreskog & Sörbom, 1984), McDonald's (1989) Fit Index (MFI), the Root Mean-Square Error of Approximation (RMSEA; Steiger & Lind, 1980), and the Standardized Root Mean-Square Residual (SRMR; Bentler, 1995). Traditionally, models with GFI, AGFI, and MFI values of .90 or greater have been considered as adequately fitting the data. Hu and Bentler (1995) proposed that SRMR values of .05 and less designate "good fit" while values between .05 and .10 designate "acceptable fit." Browne and Cudeck (1993) suggested that RMSEA values of .05 and less designate "close fit" while values between .05 and .08 designate "adequate" fit. The RMSEA is the only stand-alone fit index that is accompanied with a confidence interval and associated p-value. The null hypothesis being tested with the associated p-value is that the RMSEA is equal to or less than a value of .05, signifying "close fit." Adequate model fit is supported if a value of .05 is below the RMSEA's 90% confidence limits and the p-value is greater than .05. Adequate model fit can also be concluded if a value of .05 falls in the RMSEA's 90% confidence limits. With a p-value greater than .05, we would fail to reject the null hypothesis that the RMSEA is equal to or less than .05, which indicates that the model is "close-fitting." On the other hand, with a p-value less than .05, we would reject the null that the RMSEA is equal to or less than .05, which indicates that model is not "close-fitting."

Hu and Bentler (1999) conducted a substantial simulation study examining the performance of numerous SEM model fit indices under conditions of model complexity, model misspecification, and sample size. In general, Hu and Bentler (1999) recommended using .95 or greater as the cutoff value for the CFI, IFI, and NNFI/TLI; a cutoff value of .90 or greater for the MFI; a cutoff value of .06 or less for the RMSEA; and a cutoff value of .08 or less for the SRMR. They also suggested using joint model fit criteria when evaluating model fit in SEM. Their recommendation was offered to decrease the possibility of rejecting the "correct" model (similar to a Type II error) and the possibility of failing to reject the "incorrect" (misspecified) model (similar to a Type I error). Specifically, they suggested using a cutoff of .09 or less for the SRMR along with a cutoff value of .95 or greater for the CFI, IFI, or NNFI/TLI to support satisfactory model fit. They also suggested using a cutoff value of .09 or less for the SRMR combined with a cutoff value of .06 or less for the RMSEA to support acceptable fit.

In reaction to Hu and Bentler's (1999) article, higher cutoff values than were originally proposed for the CFI, IFI, and NNFI/TLI began to be endorsed (i.e., .95 and above instead of .90 and above). Although the use of Hu and Bentler's (1999) joint criteria helps decrease the possibility of both Type I and II errors in terms of model retention, the use of the joint criteria can also result in higher Type I and II errors than expected in some scenarios. Indeed, Hu and Bentler (1998; 1999) did note limitations about the use of their recommendations in all circumstances because the joint criteria may not generalize to all scenarios (Fan & Sivo, 2005; Yuan, 2005). Marsh, Hau, and Wen (2004) effectively summarized and discussed some concerns about model fit assessment in SEM applications. Although some continue to be advocates of using the joint criteria and/or the higher cutoff values, others continue to be advocates of using the cutoff values that were originally proposed.

Because of the different types of SEM model fit indices available, we offer some general recommendations about the assessment of model fit and reporting. According to Jackson, Gillaspy, and Purc-Stephenson (2009), a minimum set of SEM model fit indices to report would include the χ^2_{model} test statistic (with degrees of freedom and p-value); an incremental fit index (e.g., CFI, IFI, NNFI/TLI); and a residual-based absolute fit index, such as the RMSEA (with associated 90% CI and p-value) or the SRMR. M*plus* only reports a select set of fit indices, including the χ^2_{model} test statistic, the CFI, the NNFI/TLI, the RMSEA, and SRMR. Hence, many users of M*plus* report all of these indices. We recommend reporting the χ^2_{model} test statistic (with degrees of freedom and p-value); two incremental fit indices (e.g., CFI, IFI, NNFI/TLI); the RMSEA (with associated 90% CI and p-value); and the SRMR. We also endorse the use of the original model fit cutoff values that were proposed when evaluating model fit in SEM. Because the true SEM model in the population is unknown, our preference when evaluating model fit would be to fail to reject an incorrect or misspecified model as opposed to the rejection of the "true" or "correct" model if reachable.

The model fit output from LISREL and M*plus* for the union sentiment model that was introduced in Chapter 4 is presented in Tables 5.2 and 5.3, respectively. In LISREL, the χ^2 (3) = 1.251, p > .05 [labeled the "Maximum Likelihood Ratio Chi-Square (C1)"] and in M*plus*, the χ^2 (3) = 1.259, p > .05 (labeled the "Chi-square Test of Model Fit") are both non-significant (p ≈ .74), supporting good model fit to the data. The RMSEA estimate is equal to 0.0 (with a 90% confidence interval: 0.0, .09), indicating close fit of the model to the data. The p-value for the RMSEA is equal to .842, which is greater than .05, meaning that you would fail to reject the null hypothesis of "close fit" of your model to the data. The CFI and the

TLI are greater than .90, suggesting good model fit. Lastly, the SRMR is less than .05, indicating good model fit. Based on this set of model fit indices, the union sentiment path model fits the data well. When reporting model fit information in SEM, it could be presented in a table or in the text.

Table 5.2: LISREL Output with Model Fit Information for Union Sentiment Path Model

	Log-likelihood Values	
	Estimated Model	Saturated Model
Number of free parameters(t)	12	15
-2ln(L)	3079.381	3078.130
AIC (Akaike, 1974)*	3103.381	3108.130
BIC (Schwarz, 1978)*	3141.151	3155.342

*LISREL uses AIC= 2t - 2ln(L) and BIC = tln(N)- 2ln(L)

Goodness-of-Fit Statistics

Degrees of Freedom for (C1)-(C2)	3
Maximum Likelihood Ratio Chi-Square (C1)	1.251 (P = 0.7407)
Browne's (1984) ADF Chi-Square (C2_NT)	1.248 (P = 0.7415)
Estimated Non-centrality Parameter (NCP)	0.0
90 Percent Confidence Interval for NCP	(0.0 ; 4.214)
Minimum Fit Function Value	0.00728
Population Discrepancy Function Value (F0)	0.0
90 Percent Confidence Interval for F0	(0.0 ; 0.0245)
Root Mean Square Error of Approximation (RMSEA)	0.0
90 Percent Confidence Interval for RMSEA	(0.0 ; 0.0904)
P-Value for Test of Close Fit (RMSEA < 0.05)	0.842
Expected Cross-Validation Index (ECVI)	0.157
0 Percent Confidence Interval for ECVI	(0.157 ; 0.181)
ECVI for Saturated Model	0.174
ECVI for Independence Model	1.206
Chi-Square for Independence Model (10 df)	197.424
Normed Fit Index (NFI)	0.994
Non-Normed Fit Index (NNFI)	1.031
Parsimony Normed Fit Index (PNFI)	0.298
Comparative Fit Index (CFI)	1.000
Incremental Fit Index (IFI)	1.009
Relative Fit Index (RFI)	0.979
Critical N (CN)	1560.660
Root Mean Square Residual (RMR)	0.731
Standardized RMR	0.0148
Goodness of Fit Index (GFI)	0.997
Adjusted Goodness of Fit Index (AGFI)	0.986
Parsimony Goodness of Fit Index (PGFI)	0.199

Table 5.3: M*plus* Output with Model Fit Information for Union Sentiment Path Model

```
MODEL FIT INFORMATION

Number of Free Parameters                        12

Loglikelihood

            H0 Value                     -2332.069
            H1 Value                     -2331.439

Information Criteria

            Akaike (AIC)                  4688.137
            Bayesian (BIC)                4725.977
            Sample-Size Adjusted BIC      4687.978
              (n* = (n + 2) / 24)

Chi-Square Test of Model Fit

            Value                            1.259
            Degrees of Freedom                   3
            P-Value                         0.7390

RMSEA (Root Mean Square Error Of Approximation)

            Estimate                         0.000
            90 Percent C.I.                  0.000  0.090
            Probability RMSEA <= .05         0.842
CFI/TLI

            CFI                              1.000
            TLI                              1.000

Chi-Square Test of Model Fit for the Baseline Model

            Value                          153.027
            Degrees of Freedom                   9
            P-Value                         0.0000

SRMR (Standardized Root Mean Square Residual)

            Value                            0.015
```

Model fit in the SEM arena continues to be a topic of deliberation and new ways to assess model fit continue to be proposed. Various characteristics of the data and/or the model can impact model fit in SEM, such as estimation procedure, model complexity, sample size, non-normality, the calculation of the fit index, and the relationships among variables (Marsh, Balla, & McDonald, 1988; Mulaik, James, Alstine, Bennett, Lind, & Stilwell, 1989; Bollen, 1990; Marsh, Balla, & Hau, 1996; Kenny & McCoach, 2003). For this reason, model fit should also be assessed by the suitability of the parameter estimates for the relationships among variables in the theoretical model, which is discussed in the next subsection.

Parameter Fit

Even though model fit criteria indicate an acceptable model, the interpretation of parameter estimates in a model is essential. For instance, if the relationship between two variables (e.g., growth mindset and success) has been regularly demonstrated to be positively and moderately associated in the relevant literature, the estimated relationship between the same two variables in a similar study should agree with past findings. If a hypothesized relationship is not found, other problems could be happening with the variables in the theoretical model that should be investigated and/or fixed.

An examination of parameter estimates can also help in identifying a faulty or misspecified model. Sometimes parameter estimates take on impossible values, as in the case of a correlation between two variables that exceeds 1 or a negative error variance is encountered (known as *Heywood* cases). Problems with parameter estimates could be the result of suppression and/or multicollinearity. Other problems could arise from poor reliability of the surveys used to measure the variables in a model. Thus, the parameter estimates should be examined to determine whether they have the correct sign (either positive or negative) and to determine if the standardized coefficients are out of bounds or exceed an expected range of values. Although SEM software programs will most likely indicate a warning or error message when improper solutions (e.g., correlation greater than 1) occur, there could be instances in which a warning or error message may not be provided by the software program. Hence, it is recommended that parameter estimates always be evaluated for their appropriateness.

Besides issues dealing with improper solutions or estimates as mentioned above, SEM software programs may not be able to estimate model parameters for various reasons. When SEM software programs are unable to estimate model parameters, there will most likely be a warning or error message indicating the problem. Some error messages are more straightforward than others. One error message that may be encountered is that the maximum number of iterations is exceeded during the estimation process and parameter estimates could not be computed. There are default settings in SEM software programs for the maximum number of iterations permitted during the iterative estimation process. When the maximum number of iterations is exceeded without converging on a solution, it means that the change in the discrepancy value from one iteration cycle to the next did not reach the specific default value used by the SEM software program (e.g., < .000001 in LISREL; < .00005 in M*plus*) given the maximum number of iterations permitted. Users can increase the maximum number of iterations permitted in the SEM software program to see if that will produce appropriate parameter estimates. This can be done in LISREL using the OPTIONS command (e.g., **Options IT = 2000**) and in M*plus* in the ANALYSIS command (e.g., **ANALYSIS: Iterations = 2000**). In LISREL, there is also an iteration number associated with a test of the admissibility of parameter estimates. The default is 20 in LISREL and can be increased or turned off using the OPTIONS command (**Options AD = 100** or **Options AD = OFF,** respectively). If you find yourself increasing iterations to be greater than 3000 and still have convergence problems, there are probably other issues related to the data and/or the model.

Another method that may help the estimation process converge on a solution is to provide *start values* for certain model parameters. All SEM software programs have default starting values that are assigned initially to different model parameters (e.g., direct effects, variances) during the estimation process. The closer the starting values are to the final model parameter estimates, the quicker the estimation process will converge on a solution. On the other hand, as the distance between the starting values and the final model parameters increases, the more likely the maximum number of iterations will be exceeded during the estimation process. In both LISREL and M*plus*, you will be able to see the model estimates at the final iteration if it did not converge during the estimation process. These final model parameter estimates may be evaluated in terms of their appropriateness in order to possibly find the offending parameter (which has an inappropriate final estimate) and/or they may be used as starting values when trying to rerun the model and estimate model parameters. If the final model estimates used as starting values do not work, Bollen (1989) provides some recommendations, which are presented in Table 5.4.

Table 5.4: Bollen's (1989) Suggested Start Values with Observed Variables

Parameter to Estimate	Start Value	Theoretical Value of *a*
Direct effect $(X \rightarrow Y)$	$a\dfrac{s_Y}{s_X}$	$|a| = .9$ strong direct effect $|a| = .4$ moderate direct effect $|a| = .2$ weak direct effect
Covariance $(X \leftrightarrow Y)$	$a(s_Y s_X)$	$|a| = .9$ strong correlation $|a| = .4$ moderate correlation $|a| = .2$ weak correlation.
Error variance	$a\left(s_Y^2\right)$	$a = .2$ strong fit
Disturbance variance	$a\left(s_F^2\right)$	$a = .4$ moderate fit $a = .9$ weak fit

Note: the value of *a* should be selected based on theoretical knowledge; s_X is the standard deviation of variable X; s_Y is the standard deviation of variable Y; s_Y^2 is the variance of variable Y; s_F^2 is the variance of the latent factor.

Using the union sentiment path model, suppose the parameter estimate associated with the direct effect from *deference* to *sentiment* was the offending estimate. If a moderate relationship was hypothesized to exist between *deference* and *sentiment*, the start value for that parameter would be calculated as: .4*(5.65/3.82) = .59. The standard deviations of *sentiment* (5.65) and *deference* (3.82) were obtained by taking the square root of their variances from the covariance matrix, which are 31.971 and 14.610, respectively. You can assign a start value for the direct effect from *deference* to *sentiment* as follows in the SIMPLIS program:

```
Sentiment = Years (.59)*Deference Support
```

It is important that you use parenthesis around start values, otherwise the direct effect will be fixed equal to the start value. In M*plus*, you can assign start values for the direct effect as follows:

```
Model:  Sentimen on Years Deferenc*.59 Support;
```

The asterisks are important to use instead of the @ symbol in M*plus*, which will instead fix the parameter value to the value following the @ symbol.

Another error message that may be encountered when estimating models in SEM software is that a matrix is *non-positive definite* (Wothke, 1993). This error message can occur for various reasons, including a linear dependency or multicollinearity, *Heywood cases*, bad *start values*, missing data, local under-identification, and small sample sizes. Other data issues can produce improper solutions, such as outliers, highly complex models, and low factor loadings (Gerbing & Anderson, 1987; Chen, Bollen, Paxton, Curran, & Kirby, 2001; Gagné & Hancock, 2006). In the end, resolving error messages in SEM may feel like trial and error at times. Ensuring that the data meet the required assumptions in SEM and proper model specification will decrease the possibility of improper solutions and error messages.

Model Comparison

Some actually advocate for comparing theoretically plausible models instead of simply assessing the fit of one theoretical model. Model comparison involves comparing the fit between alternative models, and/or testing parameter coefficients for significance between two alternative models. These hypothesis testing methods generally involve a nested, less parameterized (or *restricted*) model and more parameterized, less restricted (*unrestricted*) model. The restricted and unrestricted models are nested if the set of parameters to be estimated in the restricted model is a subset of the parameters to be estimated in the unrestricted model. For models to be nested, the same variables must be included in the models (i.e., you cannot delete a variable in one model) and the same data set must be used.

Each model generates a χ^2 test statistic, and the difference between the models for significance testing is computed as: $\Delta\chi^2 = \chi^2_{restricted} - \chi^2_{unrestricted}$, with associated degrees of freedom for the $\Delta\chi^2$ test: $\Delta df = df_{restricted} - df_{unrestricted}$. When the $\Delta\chi^2$ test is significant $\left(\Delta\chi^2 > critical_{\alpha}\right)$, the nested or restricted model with fewer parameters has been oversimplified. Specifically, the nested or restricted model has significantly decreased in overall model fit relative to the more parameterized or unrestricted model. In this case, the unrestricted, more parameterized model would be chosen over the restricted, less parameterized model. In contrast, when the $\Delta\chi^2$ test is not significant $\left(\Delta\chi^2 < critical_{\alpha}\right)$, the two nested models are similar in terms of overall model fit. In this case, the restricted, less parameterized model would be selected over the unrestricted, more parameterized model for purposes of parsimony. It must be noted that the $\Delta\chi^2$ test is possible because it uses likelihood ratio χ^2 statistics instead of Pearson χ^2 statistics.

Using the union sentiment path models in Figure 5.2 as an example, suppose theory also supported a direct effect from years worked in the textile mill (*years*) to support for labor activism (*support*). In this comparison, the original union sentiment model would be the *restricted* model which is nested in the model with the direct effect from *years* to *support* now included, which is the *unrestricted* model. After running the alternative model with the direct effect included, the unrestricted model's chi-square from M*plus* is: $\chi^2_{unrestricted}(2) = .843, p > .05$. If you recall, the restricted model's chi-square from M*plus* is: $\chi^2_{restricted}(3) = 1.259, p > .05$ (see Table 5.3). You will notice that the restricted model has one more degree of freedom than the unrestricted model because the two models differ by one single parameter estimate. Restricted models will have larger degrees of freedom relative to their unrestricted comparison models and will generally have larger chi-square values. Hence, the more parameters that are estimated, the smaller the degrees of freedom and the chi-square value tend to become.

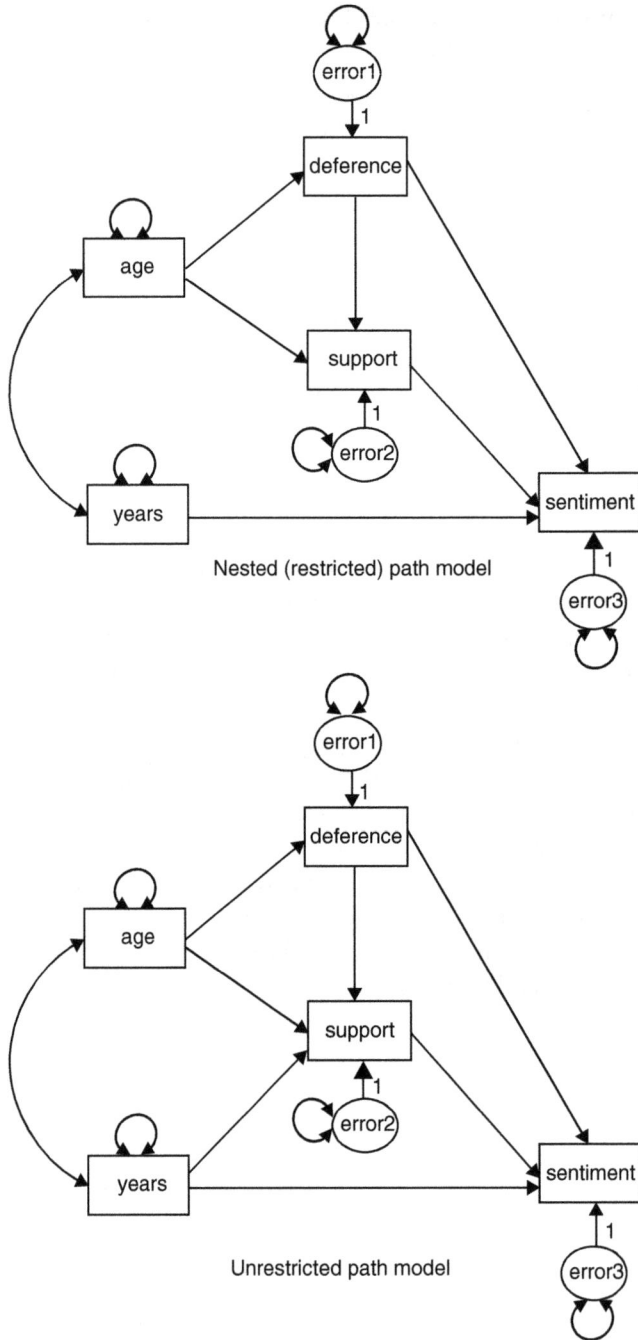

Figure 5.2: NESTED UNION SENTIMENT PATH MODELS

The $\Delta\chi^2 = 1.259 - .843 = .416$ and corresponding $\Delta df = 3 - 2 = 1$. The critical χ^2 value with 1 df and an $\alpha = .05$ is 3.84 (see Table A.4 in Appendix 2). An Excel spreadsheet is also available on the book website (**chi-square difference.xls**) for you to use and enter the $\Delta\chi^2$ value in cell A2 and the Δdf value in cell B2. It will return the one-tailed probability associated with the $\Delta\chi^2$ in cell C2. If the probability in C2 is less than .05, there is a significant difference between the two models. Figure 5.3 illustrates the chi-square difference test results in the Excel spreadsheet, which indicates that the models are not significantly different, $p > .05$. Because the models do not differ significantly, the original (restricted) model would be selected. This also suggests that the direct effect from years to support does not contribute significantly to the model.

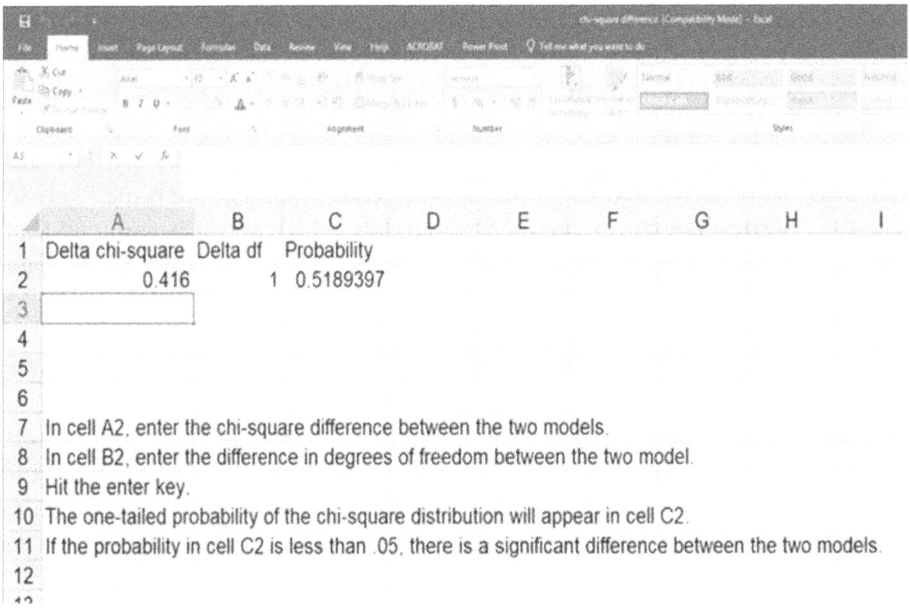

■ Figure 5.3: EXCEL SPREADSHEET WITH CHI-SQUARE DIFFERENCE TEST RESULTS FOR NESTED UNION SENTIMENT MODEL COMPARISON

It must be noted that if a robust estimator was used with the Satorra and Bentler (1994) scaling correction, the $\Delta\chi^2$ test must be modified. If you used ML robust estimation (Robust ML in LISREL; **MLM** or **MLR** in M*plus*; or **ML Robust** in EQS), a description of the Satorra-Bentler scaled chi-square difference test can be viewed here on the M*plus* website: https://www.statmodel.com/chidiff.shtml. An online calculator is available here: http://www.thestatisticalmind.com/calculat ors/SBChiSquareDifferenceTest.htm. An Excel spreadsheet was also created by Bryant and Satorra (2012) and can be retrieved here: http://econ.upf.edu/~sato rra/. If you used WLS robust estimation (**DWLS** in LISREL or **WLSMV** in

M*plus*), a test has been proposed by Asparouhov and Muthén (2010), which can be conducted using the **DIFFTEST** procedure in M*plus*.

As previously mentioned, a problem associated with the χ^2 test statistic is its sensitivity to sample size. This problem is also associated with the $\Delta\chi^2$ test, where larger sample sizes may lead to small discrepancies in fit resulting in statistically significant differences. Thus, researchers should be cautious when interpreting $\Delta\chi^2$ tests with larger sample sizes.

Models estimating relationships among the same variables that are not nested cannot be compared using the $\Delta\chi^2$ test because the nested model's parameters would not be a subset of the parameters estimated in the model in which it is nested, making a $\Delta\chi^2$ test invalid. Although the χ^2 tests for the non-nested models could be compared relatively, this would not be recommended given that more parameterized models are more inclined to be associated with smaller χ^2 test values than more parsimonious models. Model fit indices of the non-nested models involved in the comparison could also be compared relatively, but this would also not be advocated since more parameterized models are more likely to fit the data better than less parameterized models. Moreover, it is assumed that when comparing alternative models that the models to be compared fit the data adequately in the first place.

Information criteria have been proposed as indices that can be used to compare non-nested models. Information criteria are different from the SEM model fit indices mentioned previously because they do not have cutoff values. Thus, information criteria are used to compare alternative models relatively, meaning that the model with the lowest information criterion value would be selected as the model with the most predictive validity. It should be noted that information criteria can be used for nested *or* non-nested model comparisons.

Different information criteria are available, but SEM software programs tend to output a select few. For instance, LISREL prints values for two information criteria in the output whereas M*plus* prints values for three information criteria in the output. Both LISREL and M*plus* provide Akaike's (1987) Information Criterion (AIC):

$$AIC = -2LL + 2*t,$$

where $-2LL$ is the $-2*$loglikelihood and t is the number of parameters estimated in the model; and Schwarz's Bayesian information criterion (BIC; Schwarz, 1978):

$$BIC = -2LL + t[\ln(N)],$$

where ln is the natural log and N is the sample size. M*plus* also provides the sample size adjusted BIC (NBIC; Sclove, 1987):

$$NBIC = -2LL + ln\left(\frac{n+2}{24}\right)t.$$

You may see differences in the calculation of the information criteria in other references. Nevertheless, the model with the smallest information criterion value would be chosen as having the best predictive validity. Readers are encouraged to consult West et al. (2012) for more details about information criteria.

Although the models in Figure 5.2 are nested and the $\Delta\chi^2$ already supported the original union sentiment model, we can also use the information criteria to compare the two models. Table 5.5 presents the LISREL and M*plus* information criteria for the two union sentiment models. You will notice some differences in the estimates of the AIC and BIC in LISREL and M*plus*. Nonetheless, you can see that the values for each information criterion is smaller for the original nested (restricted) model than for the unrestricted model. Thus, the original model would be selected over the unrestricted model.

Table 5.5: Information Criteria for Nested Union Sentiment Path Models

Model	L_AIC	L_BIC	M_AIC	M_BIC	M_NBIC
Original nested (Restricted) model	3103.381	3141.151	4688.137	4725.977	4687.978
Unrestricted model	3104.968	3145.886	4689.722	4730.715	4689.550

Note: L_AIC = LISREL AIC; L_BIC = LISREL BIC; M_AIC = M*plus* AIC; M_BIC = M*plus* BIC; M_NBIC = M*plus* NBIC.

MODEL MODIFICATION AND RE-SPECIFICATION

When SEM models do not fit the data adequately, as suggested by the χ^2_{model} test and/or model fit indices, researchers could then modify their originally hypothesized model and retest the model fit to the data after making modifications (MacCallum, Roznowski, & Necowitz, 1992; Marcoulides & Drezner, 2001, 2003). Unacceptable model fit may be thought by some to show that the theoretical model is not reasonable because models in SEM are based on theory-driven connections among the variables, making the modification and restesting of the model inadvisable. Then again, some may believe that unacceptable model fit suggests that specification errors are in the model, indicating a difference between the correct or true model in the population and the theoretically-driven model that was tested. Specification errors mean that unimportant connections are included in the model and/or important connections are excluded in the model (MacCallum, 1986).

Unacceptable model fit is generally due to excluding important connections, which is viewed by some as worse than including unimportant connections (Saris, Satorra, & van der Veld, 2009). Hence, associations or connections among variables will most likely be added to the original theoretical model by applied researchers in order to improve model fit. Theoretically-driven models are approximations of the true population model because the correct or true model in the population is unknown to the researcher (Cudeck & Browne, 1983). For this reason, some view the process of model modification as an important part of the search for a model that adequately describes the relationships among the variables in the model (Saris et al., 2009).

Aside from the diverse perspectives about the modification and retesting of a model in SEM, it is universally acknowledged that as soon as model modifications begin, the model is no longer *a priori* and becomes *exploratory*. Thus, model modifications should be based on sound theory and the modified model should be cross-validated in a different sample to support its predictive validity in subsequent samples (MacCallum et al., 1992). The next subsections present methods available in SEM software that may aid applied researchers during the model modification process.

Modification Indices

Different statistics are available in SEM software that may be used to help assess which parameters may be freely estimated or which connections could be added to the model to improve its fit to the data. The *modification index* (MI) is computed in LISREL, M*plus*, and AMOS software packages whereas the *Lagrange Multiplier* (LM) test is computed in EQS software. Satorra (1989) and Sörbom (1989) provide more details about modification indices (MIs) or LM tests. MIs or LM tests are estimated values of how much the χ^2_{model} value would decrease if a specific connection between two variables (e.g., a direct effect or a covariance) was freely estimated in the model. Parameters or connections that are linked with an MI or LM value that is greater than 3.84, which is the χ^2 critical with 1 *df* at an $\alpha = .05$ (see Table A.4 in Appendix 2), would be evaluated in terms of its theoretical justification if included in the model. Large MI or LM values (i.e., greater than 3.84) are preferred because you want to add a parameter to the model that would significantly decrease the model's χ^2 value.

Depending upon the complexity of the original model, there could be only a few potential parameters or there could be numerous potential parameters that could be added to the model for estimation. When multiple parameters that could be added to the model are associated with large MI or LM values (i.e., greater than 3.84), we recommend that they be compared relatively. Thus, the parameters associated with the largest MI/LM value relative to other parameters would first be assessed in terms of theoretical justification. If the parameter to be added is theoretically defensible, it could be added to the model for estimation. Otherwise, the parameter associated with the next largest MI/LM value relative to other parameters should be evaluated in terms of theoretical justification. This procedure could proceed until no other parameters would significantly decrease the model's χ^2 and/or if the parameters to be included cannot be defended based on theory. We also recommend that parameters be added to the model for estimation one at a time. When adding parameters to a model, other parameter estimates could change slightly or substantially. Hence, adding parameters one at a time to a model can help detect whether changes in other parameter estimates are occurring with the addition of the parameter to the model.

There have been methodological research studies in which the performance of the MI/LM has been examined, finding disappointing results. MacCallum (1986) conducted a simulation study evaluating the accuracy of the MI/LM when reaching the correct model under different model misspecification, sample size, and search strategy conditions. MacCallum's (1986) findings revealed that the model modification process when using the MI/LM resulted in arriving at the incorrect model too frequently. Some studies have demonstrated more encouraging findings when using the MI/LM during model modifications (Chou & Bentler, 1990; Hutchinson, 1993). Nevertheless, the findings demonstrated in other studies (Kaplan, 1988; Silvia & MacCallum, 1988; MacCallum et al., 1992) were similar to those found by MacCallum (1986).

Expected Parameter Change

The *expected parameter change* (EPC) is another statistic that is available that may help during the model modification process. Saris, Satorra, and Sörbom (1987) introduced the unstandardized version of the EPC, which is an approximated value of the parameter estimate for a relationship (e.g., direct effect or covariance) if it was added to the model. As with the MI/LM, parameters with the largest EPC relative to other parameters would be evaluated in terms of theoretical justification and added to the model if theory agrees. This procedure would proceed until no other parameters have large EPC values relative to other parameters and/or if the parameter to be included cannot be defended based on theory. Standardized versions of the EPC are also available, which should also be evaluated relatively (Kaplan, 1989; Luijben, 1989; Chou & Bentler, 1993).

Saris et al. (1987) defined the four different combinations of the MI/LM and the EPC that could occur when evaluating whether a parameter should be added to a model: 1) a relatively large and statistically significant MI/LM combined with a relatively large EPC value for a specific parameter; 2) a relatively large and statistically significant MI/LM combined with a small EPC value for a specific parameter; 3) a relatively small and non-significant MI/LM combined with a relatively large EPC value for a specific parameter; and 4) a relatively small and non-significant MI/LM combined with a relatively small EPC value for a specific parameter. Combination 1 would strongly support the inclusion of the specific parameter if theory agrees. Combination 2 would not support the inclusion of the parameter because the parameter estimate would be small. Combination 3 is inconclusive and could suggest an underpowered MI/LM test. Finally, combination 4 would clearly not support the inclusion of the parameter. Methodological studies have investigated the performance of the EPC, finding encouraging results with respect to accuracy during model modification (Kaplan, 1989; Luijben & Boomsma, 1988; Whittaker, 2012).

MIs and EPCs in LISREL and M*plus*

Both LISREL and M*plus* provide modification indices and expected parameter change values (unstandardized and standardized versions). Because the union sentiment path model fits the data well, MIs and EPC values would not necessarily be considered. However, suppose that the direct effect from *deference* to *support* was removed from the original union sentiment path model as illustrated in Figure 5.4. When estimating this model, the model fit indices suggest poor fit to the data in LISREL [$\chi^2(4) = 21.303$, $p < .05$; CFI = .908; NNFI = .769; RMSEA = .159 (90% CI: .097, .228; $p < .05$); SRMR = .087] and in M*plus* [$\chi^2(4) = 21.427$, $p < .05$; CFI = .879; TLI = .728; RMSEA = .159 (90% CI: .097, .228; $p < .05$); SRMR = .086].

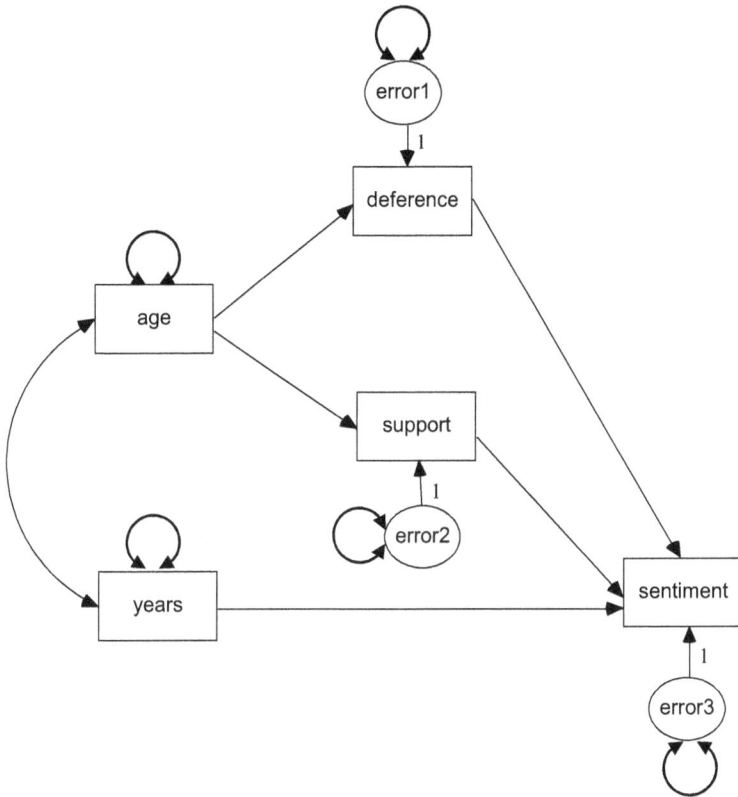

Figure 5.4: POOR FITTING MISSPECIFIED UNION SENTIMENT MODEL

To request MIs and EPCs in LISREL, you can use the following statement in LISREL: **Options: MI**. In M*plus*, you can write the following statement in the OUTPUT command: **OUTPUT: MODINDICES (3.84)**. You will notice the value of 3.84 following the request for MODINDICES in the M*plus* code. This is to request only statistically significant modification index values in the output. If you wanted to look at all of the modification index values, you can use a value of zero (0) instead of 3.84.

LISREL output with modification indices and expected change values is presented in Table 5.6. You will notice that LISREL output will suggest adding certain parameters with significant MI values (i.e., MIs > 3.84). You will also see modification indices for BETA, GAMMA, PSI, THETA-EPS, and THETA-DELTA-EPS. BETA includes direct paths among endogenous variables; GAMMA includes direct effects from exogenous variables to endogenous variables; PSI includes endogenous error covariances; THETA-EPS includes errors for endogenous variables; and THETA-DELTA-EPS includes covariances between exogenous variables and endogenous variables. As seen in Table 5.6, modification index values associated with the connection between *deference* and *support* is suggested with direct effects and error covariances. The decrease in chi-square is approximately 18.9 if a direct effect between *deference* and *support* was included. The decision now would be how to connect *deference* and *support* directly (i.e., either *deference* → *support* or *deference* ← *support*). Hence, theory would be used to support the directionality of the relationship between *deference* and *support*. Although the direct effect from *deference* to *support* would be smaller than if the direct effect from *support* to *deference* was added based on the expected change values (–.285 and –.387, respectively), theory should also help to judge the feasibility of the expected parameter estimates. LISREL provides the *Expected Change* and the *Standardized Expected Change*, which represents unstandardized and standardized EPC values, respectively. The standardization of the EPC in LISREL is sometimes referred to as a partial standardization and is explained in more detail in Chou and Bentler (1993).

Table 5.6: LISREL Output with Modification Indices and Expected Parameter Change

```
Modification Indices and Expected Change

        The Modification Indices Suggest to Add the
Path to        from                Decrease in Chi-Square    New Estimate
    Deference       Support              18.9                    -0.39
    Deference       Sentiment            11.2                    -0.31
      Support       Deference            18.9                    -0.28

        Modification Indices for BETA

                    Deference        Support        Sentiment
                  ------------     ------------    ------------
    Deference          - -            18.927          11.232
      Support        18.927            - -             2.405
    Sentiment          - -             - -             - -

        Expected Change for BETA

                    Deference        Support        Sentiment
                  ------------     ------------    ------------
    Deference          - -           -0.387          -0.306
      Support        -0.285            - -            0.207
    Sentiment          - -             - -             - -

        Standardized Expected Change for BETA

                    Deference        Support        Sentiment
                  ------------     ------------    ------------
    Deference          - -           -0.030          -0.014
      Support        -0.022            - -            0.011
    Sentiment          - -             - -             - -
```

Table 5.6: (*cont.*)

```
      Modification Indices for GAMMA

                    Years              Age
                 ------------      ------------
Deference          0.342              - -
  Support          0.169              - -
Sentiment          - -               0.517

      Expected Change for GAMMA

                    Years              Age
                 ------------      ------------
Deference          0.181              - -
  Support          0.109              - -
Sentiment          - -               0.021

Standardized Expected Change for GAMMA

                    Years              Age
                 ------------      ------------
Deference          0.048              - -
  Support          0.033              - -
Sentiment          - -               0.055

No Non-Zero Modification Indices for PHI

The Modification Indices Suggest to Add an Error Covariance
Between        and                 Decrease in Chi-Square      New Estimate
    Support      Deference                  18.9                   -3.69

      Modification Indices for PSI

                  Deference          Support         Sentiment
                 ------------      ------------      ------------
Deference          - -
  Support          18.927             - -
Sentiment          0.517             0.517             - -

      Expected Change for PSI

                  Deference          Support         Sentiment
                 ------------      ------------      ------------
Deference          - -
  Support          -3.688             - -
Sentiment          3.072            -2.387             - -

      Standardized Expected Change for PSI

                  Deference          Support         Sentiment
                 ------------      ------------      ------------
Deference          - -
  Support          -0.291             - -
Sentiment          0.145            -0.130             - -
```

Table 5.6: (cont.)

```
The Modification Indices Suggest to Add an Error Covariance
Between        and                    Decrease in Chi-Square      New Estimate
    Support        Deference                   19.4                   -3.64

        Modification Indices for THETA-EPS

                      Deference          Support          Sentiment
                    ------------      ------------      ------------
Deference             0.517
  Support            19.382              0.517
 Sentiment            0.517              0.517              - -

        Expected Change for THETA-EPS

                      Deference          Support          Sentiment
                    ------------      ------------      ------------
Deference            14.110
  Support            -3.642             2.809
 Sentiment            3.072            -2.387              - -

        Modification Indices for THETA-DELTA-EPS

                      Deference          Support          Sentiment
                    ------------      ------------      ------------
   Years              0.240             0.369              0.517
     Age              4.765             5.423              0.517

        Expected Change for THETA-DELTA-EPS

                      Deference          Support          Sentiment
                    ------------      ------------      ------------
   Years              0.118             0.121             -0.491
     Age             12.368            -8.098              2.823
```

When requesting only significant modification indices in M*plus* [i.e., MODINDICES (3.84)], the output looks like that provided in Table 5.7. The "ON Statements" represent direct effects and the "WITH Statements" represent covariances (which include error covariances). The modification indices associated with each parameter estimate is listed under the "M.I." column. Similar to LISREL, M*plus* also indicates that a connection between *deference* and *support* would decrease the chi-square value significantly and most substantially. Specifically, including a direct effect between *deference* and *support* in the model would decrease the chi-square by about 19 points. You will notice that there are three E.P.C. columns in M*plus* output, one that is unstandardized and two that are standardized. The "E.P.C." is the unstandardized value; the "Std E.P.C." uses the variances of continuous latent variables in its standardization; and "StdYX E.P.C." uses the variances of continuous latent variables and the variances of outcome and/or background variables (referred to as fully standardized; Chou & Bentler, 1993) in its standardization. These standardizations are different than those used in LISREL. Again, the E.P.C. for the direct effect from *deference* to *support* is smaller than the E.P.C.

for the direct effect from *support* to *deference*. Again, theory and the reasonableness of the expected parameter estimates should also be evaluated during the modification process.

Table 5.7: M*plus* Output with Modification Indices and Expected Parameter Change

```
MODEL MODIFICATION INDICES

NOTE: Modification indices for direct effects of observed dependent
variables regressed on covariates may not be included. To include
these, request MODINDICES (ALL).

Minimum M.I. value for printing the modification index   3.840

                           M.I.    E.P.C.    Std E.P.C.    StdYX E.P.C.

ON Statements

DEFERENC ON SUPPORT       19.036   -0.387      -0.387         -0.336
DEFERENC ON SENT          11.294   -0.306      -0.306         -0.443
SUPPORT ON DEFERENC       19.037   -0.285      -0.285         -0.328

WITH Statements

SUPPORT WITH DEFERENC     19.037   -3.667      -3.667         -0.332
```

Based on the MI and the EPC information, provided that theory also supported the relationship, the direct effect from *deference* to *support* could be added to the model. After adding the direct effect from *deference* to *support*, none of the potential model parameters were associated with a significant MI value as indicated in the M*plus* output below:

```
MODEL MODIFICATION INDICES

NOTE: Modification indices for direct effects of observed dependent
variables regressed on covariates may not be included. To include
these, request MODINDICES (ALL).

Minimum M.I. value for printing the modification index   3.840

                        M.I.    E.P.C.  Std E.P.C.  StdYX E.P.C.

No modification indices above the minimum value.
```

An important consideration when using the modification indices is that they are similar to stepwise regression techniques, which are essentially based on statistical significance. When using MIs in SEM software, the potential parameters that are listed in the relevant output are fixed and/or constrained parameters and some may not necessarily be reasonable parameters to add to the re-specified model. We strongly encourage that any potential model modifications be decided in view of theoretical plausibility. Also, the order in which parameters are added and freely estimated in a re-specified model can impact other model parameter estimates already in the model and, in turn, the MI and EPC values. We strongly recommend that re-specified models be cross-validated in a subsequent sample to provide more predictive validity for the re-specified model.

SUMMARY

This chapter explained the *basics* of structural equation modeling. The SEM modeling steps (i.e., specification, identification, estimation, testing, and modification/re-specification) were introduced. In subsequent chapters, we will present various SEM model applications that will help the five modeling steps become clearer.

The selection of the appropriate model estimation procedure is important for accurate parameter estimates, standard errors, and model fit statistics. The default estimator in SEM software packages is maximum likelihood (ML) estimation, which assumes multivariate normality and has many desirable properties. We provided recommendations about the most appropriate estimators when data are not normally distributed or categorical, including robust estimators that provide the Satorra and Bentler (1994) chi-square scaling correction.

Models can be tested using model fit information, parameter fit information, and by way of model comparisons. Model fit is a substantial area of research and topic of considerable debate in the SEM arena. We provided recommendations about evaluating model fit and model fit reporting practices. Using the union sentiment path model introduced in Chapter 4, we evaluated its fit to the data using these model fit recommendations. Once parameters in a model are estimated, it is important to evaluate the estimated parameters to make sure their strength and direction are expected and that the values are plausible. When parameters are unable to be estimated, it could be due to various data and/or model specification reasons. Thus, it is important that data are suitable for the SEM analysis and that models are specified correctly. Suggestions were provided when the estimation process does not converge on a solution, including increasing iterations and using start values. Model comparisons can involve nested and non-nested models. Depending upon the nested relationship of comparison models, they may be compared statistically and/or relatively. The union sentiment path model example was used to illustrate model comparisons.

Model modification and re-specification can occur when a theoretical model does not achieve adequate model fit to the data. Researchers will not know the true SEM population model, making the model modification and re-specification process complicated. Modification indices and expected parameter change values can be used to help during the model modification and re-specification process. Model modifications are not advisable if theory does not support the relationships in the re-specified model. When model modifications begin, the model is considered exploratory and the re-specified model should be cross-validated in another sample of data. There are also search algorithms that may be implemented during automated model specification search processes in SEM (Marcoulides & Falk, 2018), which continue to be investigated.

EXERCISES

1. Which of the following model fit indices does **NOT** take model complexity or the number of free parameters in the model into account in its calculation? Circle all that apply.

   CFI   NNFI/TLI   NFI   AGFI   IFI   RMSEA

2. Run the union sentiment path model and use a sample size of 1000 instead of 173. The LISREL and M*plus* model fit output (based on the larger sample size of 1000) are illustrated in Tables 5.8 and 5.9, respectively. Looking at the new output in Tables 5.8 and 5.9, describe what happened (if anything) to the χ^2_{model}, the CFI, NNFI/TLI, RMSEA, and SRMR as compared to the same model fit indices when using $n = 173$ (see Tables 5.2 and 5.3 for output based on $n = 173$).

Table 5.8: LISREL Output with Model Fit Information for Union Sentiment Path Model Based on $n = 1000$

```
Log-likelihood Values

                                  Estimated Model    Saturated Model
                                  ---------------    ---------------
Number of free parameters(t)                   12                 15
-2ln(L)                                 17799.924          17792.656
AIC (Akaike, 1974)*                     17823.924          17822.656
BIC (Schwarz, 1978)*                    17882.805          17896.258

*LISREL uses AIC = 2t - 2ln(L) and BIC = tln(N)- 2ln(L)


                     Goodness-of-Fit Statistics
Degrees of Freedom for (C1)-(C2)            3
Maximum Likelihood Ratio Chi-Square (C1)    7.268 (P = 0.0638)
Browne's (1984) ADF Chi-Square (C2_NT)      7.248 (P = 0.0644)

Estimated Non-centrality Parameter (NCP)    4.268
90 Percent Confidence Interval for NCP      (0.0 ; 16.276)
```

Table 5.8: (*cont.*)

Minimum Fit Function Value	0.00728
Population Discrepancy Function Value (F0)	0.00427
90 Percent Confidence Interval for F0	(0.0 ; 0.0163)
Root Mean Square Error of Approximation (RMSEA)	0.0377
90 Percent Confidence Interval for RMSEA	(0.0 ; 0.0737)
P-Value for Test of Close Fit (RMSEA < 0.05)	0.662
Expected Cross-Validation Index (ECVI)	0.0313
90 Percent Confidence Interval for ECVI	(0.0270 ; 0.0433)
ECVI for Saturated Model	0.0300
ECVI for Independence Model	1.158
Chi-Square for Independence Model (10 df)	1146.666
Normed Fit Index (NFI)	0.994
Non-Normed Fit Index (NNFI)	0.987
Parsimony Normed Fit Index (PNFI)	0.298
Comparative Fit Index (CFI)	0.996
Incremental Fit Index (IFI)	0.996
Relative Fit Index (RFI)	0.979
Critical N (CN)	1560.660
Root Mean Square Residual (RMR)	0.731
Standardized RMR	0.0148
Goodness of Fit Index (GFI)	0.997
Adjusted Goodness of Fit Index (AGFI)	0.986
Parsimony Goodness of Fit Index (PGFI)	0.199

Table 5.9: M*plus* Output with Model Fit Information for Union Sentiment Path Model Based on *n* = 1000

MODEL FIT INFORMATION

Number of Free Parameters	12
Loglikelihood	
H0 Value	-13492.157
H1 Value	-13488.520
Information Criteria	
Akaike (AIC)	27008.314
Bayesian (BIC)	27067.208
Sample-Size Adjusted BIC	27029.095
(n* = (n + 2) / 24)	
Chi-Square Test of Model Fit	
Value	7.275
Degrees of Freedom	3
P-Value	0.0636

Table 5.9: (*cont.*)

RMSEA (Root Mean Square Error Of Approximation)		
Estimate		0.038
90 Percent C.I.	0.000	0.074
Probability RMSEA <= .05		0.662
CFI/TLI		
CFI		0.995
TLI		0.985
Chi-Square Test of Model Fit for the Baseline Model		
Value		884.551
Degrees of Freedom		9
P-Value		0.0000
SRMR (Standardized Root Mean Square Residual)		
Value		0.015

3. What steps should a researcher take in examining parameter estimates in a model?
4. What steps can a researcher take when the estimation process does not converge on a solution?
5. Three observed variable path models are presented in Figure 5.5 with model fit information, including χ^2_{model}, AIC, and BIC. Variances of exogenous variables, including errors, are omitted for simplicity, but they should be considered in calculations asked below. Using the set of models, answer the questions below.
    a. What is the number of unique elements in the covariance matrix for the models?
    b. For each of the models, indicate the number of parameters to be estimated (don't forget variances of exogenous variables).
    c. Calculate the *df* associated with each model. You can check your calculations with the provided model fit information.
    d. Which models are nested and which are non-nested?
    e. For the models that are nested, conduct $\Delta\chi^2$ tests and indicate the result of those tests.
    f. For the models that are non-nested, conduct model comparisons using the appropriate criteria. What are the results of these comparisons?

MODEL A
$\chi^2(3) = 1.758, p > .05$
AIC = 3015.733
BIC = 3055.313

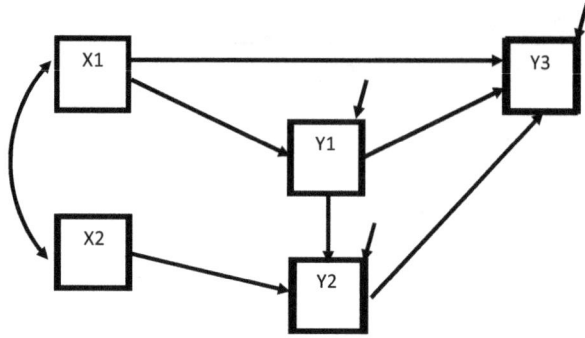

MODEL B
$\chi^2(1) = 0.093, p > .05$
AIC = 3018.067
BIC = 3064.244

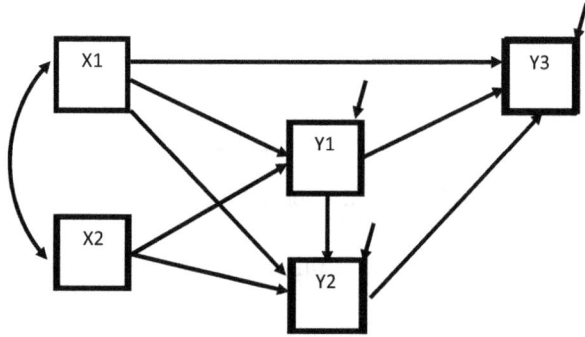

MODEL C
$\chi^2(2) = 103.241, p < .05$
AIC = 3119.216
BIC = 3162.094

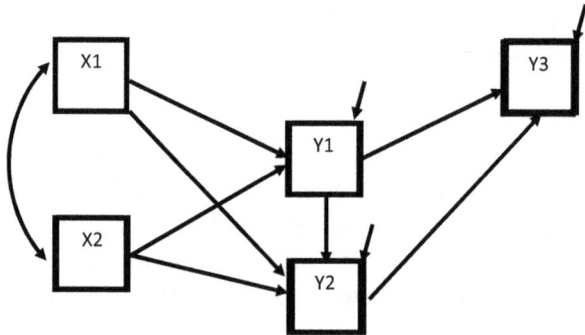

■ **Figure 5.5:** OBSERVED VARIABLE PATH MODEL COMPARISONS

6. How are modification indices used?
7. Suppose the modification index value associated with the direct effect from X1 to Y3 in Model C in Figure 5.5 was equal to **103.15.** If theory supports the direct effect, should the parameter be included (assuming alpha = .05)? Explain your answer.
8. Suppose the modification index value associated with the direct effect from X1 to Y2 in Model A in Figure 5.5 was equal to **1.32**. If theory supports the direct

effect, should the parameter be included (assuming alpha = .05)? Explain your answer.

CHAPTER FOOTNOTE

SEM software packages may output different chi-square values. For example, LISREL computes two different sets of standard errors for parameter estimates and up to four different chi-squares for testing overall fit of the model. These standard errors and chi-squares can be obtained for single-group problems as well as multiple-group problems using covariance matrices with or without means. Which standard errors and which chi-squares will be reported depends on whether an asymptotic covariance matrix is provided and which method of estimation is used to fit the model (ULS, GLS, ML, WLS, DWLS). The asymptotic covariance matrix is a consistent estimate of N times the asymptotic covariance matrix of the sample matrix being analyzed. Standard errors are estimated under non-normality if an asymptotic covariance matrix is used. Standard errors are estimated under multivariate normality if no asymptotic covariance matrix is used.

Chi-squares

Four different chi-squares are reported in LISREL and denoted below as C1, C2, C3, and C4, where the x indicates that it is reported for any of the five estimation methods.

Asymptotic covariance matrix not provided:

	ULS	GLS	ML	WLS	DWLS
C1	—	×	×	—	—
C2	×	×	×	—	—
C3	—	—	—	—	—
C4	—	—	—	—	—

Asymptotic covariance matrix provided:

	ULS	GLS	ML	WLS	DWLS
C1	—	×	×	×	—
C2	×	×	×	—	×
C3	×	×	×	—	×
C4	×	×	×	—	×

Note 1. C1 is $n - 1$ times the minimum value of the fit function; C2 is $n - 1$ times the minimum of the WLS fit function using a weight matrix estimated under multivariate normality; C3 is the Satorra–Bentler scaled chi-square statistic or its generalization to mean and covariance structures and multiple groups (Satorra & Bentler, 1994); C4 is computed from equations in Browne (1984) or Satorra (1993) using the asymptotic covariance matrix.

The corresponding chi-squares are now given in the output as follows:

C1: Minimum fit function chi-square
C2: Normal theory weighted least squares chi-square
C3: Satorra-Bentler scaled chi-square
C4: Chi-square corrected for non-normality

Note 2: Under multivariate normality of the observed variables, C1 and C2 are asymptotically equivalent and have an asymptotic chi-square distribution if the model holds exactly and an asymptotic noncentral chi-square distribution if the model holds approximately. Under normality and non-normality, C2 and C4 are correct asymptotic chi-squares, but may not be the best chi-square in small and moderate samples. Hu, Bentler, and Kano (1992) and Yuan and Bentler (1997) found that C3 performed better given different types of models, sample size, and degrees of non-normality.

REFERENCES

Akaike, H. (1987). Factor analysis and AIC. *Psychometrika*, 52, 317–332.

Asparouhov, T., & Muthén, B. (2010). Simple second-order chi-square correction. Retrieved from www.statmodel.com/download/WLSMV_new_chi21.pdf.

Bandalos, D. L. (2014). Relative performance of categorical diagonally weighted least squares and robust maximum likelihood estimation. *Structural Equation Modeling*, 21(1), 102–116.

Beauducel, A., & Herzberg, P. Y. (2006). On the performance of maximum likelihood versus means and variance adjusted weighted least square estimation in confirmatory factor analysis. *Structural Equation Modeling*, 13(2), 186–203.

Bentler, P. M. (1990). Comparative fit indexes in structural models. *Psychological Bulletin*, 107, 238–246.

Bentler, P. M. (1995). *EQS structural equations program manual*. Los Angeles, CA: BMDP Statistical Software.

Bentler, P. M., & Bonett, D. G. (1980). Significance tests and goodness-of-fit in the analysis of covariance structures. *Psychological Bulletin*, 88, 588–606.

Bollen, K. A. (1986). Sample size and Bentler and Bonnett's nonnormed fit index. *Psychometrika*, 51(3), 375–377.

Bollen, K. A. (1989). *Structural equations with latent variables*. New York: Wiley.

Bollen, K. A. (1990). Overall fit in covariance structure models: Two types of sample size effects. *Psychological Bulletin*, 107, 256–259.

Browne, M. W. (1984). Asymptotically distribution-free methods for the analysis of covariance structures. *British Journal of Mathematical and Statistical Psychology*, 37, 62–83.

Browne, M. W., & Cudeck, R. (1993). Alternative ways of assessing model fit. In K. A. Bollen & J. S. Long (Eds.), *Testing structural equation models* (pp. 132–162). Beverly Hills, CA: Sage.

Bryant, F. B., & Satorra, A. (2012). Principles and practice of scaled difference chi-square testing. *Structural Equation Modeling*, 19(3), 372–398.

Chen, F., Bollen, K. A., Paxton, P., Curran, P. J., & Kirby, J. B. (2001). Improper solutions in structural equation models: Causes, consequences, and strategies. *Sociological Methods & Research*, 29(4), 468–508.

Chou, C-P., & Bentler, P. M. (1990). Model modification in covariance structure modeling: A comparison among likelihood ratio, lagrange multiplier, and wald tests. *Multivariate Behavioral Research*, 25(1), 115–136.

Chou, C-P., & Bentler, P. M. (1993). Invariant standardized estimated parameter change for model modification in covariance structure analysis. *Multivariate Behavioral Research*, 28(1), 97–110.

Chou, C-P., Bentler, P. M., & Satorra, A. (1991). Scaled test statistics and robust standard errors for non-normal data in covariance structure analysis: A Monte Carlo study. *British Journal of Mathematical and Statistical Psychology*, 44(2), 347–357.

Cudeck, R., & Browne, M. W. (1983). Cross-validation of covariance structures. *Multivariate Behavioral Research*, 18(2), 147–167.

Curran, P. J., West, S. G., & Finch, J. F. (1996). The robustness of test statistics to nonnormality and specification error in confirmatory factor analysis. *Psychological Methods*, 1(1), 16–29.

DiStefano, C. (2002). The impact of categorization with confirmatory factor analysis. *Structural Equation Modeling*, 9(3), 327–346.

Dolan, C. V. (1994). Factor analysis of variables with 2, 3, 5 and 7 response categories: A comparison of categorical variable estimators using simulated data. *British Journal of Mathematical and Statistical Psychology*, 47(2), 309–326.

Eliason, S. R. (1993). *Maximum likelihood estimation: Logic and practice*. Thousand Oaks, CA: Sage.

Fan, X., & Sivo, S. A. (2005). Sensitivity of fit indexes to misspecified structural or measurement model components: Rationale of two-index strategy revisited. *StructuralEquation Modeling*, 12(3), 343–367.

Ferron, J. M., & Hess, M. R. (2007). Estimation in SEM: A concrete example. *Journal of Educational and Behavioral Statistics*, 32(1), 110–120.

Finney, S. J., & DiStefano, C. (2006). Non-normal and categorical data in structural equation modeling. In G. R Hancock & R. O. Mueller (Eds.), *Structural equation modeling: A second course* (pp. 269–314). Greenwich, CT: Information Age Publishing.

Flora, D. B., & Curran, P. J. (2004). An empirical evaluation of alternative methods of estimation for confirmatory factor analysis with ordinal data. *Psychological Methods*, 9(4), 466–491.

Forero, C. G., & Maydeu-Olivares, A. (2009). Estimation of IRT graded response models: Limited versus full information methods. *Psychological Methods*, 14(3), 275–299.

Gagné, P., & Hancock, G. R. (2006). Measurement model quality, sample size, and solution propriety in confirmatory factor models. *Multivariate Behavioral Research*, 41(1), 65–83.

Gerbing, D. W., & Anderson, J. C. (1987). Improper solutions in the analysis of covariance structures: Their interpretability and a comparison of alternate respecifications. *Psychometrika*, 52(1), 99–111.

Hu, L., & Bentler, P. M. (1995). Evaluating model fit. In R. H. Hoyle (Ed.), *Structural equation modeling: Concepts, issues, and applications* (pp. 76–99). Thousand Oaks, CA: Sage.

Hu, L., & Bentler, P. M. (1998). Fit indices in covariance structure modeling: Sensitivity to underparameterized model misspecification. *Psychological Methods*, 3(4), 424–453.

Hu, L., & Bentler, P. M. (1999). Cutoff criteria for fit indexes in covariance structure analysis: Conventional criteria versus new alternatives. *Structural Equation Modeling*, 6(1), 1–55.

Hu, L., Bentler, P. M., & Kano, Y. (1992). Can test statistics in covariance structure analysis be trusted? *Psychological Bulletin*, 112, 351–362.

Hutchinson, S. R. (1993). Univariate and multivariate specification search indices in covariance structure modeling. *Journal of Experimental Education*, 61(2), 171–181.

Jackson, D. L., Gillaspy, J. A., & Purc-Stephenson, R. (2009). Reporting practices in confirmatory factor analysis: An overview and some recommendations. *Psychological Methods*, 14(1), 6–23.

Jöreskog, K. G., & Sörbom, D. (1984). *LISREL VI user's guide* (3rd ed.). Mooresville, IN: Scientific Software.

Jöreskog, K. G., & Sörbom, D. (1996). *LISREL 8: User's reference guide*. Chicago, IL: Scientific Software International.

Kaplan, D. (1988). The impact of specification error on the estimation, testing, and improvement of structural equation models. *Multivariate Behavioral Research*, 23(1), 69–86.

Kaplan, D. (1989). Model modification in covariance structure analysis: Application of the expected parameter change statistic. *Multivariate Behavioral Research*, 24(3), 285–305.

Kenny, D. A., & McCoach, D. B. (2003). Effect of the number of variables on measures of fit in structural equation modeling. *Structural Equation Modeling*, 10, 333–351.

Kenny, D. A., & Milan, S. (2012). Identification: A nontechnical discussion of a technical issue. In R. H. Hoyle (Ed.), *Handbook of structural equation modeling* (pp. 145–163). New York: Guilford Press.

Loehlin, J. C. (2004). *Latent variable models* (4th ed.). Mahwah, NJ: Erlbaum.

Luijben, T. C. W. (1989). *Statistical guidance for model modification in covariance structure analysis.* Amsterdam: Sociometric Research Foundation.

Luijben, T. C., & Boomsma, A. (1988). Statistical guidance for model modification in covariance structure analysis. *Compstat*, 335–340.

MacCallum, R.C. (1986). Specification searches in covariance structure analysis. *Psychological Bulletin*, 100(1), 107–120.

MacCallum, R. C., Roznowski, M., & Necowitz, L. B. (1992). Model modifications in covariance structure analysis: The problem of capitalization on chance. *Psychological Bulletin*, 111(3), 490–504.

Marcoulides, G. A., & Drezner, Z. (2001). Specification searches in structural equation modeling with a genetic algorithm. In G. A. Marcoulides & R. E. Schumacker (Eds.), *New developments and techniques in structural equation modeling* (pp. 247–268). Mahwah, NJ: Erlbaum.

Marcoulides, G. A., & Drezner, Z. (2003). Model specification searches using ant colony optimization algorithms. *Structural Equation Modeling*, 10, 154–164.

Marcoulides, K. M., & Falk, C. F. (2018). Model specification searches in structural equation modeling. *Structural Equation Modeling*, 25(3), 1–8.

Marsh, H. W., Balla, J. R., & Hau, K.-T. (1996). An evaluation of incremental fit indices: A clarification of mathematical and empirical properties. In G. A. Marcoulides & R. E. Schumacker (Eds.), *Advanced structural equation modeling: Issues and techniques* (pp. 315–353). Mahwah, NJ: Lawrence Erlbaum.

Marsh, H. W., Balla, J. R., & McDonald, R. P. (1988). Goodness-of-fit indexes in confirmatory factor analysis: The effect of sample size. *Psychological Bulletin*, 103, 391–410.

Marsh, H. W., Hau, K.-T., & Wen, Z. (2004). In search of golden rules: Comment on hypothesis-testing approaches to setting cutoff values for fit indexes and dangers in overgeneralizing Hu and Bentler's findings. *Structural Equation Modeling*, 11(3), 320–341.

McDonald, R. P. (1989). An index of goodness-of-fit based on noncentrality. *Journal of Classification*, 6, 97–103.

Mulaik, S. A., James, L. R., Alstine, J. V., Bennett, N., Lind, S., & Stilwell, C. D. (1989). Evaluation of goodness-of-fit indices for structural equation models. *Psychological Bulletin*, 105, 430–445.

Muthén, B., du Toit, S.H.C. & Spisic, D. (1997). Robust inference using weighted least squares and quadratic estimating equations in latent variable modeling with categorical and continuous outcomes. Retrieved from http://pages.gseis.ucla.edu/faculty/ uthen/articles/Article_075.pdf.

Muthén, B., & Kaplan, D. (1985). A comparison of some methodologies for the factor analysis of non-normal Likert variables. *British Journal of Mathematical and Statistical Psychology*, 38(2), 171–189.

Muthén, B. & Kaplan D. (1992). A comparison of some methodologies for the factor analysis of non-normal Likert variables: A note on the size of the model. *British Journal of Mathematical and Statistical Psychology*, 45(1), 19–30.

Olsson, U. H., Foss, T., Troye, S. V., & Howell, R. D. (2000). The performance of ML, GLS, and WLS estimation in structural equation modeling under conditions of misspecification and nonnormality. *Structural Equation Modeling*, 7(4), 557–595.

Rhemtulla, M., Brosseau-Liard, P. E., & Savalei, V. (2012). When can categorical variables be treated as continuous? A comparison of robust continuous and categorical SEM estimation methods under suboptimal conditions. *Psychological Methods*, 17(3), 354–373.

Saris, W. E., Satorra, A., & Sörbom, D. (1987). The detection and correction of specification errors in structural equation models. In C. C. Clogg (Ed.), *Sociological methodology* (pp. 105–129). San Francisco, CA: Jossey-Bass.

Saris, W. E., Satorra, A., & van der Veld, W. M. (2009). Testing structural equation models or detection of misspecifications? *Structural Equation Modeling,* 16(4), 561–582.

Satorra, A. (1989). Alternative test criteria in covariance structure analysis: A unified approach. *Psychometrika*, 54(1), 131–151.

Satorra, A. (1993). Multi-sample analysis of moment structures: Asymptotic validity of inferences based on second-order moments. In K. Haagen, D. J. Bartholomew, & M. Deistler (Eds.), *Statistical modeling and latent variables* (pp. 283–298). Amsterdam: Elsevier.

Satorra, A., & Bentler, P. M. (1994). Corrections for test statistics and standard errors in covariance structure analysis. In A. Von Eye & C. C. Clogg (Eds.), *Latent variable analysis: Applications for developmental research* (pp. 399–419). Thousand Oaks, CA: Sage.

Schwarz, G. (1978). Estimating the dimension of a model. *The Annals of Statistics*, 6(2), 461–464.

Sclove, S. L. (1987). Application of model-selection criteria to some problems in multivariate analysis. *Psychometrika*, 52(3), 333–343.

Silvia, E. M., & MacCallum, R. C. (1988). Some factors affecting the success of specification searches in covariance structure modeling. *Multivariate Behavioral Research*, 23(3), 297–326.

Sörbom, D. (1989). Model modification. *Psychometrika*, 54(3), 371–384.

Steiger, J. H., & Lind, J. M. (1980, May). *Statistically-based tests for the number of common factors*. Paper presented at Psychometric Society Meeting, Iowa City, IA.

Tucker, L. R., & Lewis, C. (1973). The reliability coefficient for maximum likelihood factor analysis. *Psychometrika*, 38, 1–10.

West, S. G., Taylor, A. B., & Wu, W. (2012). Model fit and model selection in structural equation modeling. In R. H. Hoyle (Ed.), *Handbook of structural equation modeling* (pp. 209–231). New York: Guilford Press.

Whittaker, T. A. (2012). Using the modification index and standardized expected parameter change for model modification. *Journal of Experimental Education*, 80(1), 26–44.

Wothke, W. (1993). Nonpositive definite matrices in structural modeling. In K. A. Bollen & J. S. Long (Eds.), *Testing structural equation models* (pp. 256–293). Newbury Park, CA: Sage.

Yuan, K.-H. (2005). Fit indices versus test statistics. *Multivariate Behavioral Research*, 40(1), 115–148.

Yuan, K.-H., & Bentler, P. M. (1997). Mean and covariance structure analysis: Theoretical and practical improvements. *Journal of the American Statistical Association*, 92, 767–774.

Chapter 6

FACTOR ANALYSIS

EXPLORATORY FACTOR ANALYSIS

Exploratory factor analysis (EFA) is a statistical method that is used to find a small set of latent constructs or factors from a larger number of survey items or observed variables. There are important considerations when conducting EFA, including sample size, number of factors, factor rotation, factor scores, and alternative data reduction techniques. These topics are subsequently discussed with recommendations for best practices in EFA.

Sample Size

Basically, it takes only two pairs of scores to compute a correlation coefficient. Would the correlation coefficient be a good sample estimate of the population correlation, rho (ρ)? The issues become those of sampling, inference, and validity. The more stable the sample correlations, the more valid the inference of the scores to the population. Sample size in EFA is an important issue and is a topic of considerable debate and study. There have been minimum sample size recommendations for EFA (e.g., 300; Comrey, 1973) as well as suggested ratios of the number of participants to the number of items (e.g., 5:1; Gorsuch, 1983). Nonetheless, the sample size issue is more complicated because the accuracy of EFA is dependent upon interactions between sample size and other conditions, such as the number of items per factor and item communalities (Guadagnoli & Velicer, 1988; MacCallum, Widaman, Zhang, & Hong, 1999). An item communality is the proportion of variance in the item that is explained by all the latent factors in the model. Items with low communalities are not desired because it means that the item would probably not load highly on a factor.

Mundfrom, Shaw, and Ke (2005) conducted a simulation study to determine the minimum sample size under different conditions, including number of factors,

DOI: 10.4324/9781003044017-6

number of items per factor, and communality levels. The findings from their study generally indicated that if communalities are high (.60 or greater) and there are several items per factor (5 or more), smaller sample sizes can be used with EFA (e.g., less than 200; Mundfrom et al., 2005). The problem will be that researchers will not know the values of communalities when collecting data to be used in an EFA. Thus, researchers may need to collect larger sets of data to ensure the adequacy of results from EFA. Based on the findings from Mundfrom et al. (2005), we offer general guidelines concerning sample size: sample sizes of approximately 300 should be satisfactory if the number of items per factor is at least four with low to high ranging item communalities and less than five factors extracted. The number of items to factor ratio of at least 7 can reduce the negative impact of low item communalities with sample sizes as low as 150. Under more ideal scenarios with high communalities, five items per factor, and only two factors extracted, sample size could be much lower (e.g., less than 50).

Number of Factors

Because EFA is exploratory, researchers will not know how many factors exist when conducting the EFA. There are different criteria available to help researchers when determining the number of factors underlying the data. The determination of how many factors is commonly done using a scree plot with the eigenvalues plotted against the number of factors (Cattell, 1966) and/or the eigenvalues greater than one rule (Kaiser, 1958). The estimation process during factor analysis will produce less valuable explanation of variance in the set of items as each new sequential factor is extracted. This is because each new factor extracted after the last is based on unexplained variance in the set of items with previous factors already extracted. So, the first factor will have the largest eigenvalue (largest variance explained), the second factor a lower eigenvalue (next largest variance explained), and so forth. When using a scree plot, Cattell (1966) recommended selecting the factor(s) just before the plotted eigenvalues level off horizontally on the plot. Kaiser (1958) recommended against selecting factors with eigenvalues less than one because the factors associated with small eigenvalues will most likely not contribute meaningfully to the explanation of variance in the set of items. The goal in EFA is to find the fewest factors that have the largest amount of variance accounted for and provide a meaningful interpretation (Schumacker, 2015). The scree plot can be requested in M*plus*, but is not available in LISREL.

Parallel analysis (Horn, 1965) may also be used to help researchers when determining the number of factors underlying the data. With parallel analysis, in general, the item-level data are randomly arranged repeatedly in order to create several random data sets. A minimum of 100 random data sets is recommended. Factor analyses are performed on the original data set and on the 100 random data sets.

The eigenvalues across the 100 random data sets are then averaged. The number of factors to select is decided by comparing the eigenvalues from the original sample data set to the averaged eigenvalues from the random data sets. A factor is selected if the eigenvalue for a certain factor using the original data set is greater than the same factor's averaged eigenvalue using the random data sets. Parallel analysis has been found to work well when identifying the number of factors underlying a set of data (Fabrigar, Wegener, MacCallum, & Strahan, 1999; Preacher & MacCallum, 2003). Parallel analysis is available in M*plus*, but it is not currently available in LISREL. O'Connor (2000) has developed programs that can conduct parallel analysis in SPSS, SAS, MATLAB, and R.

When deciding on the number of factors to keep in EFA, another consideration is what is called *approximate simple structure*. McDonald (1985) defines *approximate simple structure* as a factor pattern with several items loading strongly on one single factor while loading weakly on other factors. Many consider approximate simple structure an important issue when deciding on the number of factors as well as items to keep in a factor solution. According to Worthington and Whittaker (2006, p. 821), "if factors share items that cross-load too highly on more than one factor (e.g., > .32), the items are considered complex because they reflect the influence of more than one factor." This does not mean that you cannot have items load on more than one factor. However, many prefer approximate simple structure for easier interpretation of the factors and calculations of subscale scores. Approximate simple structure can be accomplished by deleting factors and/or deleting items.

Both LISREL and M*plus* also provide fit statistics to help determine the number of factors underlying the data. In addition to the criteria available to help when selecting the number of factors, theory is an important consideration during the factor selection process. Factor analysis tends to be an iterative process in which decisions are made and models are re-estimated in the process. In the end, the factor structure selected should agree with theoretical explanations.

Rotation Methods

Factor analysis is invariant within rotations. That is, the initial factor pattern matrix is not unique. We can get an infinite number of solutions that produce the same correlation matrix by rotating the reference axes of the factor solution to simplify the factor structure and to achieve a more meaningful and interpretable solution. Rotation methods provide a different orientation of data points to factors.

There are two rotation approaches available with factor analyses: orthogonal and oblique. Orthogonal rotations do not allow the factors to correlate whereas oblique rotations allow the factors to be intercorrelated. Various types of orthogonal and oblique rotations are available (see Gorsuch, 1983 for more details about the different rotation methods and their purpose). The most commonly used orthogonal rotation is the varimax rotation (Kaiser, 1958) and the most commonly used oblique rotations include the oblimin rotation (Jenrich & Sampson, 1966) and the promax rotation (Hendrickson & White, 1964). These rotations are all available in M*plus* whereas LISREL provides varimax and promax rotations.

Determining whether factors are intercorrelated or not should be mainly grounded in theory. This does not mean that a researcher could not examine both orthogonal and oblique rotated solutions to examine the factor structure. Hence, if a researcher finds that factors are intercorrelated when it was originally hypothesized that the factors would not correlate, it would be suitable to use an oblique rotation given the results.

Factor Scores

Exploratory factory analysis permits the computation of factor scores. Factor scores are usually computed using a multiple regression equation where the factor loadings are multiplied by their respective variables. Factor scores are typically output as standardized scores, but can be converted to a scaled score using a linear transformation. There are three methods used to compute factor scores: Regression (Thurstone, 1935); Bartlett (Bartlett, 1937); and Anderson–Rubin (Anderson & Rubin, 1956). The regression method maximizes the validity of the construct (factor variance explained). The Bartlett method computes factor scores keeping the factors orthogonal (uncorrelated). The Anderson–Rubin method computes factor scores that have correlations which match the correlations among the factors. SEM software uses the regression method where the standardized factor loadings are used as regression weights to compute a score. Factor scores suffer from indeterminacy (Mulaik, 2005; Bartholomew, Deary, & Lawn, 2009), meaning that there are infinite sets of factor scores that could be calculated to explain the association between factors and items. Factor scores can be saved in both LISREL and M*plus* during factor analysis.

EFA Versus PCA

Exploratory factor analysis (EFA) is often confused with principal components analysis (PCA). Both methods extract variance from the correlation matrix. Factor analysis extracts variable variance based on the squared multiple correlation in the diagonal of the matrix. Principal components analysis extracts variable variance, but from the diagonal of the correlation matrix where each variable has a variance = 1. Since each variable has variance = 1, the sum of the diagonal is the number of variables in the correlation matrix. PCA *components* (as opposed to factors) reproduce all of the variable variance. Principal components analysis is designed to account for all of the variable variance, as indicated by the 1s in the diagonal of the matrix. Factor analysis is designed to account for the shared variance in the correlation matrix, so it does not include all the variable variance. This distinction is not always made clear. Thus, principal components analysis may inadvertently be conducted when the researcher intended to conduct a factor analysis (Schumacker, 2015). Readers are encouraged to consult *Multivariate Behavioral Research*, 1990, Volume 25, Issue 1 for more details about the differences between PCA and factor analysis.

It must be noted that there are different factor analysis extraction methods, including unweighted and generalized least squares, alpha factoring, image factoring, principal axis factoring, and maximum likelihood. In LISREL and M*plus*, maximum likelihood is available. Both LISREL and M*plus* can also perform EFA with ordinal data. Regardless of extraction method used, factor analysis attempts to determine which sets of observed variables share common variance–covariance characteristics that define theoretical constructs or factors (latent variables). Factor analysis presumes that the factors are smaller in number than the number of observed variables and are responsible for the shared variance–covariance among the observed variables. We present an example using EFA next.

EFA EXAMPLE

For the EFA example, we are using the 11 subscales on the Wechsler Intelligence Scale for Children from a sample of 175 children (Tabachnick & Fidell, 2007). The 11 subscales measure the following: *info* (Information), *comp* (Comprehension), *arith* (Arithmetic), *simil* (Similarities), *vocab* (Vocabulary), *digit* (Digit Span), *pictcomp* (Picture Completion), *parang* (Paragraph Arrangement), *block* (Block Design), *object* (Object Assembly), and *coding* (Coding). There are no missing data. The data set and programs are on the book website.

The easiest way to run an EFA in LISREL is to use the ***WISC.LSF*** file with the **Exploratory Factor Analysis** option in the **Statistics** menu on the toolbar. After opening the ***WISC.LSF*** file in LISREL, click on the **Statistics** menu and scroll down to click on the **Exploratory Factor Analysis** option. You will need to select variables from the *Variable List* and put them in the box on the right of the *Select>>* button in the dialog box. The *client* number and *agemate* variables should not be included in the subset of variables used for the EFA. ML Factor Analysis is the default estimator. There is an option to select the number of factors. If you enter the number of factors to be extracted, model fit information for different factor structures will not be provided. For this example, we did not enter the number of factors to be extracted, which allows LISREL to decide on the number of factors based on model fit information. You will also see an option to click if you would like to save Factor Scores. The dialog box should look like the one in Figure 6.1. You can then click the *Run* button.

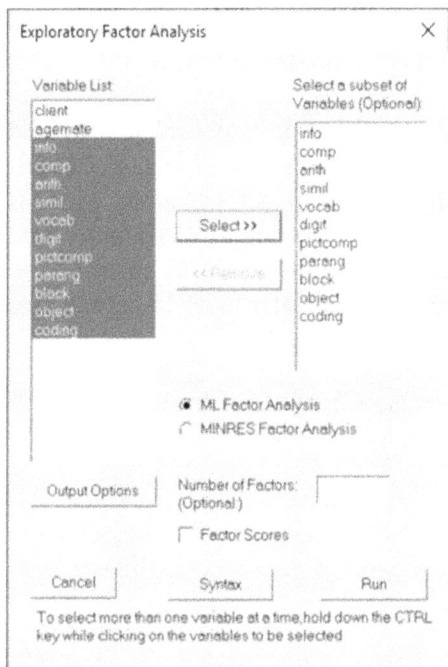

■ **Figure 6.1:** LISREL EFA DIALOG BOX

The "Decision Table for Number of Factors" in LISREL output is presented below and can be used to help determine the number of factors to retain.

```
Decision Table for Number of Factors

Factors   Chi2     df    P      DChi2    Df    PD     RMSEA
-------   ----     --    -      -----    --    --     -----
   0      502.89   55    0.000                        0.216
   1      112.76   44    0.000  390.12   11    0.000  0.095
   2       47.21   34    0.065   65.55   10    0.000  0.047
   3       30.58   25    0.203   16.63    9    0.055  0.036
```

The "Chi2" column represents the model chi-square statistic with corresponding df and p-value found in the "df" and "P" columns, respectively. You can see that the chi-square test of model fit is significant with a one-factor solution, but not for the two- and three-factor solutions. The "DChi2" column represents the $\Delta\chi^2$ between that specific factor model and the model with one less factor, with corresponding Δdf in the "Df" column and p-value in the "PD" column. There is a significant difference between the zero- and one-factor models [$\Delta\chi^2$ (11) = 390.12, $p < .05$] and between the one- and two-factor models [$\Delta\chi^2$ (10) = 65.55, $p < .05$]. There is no significant difference between the two- and three-factor models [$\Delta\chi^2$ (9) = 16.63, $p > .05$]. The RMSEA is below the .05 cutoff for the two- and three-factor models. Although no significant difference was found between the two- and three-factor models, LISREL provided the final solution for a three-factor model.

LISREL provides both the varimax (orthogonal) and promax (oblique) rotated factor loadings with factor correlations, which are presented in Table 6.1. A variable is determined to load on a factor with the highest factor loading, which is based on comparing their rotated factor loadings. According to Worthington and Whittaker (2006), it is recommended that factor loadings be greater than .32 and if items cross-load, the absolute difference should be greater than .15 between the highest and next highest factor loadings.

Table 6.1: LISREL Rotated Factor Loadings, Factor Correlations, and Residual Variances for Initial EFA Analysis

Varimax-Rotated Factor Loadings

	Factor 1	Factor 2	Factor 3	Unique Var
info	0.779	0.156	0.074	0.363
comp	0.551	0.448	−0.032	0.494
arith	0.556	0.140	0.269	0.599
simil	0.620	0.366	−0.160	0.455
vocab	0.721	0.252	0.035	0.415
digit	0.431	−0.003	0.134	0.797
pictcomp	0.202	0.605	−0.194	0.556
parang	0.154	0.392	0.135	0.805
block	0.117	0.714	0.380	0.332
object	0.084	0.573	−0.051	0.662
coding	0.054	0.004	0.290	0.913

Promax-Rotated Factor Loadings

	Factor 1	Factor 2	Factor 3	Unique Var
info	0.819	−0.057	0.018	0.363
comp	0.509	0.328	0.005	0.494
arith	0.549	0.064	0.260	0.599
simil	0.622	0.167	−0.166	0.455
vocab	0.736	0.062	0.007	0.415
digit	0.462	−0.105	0.091	0.797
pictcomp	0.104	0.577	−0.088	0.556

Table 6.1: (*cont.*)

parang	0.059	0.451	0.228	0.805
block	-0.090	0.920	0.000	0.332
object	-0.038	0.626	0.078	0.662
coding	0.022	0.081	0.316	0.913

Factor Correlations

	Factor 1	Factor 2	Factor 3
Factor 1	1.000		
Factor 2	0.421	1.000	
Factor 3	0.042	-0.423	1.000

Because the factor correlations are estimated to be moderate between Factors 1 and 2 (.421) and between Factors 2 and 3 (−.423), the promax rotated factor loadings should be examined to determine the number of factors and items to retain. As seen in Table 6.1, *info* has the highest factor loading on Factor 1. It appears that the first factor contains *info*, *comp*, *arith*, *simil*, *vocab*, and *digit*. *Comp* cross-loads on Factors 1 and 2 with a difference greater than .15 (.18). The second factor contains *pictcomp*, *parang*, *block*, and *object*. *Coding* loads the highest on Factor 3, but the value is not very large.

You will notice the column in the output labeled "Unique Var," which is the residual variance associated with the item. Thus, approximately 36.3% of the variance in *info* is not explained by the factors. The communality is calculated by subtracting the residual variance from 1. Hence, approximately 63.7% (1 − .363 = .637) of the variance in *info* is explained by the factors. You will notice that the residual variances are the same in the varimax and promax rotations. This is because item communalities do not change with factor rotations.

The three-factor solution is not desirable because Factor 3 did not contain many high-loading items and *coding* did not load strongly (> .32) on any of the factors. Hence, we ran a two-factor solution next with all of the subscales. Although not presented, the promax solution with two factors showed that *comp* cross-loaded on both factors with a difference less than .15 between loadings and *coding* did not load on either factor. We subsequently ran a two-factor solution without the *coding* and *comp* variables. The dialog box should look like the one presented in Figure 6.2.

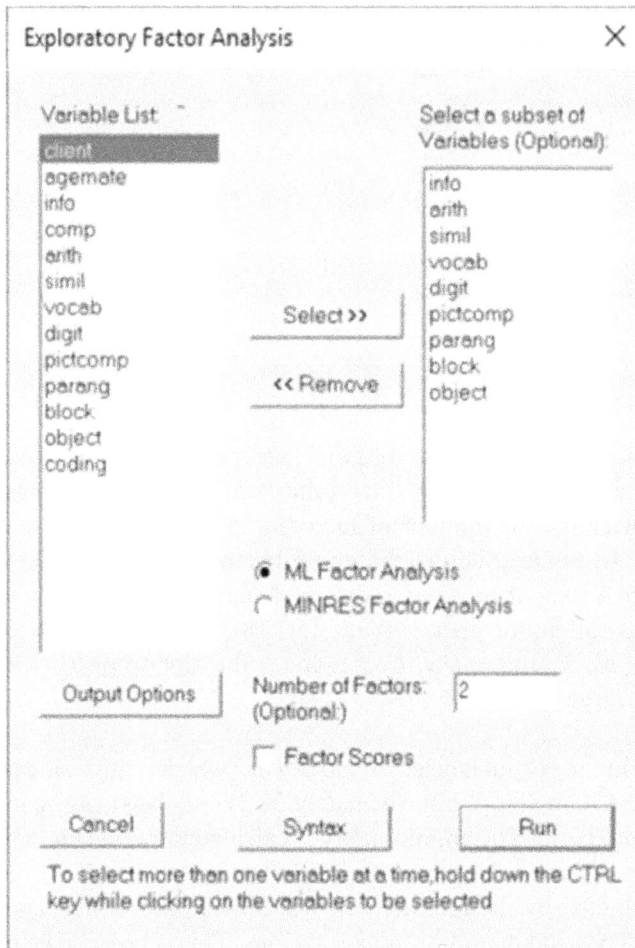

■ **Figure 6.2:**   LISREL EFA DIALOG BOX FOR THE TWO-FACTOR MODEL

The promax rotated loadings and factor correlations in LISREL with the two-factor solution are presented in Table 6.2. The primary loadings, or the highest loading associated with an item, are bolded to help illustrate the approximate simple structure. As shown in Table 6.2, all primary factor loadings are greater than .32 and the factor pattern is demonstrating approximate simple structure. Although *simil* loads on both Factors 1 and 2, the difference between the respective loadings is greater than .15 (.535 – .255 = .28). The two factors are correlated moderately and positively, *r* = .460.

Table 6.2: EFA Promax Two-Factor Solution in LISREL

| | FACTOR STRUCTURE | |
	1	2
INFO	0.842	−0.034
ARITH	0.566	0.019
SIMIL	0.535	0.255
VOCAB	0.716	0.077
DIGIT	0.458	−0.086
PICTCOMP	0.044	0.584
PARANG	0.056	0.439
BLOCK	0.016	0.645
OBJECT	−0.088	0.639
FACTOR 1	1.000	
FACTOR 2	0.460	1.000

Mplus software was used to read in the raw data from the ASCII text file, ***WISC. txt***. The *Mplus* program code is provided below:

```
TITLE:       Exploratory Factor Analysis

DATA:        FILE is "C:\WISC.txt";

VARIABLE:    NAMES are client agemate info comp arith simil
             vocab digit pictcomp parang block
             object coding;

USEVAR       are info comp arith simil
             vocab digit pictcomp parang block
             object coding;

ANALYSIS:    TYPE = EFA 1 3;
             PARALLEL = 100;
             ! ROTATION = VARIMAX;

PLOT:        TYPE = PLOT2;
```

In M*plus*, the *client* number and *agemate* variables were not included so the **USEVAR** (short for USEVARIABLES) option was used in the VARIABLE command to indicate which variables will be included in the analysis. The **ANALYSIS** command specifies the **TYPE** of analysis as an **EFA** analysis with a minimum of 1 factor and a maximum of 3 factors. The **PARALLEL = 100** option requests that parallel analysis be conducted using 100 random data sets. In the code, we commented out a request for VARIMAX rotation using the exclamation point, which users should notice the font for the code that is commented out turn green. This was done because the default rotation method in M*plus* is an oblique rotation (i.e., Geomin), but users could request Varimax instead by removing the exclamation point in the code. You can also use the same command to request a PROMAX rotation, which is used in LISREL. The **PLOT** command with the **TYPE = PLOT2** option produces the scree plot. The *Mplus* EFA program by default uses the maximum likelihood estimation method.

The abbreviated M*plus* output below indicates defaults in M*plus*, including that the estimation method was maximum likelihood (ML) estimation and the rotation method was GEOMIN, which is an OBLIQUE rotation.

```
Estimator                                          ML
Rotation                                       GEOMIN
Row standardization                       CORRELATION
Type of rotation                              OBLIQUE
```

The output also provides the eigenvalues for each of the 11 variables using the sample correlation matrix, the average eigenvalues from parallel analysis, and the 95th percentile eigenvalues from parallel analysis. The 95th percentile eigenvalue has been recommended by some because it resembles a hypothesis testing scenario (Green, Levy, Thompson, Lu, & Lo, 2012). The 95th percentile eigenvalue can be used similarly to that of the mean eigenvalue from parallel analysis. The M*plus* output with these eigenvalues is provided below.

```
EIGENVALUES FOR SAMPLE CORRELATION MATRIX
              1          2          3          4          5

            3.829      1.442      1.116      0.890      0.768

EIGENVALUES FOR SAMPLE CORRELATION MATRIX
              6          7          8          9         10

            0.633      0.595      0.522      0.471      0.419

EIGENVALUES FOR SAMPLE CORRELATION MATRIX
             11

            0.315

AVERAGE EIGENVALUES FROM PARALLEL ANALYSIS
              1          2          3          4          5

            1.427      1.302      1.207      1.123      1.049

AVERAGE EIGENVALUES FROM PARALLEL ANALYSIS
              6          7          8          9         10
```

| | 0.982 | 0.923 | 0.852 | 0.781 | 0.714 |

AVERAGE EIGENVALUES FROM PARALLEL ANALYSIS
       11

| | 0.640 |

95 PERCENTILE EIGENVALUES FROM PARALLEL ANALYSIS

1	2	3	4	5
1.556	1.366	1.277	1.183	1.103

95 PERCENTILE EIGENVALUES FROM PARALLEL ANALYSIS

6	7	8	9	10
1.026	0.973	0.892	0.840	0.757

95 PERCENTILE EIGENVALUES FROM PARALLEL ANALYSIS
       11

| | 0.696 |

Using Kaiser's (1958) criterion for selecting the number of factors based on the number of eigenvalues greater than one rule, three factors have eigenvalues greater than one based on the sample correlation matrix. Using the parallel analysis results, the first two eigenvalues for the sample correlation matrix (3.829 and 1.442, respectively) are larger than the first two average eigenvalues from parallel analysis (1.427 and 1.302, respectively) and the first two 95th percentile eigenvalues from parallel analysis (1.556 and 1.366, respectively). However, you will notice that the third eigenvalue for the sample correlation matrix (1.116) is smaller than the third average eigenvalue from parallel analysis (1.207) and the third 95th percentile eigenvalue from parallel analysis (1.277). The parallel analysis suggests selecting two factors.

To see the scree plot in M*plus*, you must click on the **Plot** menu on the toolbar after analyzing the EFA model followed by clicking the **View Plots** option. Users must then click on the **Eigenvalues for exploratory factor analysis** option and click the **View** button. The scree plot in Figure 6.3 provides a visual representation of the eigenvalues by the number of factors. The eigenvalues are on the Y axis and the number of factors on the X axis. You will notice that the scree plot has the original eigenvalues as well as the eigenvalues (average and 95th percentile) from the parallel analysis. The sample eigenvalues plotted in the scree plot appears to level off substantially after one factor.

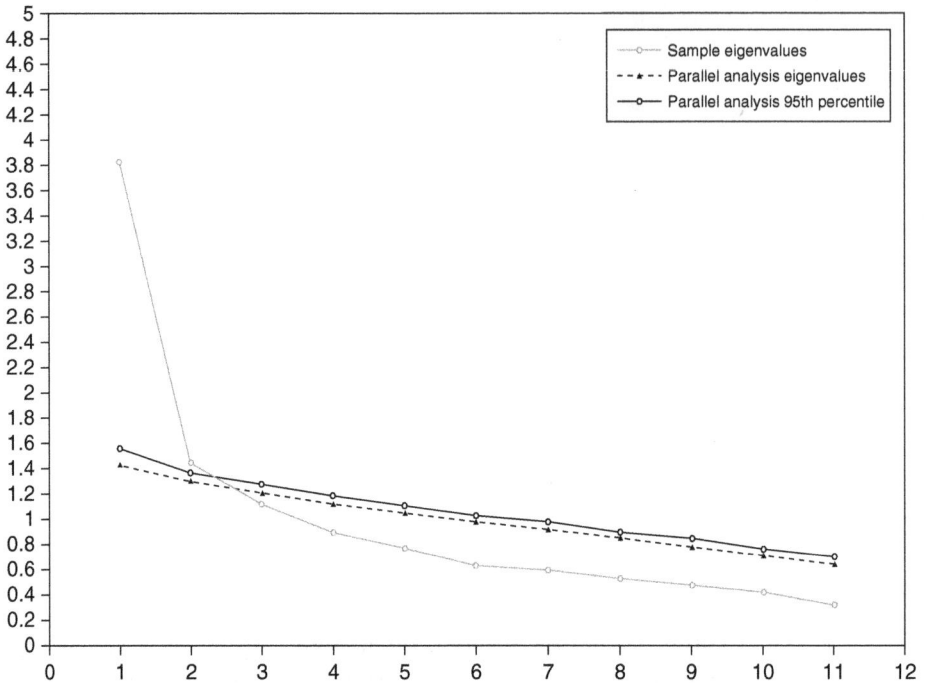

■ **Figure 6.3:** SCREE PLOT IN M*PLUS*

M*plus* also provides the following summary of model fit information.

```
SUMMARY OF MODEL FIT INFORMATION
                Number of                  Degrees of
Model           Parameters    Chi-Square   Freedom     P-Value
1-factor            33        116.882        44        0.0000
2-factor            43         49.133        34        0.0450
3-factor            52         31.953        25        0.1594
                                            Degrees of
Models Compared               Chi-Square   Freedom     P-Value
1-factor against 2-factor       67.749       10        0.0000
2-factor against 3-factor       17.180        9        0.0460
```

The chi-square test of model fit becomes non-significant with a three-factor solution. Further, the $\Delta\chi^2$ tests suggest selecting a three-factor solution over the two-factor solution since it resulted in a significant $\Delta\chi^2$ ($p < .05$). M*plus* also prints out all model fit information (CFI, TLI, RMSEA, and SRMR) for each of the three EFA models.

The rotated factor loadings, factor correlations, and residual variances for the three-factor solution are presented in Table 6.3. Similar to the LISREL output, *info* has the highest factor loading on Factor 1. Also, the first factor contains *info, comp, arith, simil, vocab,* and *digit. Comp* cross-loads on Factors 1 and 2 with a difference less than .15 (.11). The second factor contains *pictcomp, parang, block,* and *object. Coding* loads the highest on Factor 3, but the value is not very large. Also, you will notice a moderately large correlation between Factors 1 and 2, but neither factors correlate moderately with Factor 3. You will notice that the residual variances are the same as those in LISREL. Again, once factors are extracted, factor rotations do not change the values of item communalities. M*plus* also provides standard errors and *z*-tests for each of the loadings and residual variances in the output.

Table 6.3: M*plus* GEOMIN Rotated Factor Loadings, Factor Correlations, and Residual Variances for Initial EFA

GEOMIN ROTATED LOADINGS (* significant at 5% level)

	1	2	3
INFO	0.810*	-0.025	0.018
COMP	0.468*	0.361*	-0.011
ARITH	0.607*	0.004	0.234
SIMIL	0.543*	0.259	-0.159
VOCAB	0.716*	0.097	0.003
DIGIT	0.488*	-0.117	0.087
PICTCOMP	0.021	0.630*	-0.106
PARANG	0.082	0.390*	0.187
BLOCK	-0.003	0.746*	0.492
OBJECT	-0.073	0.617*	0.043
CODING	0.106	-0.022	0.283

GEOMIN FACTOR CORRELATIONS (* significant at 5% level)

	1	2	3
1	1.000		
2	0.457*	1.000	
3	-0.084	-0.175	1.000

ESTIMATED RESIDUAL VARIANCES

INFO	COMP	ARITH	SIMIL	VOCAB
0.363	0.494	0.599	0.455	0.415

ESTIMATED RESIDUAL VARIANCES

DIGIT	PICTCOMP	PARANG	BLOCK	OBJECT
0.797	0.556	0.805	0.332	0.662

ESTIMATED RESIDUAL VARIANCES

CODING
0.913

Similar to the LISREL output, the three-factor solution is not desirable because Factor 3 did not contain many high-loading items and *coding* did not load strongly (> .32) on any of the factors. Moreover, *comp* loaded on both Factors 1 and 2 with a small difference between the respective loadings (< .15). Thus, we ran a two-factor solution next without the *coding* and *comp* variables in M*plus* with the following code.

```
TITLE:      Exploratory Factor Analysis
            2-Factor Solution

DATA:       FILE is "C:\WISC.txt";

VARIABLE:   NAMES are client agemate info comp arith simil
            vocab digit pictcomp parang block
            object coding;

USEVAR      are info arith simil
            vocab digit pictcomp parang block
            object;

ANALYSIS:   TYPE = EFA 2 2;
```

The rotated factor loadings, factor correlations, and communalities ($h^2 = 1 -$ residual variance) for the two-factor solution are shown in Table 6.4. The primary loadings are bolded. All primary factor loadings are greater than .32 and the factor pattern is demonstrating approximate simple structure. Although *simil* loads on both Factors 1 and 2, the difference between the respective loadings is greater than .15 (.528 − .260 = .268). The two factors are correlated moderately, positively, and significantly, $r = .475$, $p < .05$. The chi-square test of model fit was non-significant [$\chi^2(19) = 26.442$, $p > .05$] and other model fit indices also support good model fit [CFI = .98; TLI = .961; SRMR = .039; RMSEA = .047, $p > .05$ (90% CI: .000, .087)].

Table 6.4: EFA GEOMIN Two-Factor Solution in M*plus*

	FACTOR STRUCTURE		h^2
	1	2	
INFO	**0.842***	−0.031	0.684
ARITH	**0.565***	0.021	0.331
SIMIL	**0.528***	0.260*	0.477
VOCAB	**0.713***	0.080	0.568
DIGIT	**0.459***	−0.085	0.181
PICTCOMP	0.030	**0.591***	0.367
PARANG	0.045	**0.444***	0.219
BLOCK	0.000	**0.652***	0.425
OBJECT	−0.103	**0.646***	0.365
FACTOR 1	1.000		--
FACTOR 2	0.475*	1.000	--

Note: *$p < .05$. h^2 = Communality.

The researcher is now faced with the task of naming or labeling these factors. The first factor might be called *verbal ability* and the second factor *spatial ability*? According to Worthington and Whittaker (2006):

> Conceptual interpretability is the definitive factor-retention criterion. In the end, researchers should retain a factor only if they can interpret it in a meaningful way no matter how solid the evidence for its retention based on the empirical criteria earlier described. EFA is ultimately a combination of empirical and subjective approaches to data analysis because the job is not complete until the solution makes sense.

p. 822

PATTERN AND STRUCTURE MATRICES

When factor rotations are orthogonal, the factor loadings represent the correlation between the items and the factors. Communalities, which represent the proportion of variance in the item that is accounted for by the extracted factors, can be calculated by squaring and summing the factor loadings for the respective item. When factor rotations are oblique, however, two different factor loadings are provided, which are called *pattern coefficients* or *structure coefficients*. These coefficients are provided as elements in the *pattern matrix* and the *structure matrix*, respectively. The *structure* matrix holds the correlations between the items and the factors. The *pattern* matrix holds the factor loadings. Each row of the *pattern* matrix is essentially a regression equation where the standardized observed item is expressed as a function of the factors. The pattern matrix has factor loadings similar to partial standardized regression coefficients, which indicates the effect of a given factor for a specific item while controlling for the other factors. The pattern coefficients are typically used to determine which items load on which factor.

When two or more factors are present, the pattern and structure matrices will be different with oblique rotations. The size of the differences between respective pattern and structure coefficients depends upon the magnitude of the factor correlations. The more correlated the factors, the greater the difference between the factor pattern loadings and structure matrix elements. When factors are correlated, the communalities for items are not as easily computed as with orthogonal factor rotations because the factor intercorrelations must be included in the calculation. Consequently, path tracing must be used to calculate the communalities for items involved in oblique factor rotations.

The factor pattern matrix is used when obtaining factor scores and reproducing the correlation matrix (Pett, Lackey, & Sullivan, 2003). The factor structure matrix is $P\phi$, or the pattern matrix times the covariance (correlation) matrix. When factors are orthogonal, $\phi = I$ (an identity matrix), so $PI = P$ (the pattern and structure matrices are identical). It must be noted that both LISREL and M*plus* only provide pattern coefficients with promax rotation. M*plus* does provide both pattern and structure coefficients with geomin rotation.

CONFIRMATORY FACTOR ANALYSIS

The validity and reliability issues in measurement have traditionally been handled by first examining the validity and reliability of scores on instruments used in a particular design context. Given an acceptable level of validity and reliability, the scores are then used in a statistical analysis. However, the traditional statistical analysis of these scores, such as multiple regression and path analysis, does not account for measurement error. The impact of measurement error has been investigated and found to have serious consequences, such as producing biased parameter estimates (Cochran, 1968; Fuller, 1987). A general approach to confirmatory factor analysis using maximum likelihood estimation was therefore developed by Jöreskog (1969) to counter the negative impact of measurement error.

Confirmatory factor analysis (CFA) tests a hypothesized theoretical measurement model. CFA models are commonly used by researchers to test the construct validity of latent factors underlying the items on surveys or instruments. In contrast to EFA, a researcher specifies which variables go together and are assigned to a specific factor in the CFA model. CFA is commonly conducted after an EFA (or multiple EFAs) using a different sample of data in order to provide additional support for the factor structure that was retained by the EFA. Another distinguishing characteristic of CFA is that not all items load on all factors in the model as is done in EFA. That is, EFA allows all items to load on all factors, with the goal of reaching an approximate simple structure. Recall that the rotated factor loadings were large for primary factors, yet smaller non-zero loadings were associated with other non-primary factors. In CFA, researchers typically only allow an item to load on one factor, unless cross-loadings were found in previous EFA models and/or theoretically the item was thought to cross-load. Readers are encouraged to read Brown's (2015) book for more details about CFA. We present an example of a CFA analysis next with considerations of specification, identification, testing, and modification/re-specification.

CFA EXAMPLE

Holzinger and Swineford (1939) collected data on 26 psychological tests from seventh- and eighth-grade children in a suburban school district of Chicago.

Over the years, different subsamples of the children and different subsets of the variables in this dataset have been analyzed and presented in various multivariate statistics textbooks, for example, Harmon (1976), Gorsuch (1983), and Jöreskog & Sörbom (1993, example 5, pp. 23–28).

The data analyzed here are the first six psychological variables for all 301 subjects. The CFA model consists of the following six observed variables: *Visual Perception, Cubes, Lozenges, Paragraph Comprehension, Sentence Completion,* and *Word Meaning.* The first three variables were hypothesized to measure a *Spatial* ability factor and the second three variables to measure a *Verbal* ability factor.

The CFA diagram of the theoretical proposed model is shown in Figure 6.4. The observed variables are enclosed by boxes or rectangles, and the factors or latent variables (*Spatial* and *Verbal*) are enclosed by circles. Conceptually, a factor represents the common variation among a set of observed variables. Thus, for example, the *Spatial* ability factor represents the common variation among the *Visual Perception, Cubes,* and *Lozenges* tasks. That is, spatial ability is the hypothesized reason for scores on *Visual Perception, Cubes,* and *Lozenges* tasks. Lines directed from a factor to a particular observed variable denote the relations between that factor and that variable, which are the factor loadings. Factor loadings are similar to traditional regression coefficients in one-factor models and partial regression coefficients in correlated multi-factor models.

The measurement errors are enclosed by smaller ovals and indicate that some portion of each observed variable is not explained by the hypothesized factor. Conceptually, a measurement error represents the unique variation for a particular observed variable beyond the variation due to the relevant factor. For example, the *Cubes* task is largely a measure of *Spatial* ability, but may also be assessing other characteristics, such as a different common factor or unreliability. To assess measurement error, the variance of each measurement error is estimated (known as *measurement error variance*). Hence, you will notice the circular two-headed arrows associated with each indicator's error.

You will also notice the circular two-headed arrows associated with each factor. Remember that the variances of exogenous variables are estimated in SEM models. Just like errors in the model, neither of the two factors (*Spatial* and *Verbal*) have a scale of measurement because they are unobserved. Because the latent factors do not have an observed metric, the estimation of all of the model parameters would be prohibited unless we set their scale of measurement. There are two ways to set the scale of latent factors in SEM models. One way is to scale the factor using the variance of one of the indicator variables, which is commonly done by setting the direct path or factor loading from a factor to an indicator variable equal to 1. Indicators with fixed loadings equal to 1 are called *reference indicators*. In Figure 6.4, you will notice that *Visual Perception* and *Paragraph Comprehension* are serving as reference indicators for the *Spatial* and *Verbal* factors, respectively. The second way of setting the scale of a factor is to fix the factor variance to a value of 1. The default in LISREL is to set the factor variance equal to 1 whereas the default in M*plus* is to use the reference indicator scaling method.

Regardless of the scaling method used, the standardized estimates and model fit will not be impacted. One consideration when using the reference indicator scaling approach is the selection of the indicator to serve as the reference indicator. Although the selection of a reference indicator is somewhat arbitrary in CFA models, the selection can impact the results for other SEM techniques (e.g., multiple-group modeling; see Cheung & Rensvold, 1999; Yoon & Millsap, 2007; Whittaker & Khojasteh, 2013). In the context of CFA models, the reference indicator selected should correlate positively with its respective factor given that its loading will be constrained to equal a positive value of 1. If not, "estimation problems may be encountered because you are fixing its relationship with the factor (and, in turn, its relationship with the other variables loading on the same factor) to a relationship that contradicts the original data" (Whittaker, 2016, p. 691). Readers should consult Hancock, Stapleton, and Arnold-Berkovits, 2009 for more information about reference indicators in SEM techniques.

A curved arrow between the two factors indicates that they have shared variance or are correlated. In this example, *Spatial* ability and *Verbal* ability are specified to covary or correlate. The rationale for this particular factor correlation is that *Spatial* ability and *Verbal* ability are related to a more general ability factor and thus should be theoretically related.

A curved arrow between two measurement error variances indicates that they also have shared variance or are correlated. Although not shown in this example, two measurement error variances could be correlated if they shared something in common such as: (a) common method variance where the method of measurement is the same (such as the same scale of measurement) or they are both part of the same global instrument; or (b) the same measure is being used at different points in time; that is, the *Cubes* task is measured at Time 1 and again at Time 2.

The *model specification* is a first step in confirmatory factor analysis, just as it was for multiple regression and path models. Model specification is necessary because many different relations among a set of variables can be postulated with many different parameters being estimated. Thus, many different factor models can be postulated on the basis of different hypothesized relationships between the observed variables and the factors.

Our CFA model is specified to have six observed variables with two different correlated latent variables (factors) being hypothesized. In Figure 6.4, each observed variable is hypothesized to load on only a single factor (thus, there are three observed variables per factor); the factors are believed to be correlated (a single factor correlation); and the measurement error variances are not related (zero correlated measurement errors). How does the researcher determine which factor model is correct? We already know that *model specification* is important in this process and indicates the important role that theory and prior research play in justifying a specified model. Confirmatory factor analysis does not tell us how to specify the model, but rather estimates the parameters of the model once the model has been specified by the researcher.

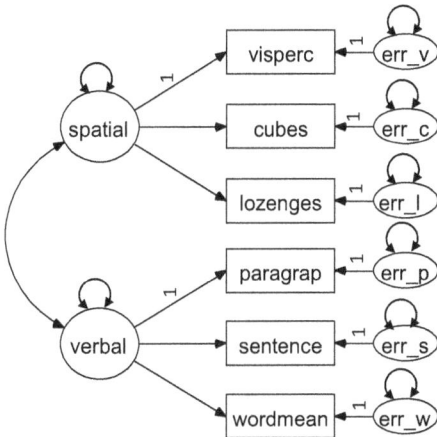

■ **Figure 6.4:** TWO-FACTOR CORRELATED CONFIRMATORY FACTOR MODEL

The CFA model in Figure 6.4 contains six measurement equations in the model, one for each of the six observed variables. Substantive theory and prior research suggest that items only load on their respective factor and that other possible factor loadings (e.g., *visperc* also loading on *verbal*) should not be included in the confirmatory factor model. Using the variable names from Figure 6.4, the measurement equations are as follows:

```
visperc  = 1*spatial + err_v
cubes    = λ₁₂ spatial + err_c
lozenges = λ₁₃ spatial + err_l
paragrap = 1* verbal + err_p
sentence = λ₂₂ verbal + err_s
wordmean = λ₂₃ verbal + err_w
```

Model identification is the next step in the confirmatory factor analysis. It is crucial that the researcher check for any *identification problem* prior to the estimation of parameters. In our confirmatory factor model, some parameters are fixed and others are free. An example of a *fixed parameter* is that *cubes* is not allowed to load on *verbal*. An example of a *free parameter* is that *cubes* is allowed to load on *spatial*.

In determining identification, we first assess the number of distinct or non-redundant values in the sample covariance matrix: $p(p + 1)/2 = 6(6 + 1)/2 = 21$, where p is the number of observed variables (i.e., 6 indicators). Next, we need to determine the number of free parameters to be estimated in the model. Using the reference indicator scaling method, our CFA model includes four factor loadings; six measurement error variances, one for each observed variable; two factor variances (one for *spatial* ability and one for *verbal* ability); and one covariance between the *spatial* ability factor and the *verbal* ability factor. A count of the free parameters is as follows (with the *reference indicator* scaling method used):

  4 factor loadings
  6 measurement error variances
  2 factor variances
  1 covariance among the latent variables

Thus, there are a total of 13 free parameters that we wish to estimate. The number of distinct or non-redundant elements in our sample covariance matrix (**S**) is 21, which is greater than the number of free parameters (13), with the difference being the degrees of freedom for the specified model, $df = 21 - 13 = 8$. According to the counting rule, this model is *over-identified* because there are more non-redundant elements in **S** than parameters to be estimated. That is, our degrees of freedom are positive and not zero (*just-identified*) or negative (*under-identified*). This does not guarantee that the model is identified. For instance, if we did not set the scale of the latent factors somehow, the model would be *locally under-identified* even though it is *globally* identified. The reason for this is that we could not estimate parameters in the model without some scale of measurement associated with latent variables because the variances of the latent factors are included in each of the decomposition equations.

The next step is to estimate the parameters for the hypothesized factor model. In CFA, the goal is still to minimize the difference between the sample covariance matrix and the model-implied covariance matrix that is based on the relationships in the model. The complete decomposition of the covariance matrix is possible when all variable relations are accounted for by the factors in a specified factor model. When an over-identified factor model is hypothesized, the covariance matrix will not be completely reproduced. The goal is to have a hypothesized factor model that closely reproduces the original sample covariance matrix.

The model parameters can be estimated by different estimation procedures, such as maximum likelihood (ML) and robust weighted least squares (DWLS or WLSMV). We analyzed the confirmatory factor model using maximum likelihood estimation, requesting the standardized solution to report our statistical estimates of the factor model parameters. The following SIMPLIS program was used to specify the CFA model:

```
LISREL-SIMPLIS Confirmatory Factor Model Program
Confirmatory Factor Model in Figure 6.4
Observed Variables:
VISPERC CUBES LOZENGES PARCOMP SENCOMP WORDMEAN
Sample Size: 301
Covariance Matrix
49.064
9.810 22.182
27.928 14.482 81.863
9.117 2.515 5.013 12.196
10.610 3.389 3.605 13.217 26.645
19.166 6.954 13.716 18.868 28.502 58.817
Latent Variables: Spatial Verbal
Relationships:
VISPERC CUBES LOZENGES = Spatial
PARCOMP SENCOMP WORDMEAN = Verbal
Print Residuals
Number of Decimals = 3
OPTIONS: SC MI
Path Diagram
End of problem
```

The program syntax is case sensitive, so capitalized variable names must also be used in the *Relationships* commands, which specifies which observed variables go with the latent variables. The latent variables are named using the *Latent Variable* command. The following code: VISPERC CUBES LOZENGES = Spatial indicates that these three indicators load on the *Spatial* factor. Likewise: PARCOMP SENCOMP WORDMEAN = Verbal indicates that these three indicators load on the *Verbal* factor. The **Print Residuals** command will output the residual values, **SC** and **MI** in the OPTIONS command will print out the standardized solution and modification indices, respectively. The default in LISREL is to set the factor variances equal to 1 unless otherwise specified by the user to use the reference indicator approach. If using the *reference indicator* method is preferred, the SIMPLIS code would look as follows:

```
LISREL-SIMPLIS Confirmatory Factor Model Program
Reference Indicators Used
Observed Variables:
VISPERC CUBES LOZENGES PARCOMP SENCOMP WORDMEAN
Sample Size: 301
Covariance Matrix
49.064
9.810 22.182
27.928 14.482 81.863
9.117 2.515 5.013 12.196
10.610 3.389 3.605 13.217 26.645
19.166 6.954 13.716 18.868 28.502 58.817
Latent Variables: Spatial Verbal
Relationships:
VISPERC = 1*Spatial
CUBES LOZENGES = Spatial
PARCOMP = 1*Verbal
SENCOMP WORDMEAN = Verbal
Print Residuals
Number of Decimals = 3
OPTIONS: SC MI
Path Diagram
End of problem
```

You will notice that the reference indicators (*VISPERC* and *PARCOMP*) are equal to the product of the fixed value of 1 and their respective factor (i.e., 1*Spatial and 1*Verbal, respectively). The factors are also correlated by default in LISREL. If you thought the factors were not correlated or orthogonal, you can write the following statement prior to the **End of Problem** command:

```
Set the Covariance Between Verbal and Spatial Equal to 0
```

The **Path Diagram** command requests a diagram of the CFA model, which is presented in Figure 6.5 with the standardized solution (which was chosen from the *Estimates* or the *View* pull-down menu).

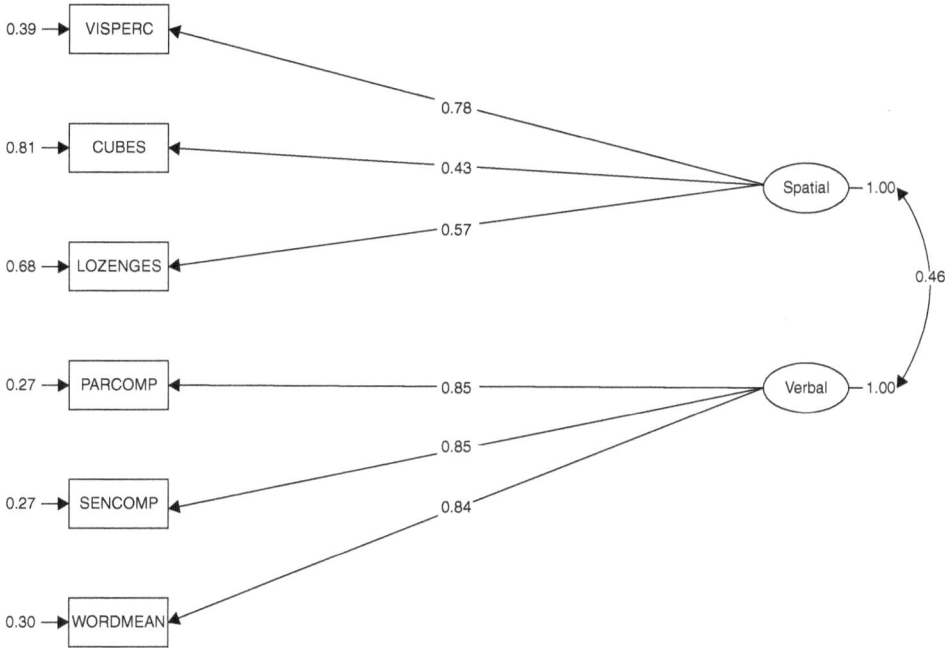

■**Figure 6.5:** LISREL PATH DIAGRAM OUTPUT FOR TWO-FACTOR CORRELATED CFA MODEL (STANDARDIZED SOLUTION)

The LISREL output with standardized factor loadings (LAMBDA-X), factor correlation between *spatial* and *verbal* factors (PHI), and residual variances or unexplained variance for each indicator (THETA-DELTA) is reported in Table 6.5. The model fit of the CFA model is summarized below:

LISREL Model Fit of Two-Factor Correlated CFA Model

χ^2	df	CFI	NNFI/TLI	RMSEA (90% CI)	SRMR
24.284*	8	.975	.953	.082 (.046, .121)	.047

Note: $^*p < .05$.

The chi-square model fit value was statistically significant and the RMSEA was greater than .08, whereas the CFI and NNFI/TLI were above and the SRMR was below their suggested cutoff values for adequate model fit. Thus, the model is demonstrating marginal model fit.

Table 6.5: Two-factor Correlated CFA Model Results from LISREL – Standardized Solution

```
Completely Standardized Solution

        LAMBDA-X
                Spatial     Verbal
                --------    --------
   VISPERC       0.778       - -
     CUBES       0.431       - -
  LOZENGES       0.568       - -
   PARCOMP       - -         0.852
   SENCOMP       - -         0.854
  WORDMEAN       - -         0.838

        PHI
                Spatial     Verbal
                --------    --------
   Spatial       1.000
    Verbal       0.461       1.000


        THETA-DELTA

              VISPERC     CUBES    LOZENGES    PARCOMP    SENCOMP    WORDMEAN
              --------   --------  --------   --------   --------   --------
               0.395      0.814     0.677      0.274      0.270      0.298
```

When the CFA model does not fit the data adequately, the researcher has several options to determine whether model misspecification may be occurring in the model and/or how to improve model fit. First, the normalized and/or standardized values in the residual matrix could be examined. If any values are greater than $z = 1.96$, that could suggest that a relationship between two variables is not explained well and a possible additional connection in the model could be added. Second, the modification indices (MIs) could be examined. Third, the unstandardized and standardized versions of the expected parameter change (EPC) values could be examined. Fourth, the structure matrix could be computed to determine if the CFA model was correctly specified.

In CFA, the items that are not loading on other factors have values of zero in the pattern coefficient matrix. According to Graham, Guthrie, and Thompson (2003), values of zero in the pattern coefficient matrix does not directly translate into values of zero in the structure coefficient matrix. When factors are correlated in CFA models, it suggests that the items with loadings of zero on non-primary factors actually have non-zero correlations with the non-primary factors. Nevertheless, the goal is still to have items correlate strongly with one single factor while correlating weakly with other factors in the structure coefficient matrix (Graham et al., 2003).

It is possible to obtain the structure coefficient matrix from the product of the LAMBDA-X pattern matrix times the PHI matrix: $\psi_s = \Lambda_x \phi$. You can use R, SPSS, SAS, or other software to multiply the matrices. For the CFA results in Table 6.5, R was used to calculate the structure coefficients as follows:

$$\Psi_s = \Lambda_x \Phi$$

$$\Psi_s = \begin{bmatrix} .78 & 0 \\ .43 & 0 \\ .57 & 0 \\ 0 & .85 \\ 0 & .85 \\ 0 & .84 \end{bmatrix} \begin{bmatrix} 1 & .461 \\ .461 & 1 \end{bmatrix}$$

$$\Psi_s = \begin{bmatrix} .78 & .36 \\ .43 & .20 \\ .57 & .26 \\ .39 & .85 \\ .39 & .85 \\ .39 & .84 \end{bmatrix}$$

The structure coefficients $\left(\Psi_s\right)$ indicate that the pattern coefficients $\left(\Lambda_x\right)$ are specified correctly. That is, the observed variables are indicators of factors where they had the highest factor loading. We therefore turn our attention to the MIs in LISREL. The SIMPLIS program provides MIs for cross-loadings and for error covariances as follows, respectively:

```
The Modification Indices Suggest to Add the
Path to          from            Decrease in Chi-Square    New Estimate
    VISPERC          Verbal             10.4                    2.62
    LOZENGES         Verbal              9.2                   -2.32
    SENCOMP          Spatial             7.9                   -0.79

The Modification Indices Suggest to Add an Error Covariance
Between          and             Decrease in Chi-Square    New Estimate
    CUBES            VISPERC             9.2                   -8.53
    LOZENGES         CUBES              10.4                    8.59
    WORDMEAN         ARCOMP              7.9                   -5.87
```

A researcher would generally pick the MI with the largest value. However, not all of the MIs will make sense given the theoretical CFA model. For example, the cross-loading of *VISPERC* on the *Verbal* factor would not necessarily make theoretical sense given that visual perception would not be thought to be strongly related to verbal ability. Error covariances among indicators in a CFA model represent that the relationship between the respective indicators is being explained by something in addition to the latent factor through which they are spuriously associated. Thus, strong theoretical justifications are needed when connecting error covariances. When looking at the modification indices for the error covariances, the largest drop in chi-square would be the result of adding the error covariance between *LOZENGES* and *CUBES*, which are both loading on the same factor, *Spatial* ability. We could substantiate the added error covariance between these observed variables because the subtests both have visual pictures that involve the conjectural manipulation of pictures to understand the next picture that would be reasonable in the sequence.

To modify the SIMPLIS program with this error covariance, we added the following command line after our CFA model statements on the *Relationships* command:

```
Let the error covariances between LOZENGES and CUBES correlate
```

The LISREL diagram for the modified model is presented in Figure 6.6 and the results from the model with the added error covariance (THETA-DELTA) are reported in Table 6.6. The path diagram now includes a curved arrow between *CUBES* and *LOZENGES*, which indicates the coefficient for the error correlation and is estimated to be equal to .196.

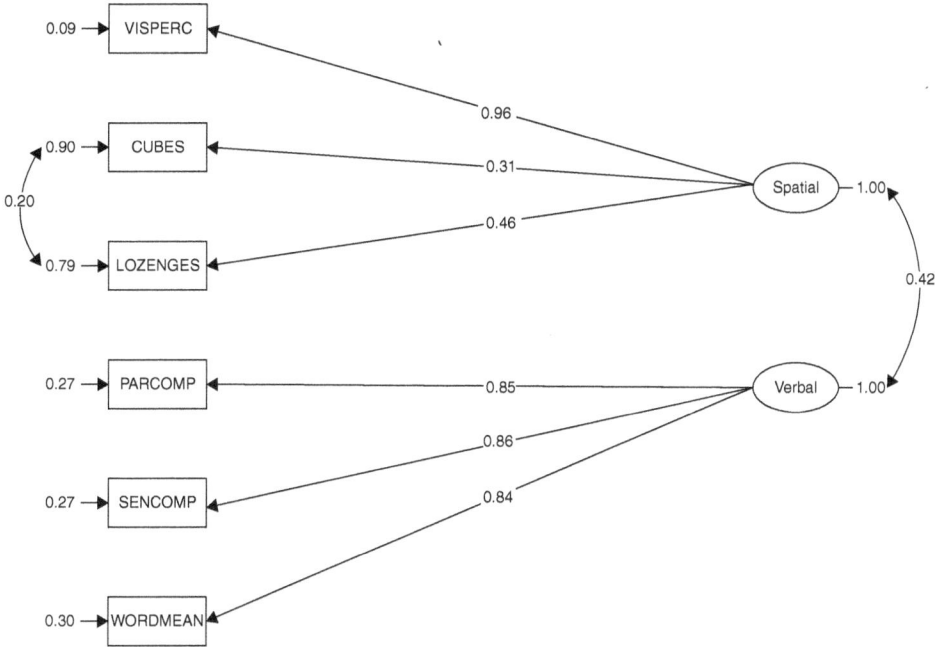

■ **Figure 6.6:** MODIFIED TWO-FACTOR CORRELATED CFA MODEL DIAGRAM IN LISREL

Table 6.6: LISREL Results for Modified Two-factor Correlated CFA Model – Standardized Solution

```
Completely Standardized Solution

        LAMBDA-X
              Spatial      Verbal
              --------    --------
VISPERC        0.956        - -
  CUBES        0.314        - -
LOZENGES       0.460        - -
PARCOMP         - -        0.854
SENCOMP         - -        0.855
WORDMEAN        - -        0.836

        PHI
              Spatial      Verbal
              --------    --------
Spatial        1.000
 Verbal        0.419       1.000
```

Table 6.6: (*cont.*)

THETA-DELTA	VISPERC	CUBES	LOZENGES	PARCOMP	SENCOMP	WORDMEAN
VISPERC	0.087					
CUBES	- -	0.902				
LOZENGES	- -	0.196	0.788			
PARCOMP	- -	- -	- -	0.271		
SENCOMP	- -	- -	- -	- -	0.269	
WORDMEAN	- -	- -	- -	- -	- -	0.301

Below is the summary of model fit for the modified model.

LISREL Model Fit of Modified Two-Factor Correlated CFA Model

χ^2	df	CFI	NNFI/TLI	RMSEA (90% CI)	SRMR
13.929	7	.989	.977	.057 (.000, .101)	.032

The chi-square test of model fit is not significant and all of the model fit indices are in the acceptable range, suggesting good fit of the model to the data. With the added error covariance, the model *df* decreased by 1 because we estimated an additional parameter.

The M*plus* code to run the same original CFA model with the summary data in the **HS_Data. txt** file is provided below.

```
TITLE:      Confirmatory Factor Analysis
            CFA Model in Figure 6.4

DATA:       FILE is "C:\HS_Data.txt";
            TYPE IS COVARIANCE;
            NOBS = 301;

VARIABLE:   NAMES are VISPERC CUBES LOZENGES
            PARCOMP SENCOMP WORDMEAN;

MODEL:      SPATIAL BY VISPERC* CUBES LOZENGES;
            VERBAL  BY PARCOMP* SENCOMP WORDMEAN;
            SPATIAL WITH VERBAL;
            SPATIAL@1;
            VERBAL@1;

OUTPUT: STDYX RESIDUAL MODINDICES(3.84);
```

In M*plus*: SPATIAL BY VISPERC* CUBES LOZENGES in the MODEL command indicates that the *Spatial* factor is indicated **BY** *VISPERC, CUBES,* and *LOZENGES*. Likewise: VERBAL BY PARCOMP* SENCOMP WORDMEAN in the MODEL command indicates that the *Verbal* factor is indicated **BY** *PARCOMP, SENCOMP,* and *WORDMEAN*. In M*plus*, the reference indicator approach of scaling the factor is the default and will automatically set the loading of the first variable for each factor following the **BY** statement in the MODEL command (i.e., *VISPERC* and *PARCOMP*) equal to 1 unless otherwise specified. To avoid this default, you can type an asterisk (*) immediately after the first variable for each factor following the **BY** statement in order to estimate the loading. You will still need to set the scale of the factor. This is done by setting the factor variances equal to 1 using the following statements:

```
SPATIAL@1;
VERBAL@1;
```

Although the default in M*plus* is to covary exogenous factors, the following statement requested that the factors covary:

```
SPATIAL WITH VERBAL;
```

To request no covariance between factors, you would write the following to set the covariance to zero:

```
SPATIAL WITH VERBAL@0;
```

The standardized estimates are requested with **STDYX** in the OUTPUT command. Further, statistically significant modification indices are requested with **MODINDICES(3.84)** in the OUTPUT command. If the reference indicator approach was preferred, the M*plus* code would look as follows:

```
TITLE:      Confirmatory Factor Analysis
            Reference Indicators Used

DATA:       FILE is "C:\HS_Data.txt";
            TYPE IS COVARIANCE;
            NOBS = 301;

VARIABLE:   NAMES are VISPERC CUBES LOZENGES
            PARCOMP SENCOMP WORDMEAN;

MODEL:      SPATIAL BY VISPERC CUBES LOZENGES;
            VERBAL  BY PARCOMP SENCOMP WORDMEAN;
            SPATIAL WITH VERBAL;

OUTPUT: STDYX RESIDUAL MODINDICES(3.84);
```

VISPERC and *PARCOMP* would then serve as reference indicators for the *Spatial* and *Verbal* factors, respectively.

The standardized results from M*plus* are presented in Table 6.7, which match the standardized output in LISREL (see Table 6.5).

Table 6.7: Two-factor Correlated CFA Model – Standardized Estimates from M*plus*

```
STANDARDIZED MODEL RESULTS

STDYX Standardization
```

	Estimate	S.E.	Est./S.E.	Two-Tailed P-Value
SPATIAL BY				
VISPERC	0.778	0.062	12.583	0.000
CUBES	0.431	0.061	7.094	0.000
LOZENGES	0.568	0.059	9.663	0.000
VERBAL BY				
PARCOMP	0.852	0.023	37.826	0.000
SENCOMP	0.854	0.022	38.094	0.000
WORDMEAN	0.838	0.023	35.845	0.000
SPATIAL WITH				
VERBAL	0.461	0.064	7.194	0.000
Variances				
SPATIAL	1.000	0.000	999.000	999.000
VERBAL	1.000	0.000	999.000	999.000
Residual Variances				
VISPERC	0.395	0.096	4.107	0.000
CUBES	0.814	0.052	15.533	0.000
LOZENGES	0.677	0.067	10.119	0.000
PARCOMP	0.274	0.038	7.121	0.000
SENCOMP	0.270	0.038	7.055	0.000
WORDMEAN	0.298	0.039	7.595	0.000

```
R-SQUARE
```

Observed Variable	Estimate	S.E.	Est./S.E.	Two-Tailed P-Value
VISPERC	0.605	0.096	6.292	0.000
CUBES	0.186	0.052	3.547	0.000
LOZENGES	0.323	0.067	4.831	0.000
PARCOMP	0.726	0.038	18.913	0.000
SENCOMP	0.730	0.038	19.047	0.000
WORDMEAN	0.702	0.039	17.922	0.000

You will notice that M*plus* printed out the R^2 estimates for each indicator variable with the standardized solution, which are calculated as 1 minus the standardized residual variance (e.g., $1 - .395 = .605$ for *VISPERC*). This indicates that approximately 61% of the variance in *VISPERC* is explained by the factor model. The R^2 values in LISREL can be found in the *Measurement Equations* output.

The model fit information from M*plus* for the CFA model is presented below, and is analogous to the model fit information provided in LISREL (with the exception of the subtle chi-square value difference).

M*plus* Model Fit of Two-Factor Correlated CFA Model

χ^2	df	CFI	NNFI/TLI	RMSEA (90% CI)	SRMR
24.365*	8	.975	.953	.082 (.046, .121)	.047

Note: * $p < .05$.

The M*plus* output from requesting statistically significant modification indices is provided below:

```
MODEL MODIFICATION INDICES

NOTE: Modification indices for direct effects of observed dependent
variables regressed on covariates may not be included.  To include
these, request MODINDICES (ALL).

Minimum M.I. value for printing the modification index     3.840

                        M.I.    E.P.C.   Std E.P.C.   StdYX E.P.C.
BY Statements
SPATIAL BY SENCOMP       7.966   -0.789      -0.789        -0.153
VERBAL BY VISPERC       10.439    2.615       2.615         0.374
VERBAL BY LOZENGES       9.201   -2.315      -2.315        -0.256

WITH Statements
CUBES WITH VISPERC       9.206   -8.507      -8.507        -0.456
LOZENGES WITH CUBES     10.430    8.558       8.558         0.271
SENCOMP WITH LOZENGES    7.349   -4.080      -4.080        -0.205
WORDMEAN WITH PARCOMP    7.964   -5.845      -5.8451       -0.767
```

The BY Statements in the output for the modification indices are suggesting cross-loadings. These are the same cross-loadings that were suggested in LISREL. The WITH Statements in the output for the modification indices are suggesting error covariances. Similar to the LISREL output, the largest MI value for the error covariances (WITH Statements) is associated with the error covariance between *LOZENGES* and *CUBES*.

To add the error covariance between *LOZENGES* and *CUBES*, you can write the following statement after setting the factor variances equal to one in the M*plus* code:

```
LOZENGES WITH CUBES
```

The standardized estimates from M*plus* for the modified CFA model with the added error covariance are presented in Table 6.8, which look very similar to the LISREL estimates. The estimated correlation between the errors for *LOZENGES* and *CUBES* is slightly larger in M*plus* than in LISREL (.232 and .196, respectively). Table 6.9 illustrates the final model estimates from M*plus* similarly to what would be presented in a manuscript. The M*plus* model fit information for the modified CFA model is presented below, which is also similar to the estimates in LISREL.

M*plus* Model Fit of Modified Two-Factor Correlated CFA Model

χ^2	df	CFI	NNFI/TLI	RMSEA (90% CI)	SRMR
13.976	7	.989	.977	.058 (.000, .101)	.033

Table 6.8: Modified Two-factor CFA Model – Standardized Estimates from M*plus*

```
STANDARDIZED MODEL RESULTS

STDYX Standardization
```

	Estimate	S.E.	Est./S.E.	Two-Tailed P-Value
SPATIAL BY				
VISPERC	0.956	0.117	8.163	0.000
CUBES	0.314	0.066	4.750	0.000
LOZENGES	0.460	0.073	6.336	0.000
VERBAL BY				
PARAGRAP	0.854	0.022	38.065	0.000
SENTENCE	0.855	0.022	38.297	0.000
WORDMEAN	0.836	0.023	35.593	0.000
SPATIAL WITH				
VERBAL	0.419	0.072	5.786	0.000
LOZENGES WITH				
CUBES	0.232	0.063	3.683	0.000
Variances				
SPATIAL	1.000	0.000	999.000	999.000
VERBAL	1.000	0.000	999.000	999.000

Table 6.8: (*cont.*)

```
Residual Variances
   VISPERC              0.087      0.224       0.387      0.699
   CUBES                0.902      0.041      21.790      0.000
   LOZENGES             0.788      0.067      11.804      0.000
   PARAGRAP             0.271      0.038       7.092      0.000
   SENTENCE             0.269      0.038       7.035      0.000
   WORDMEAN             0.301      0.039       7.676      0.000

R-SQUARE
   Observed                                            Two-Tailed
   Variable          Estimate       S.E.   Est./S.E.    P-Value

   VISPERC              0.913      0.224       4.082      0.000
   CUBES                0.098      0.041       2.375      0.018
   LOZENGES             0.212      0.067       3.168      0.002
   PARAGRAP             0.729      0.038      19.033      0.000
   SENTENCE             0.731      0.038      19.149      0.000
   WORDMEAN             0.699      0.039      17.797      0.000
```

Table 6.9: Unstandardized Estimates, Standard Errors, Standardized Estimates, and *R* Square Values for the Final Two-factor CFA Model

Item	Unstandardized Estimates	S.E.	Standardized Estimates	*R* Square
Spatial Ability Factor Loadings				
visperc	6.68*	.87	.96	.91
cubes	1.47*	.33	.31	.10
lozenges	4.16*	.71	.46	.21
Verbal Ability Factor Loadings				
paragraph	2.98*	.17	.85	.73
sentence	4.41*	.25	.86	.73
wordmean	6.40*	.38	.84	70
Factor Covariance				
Spatial ↔ Verbal	0.42*	.07	.42	--
Error Covariances				
lozenges ↔ cubes	8.31*	2.60	.23	--

Note: * *p* < .05.

CFA WITH MISSING CONTINUOUS DATA

When continuous data are missing, both LISREL and M*plus* can perform full information maximum likelihood (FIML) estimation (see Chapter 2). When using FIML, the assumption is that the data are missing at random (MAR). Thus, the reason for missing data should be investigated prior to estimating model parameters. For this example, we are assuming that the data are MAR. The example data for this demonstration are simulated data for 1,250 girls based on the same six psychological variables from Holzinger and Swineford (1939) used previously. Specifically, *Visual Perception*, *Cubes*, and *Lozenges* will load on the *Spatial* ability factor and *Paragraph Comprehension*, *Sentence Completion*, and *Word Meaning* will load on the *Verbal* ability factor.

The raw data are necessary when estimating models with missing data. In LISREL, the ***MGGIRLS.LSF*** was used and is available on the book website. The missing data in the .LSF file are indicated as –9.00. Thus, it is important that the user first designate the 'Global missing value' in the .LSF file prior to estimating model parameters. This can be done with the .LSF file open and clicking on the **Data** menu and scrolling down to click the **Define Variables** option. Screenshots of this are presented in Figure 6.7.

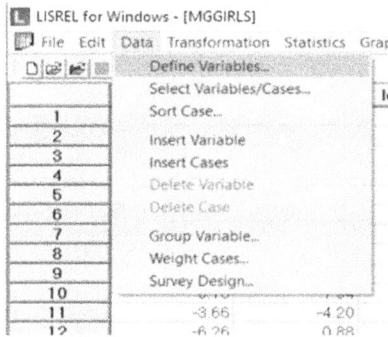

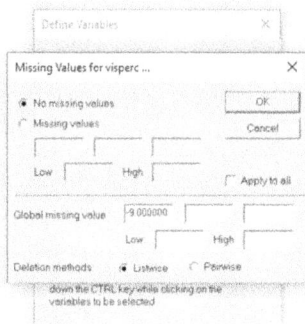

▨ **Figure 6.7:** SCREENSHOTS IN LISREL DESIGNATING MISSING VALUES IN MGGIRLS.LSF

Once the missing values have been designated, the .LSF file should be saved. The following SIMPLIS code can be used to estimate the CFA with missing data:

```
CFA WITH MISSING DATA
Observed Variables:
VISPERC CUBES LOZENGES PARCOMP SENCOMP WORDMEAN
RAW DATA FROM FILE C:\MGGIRLS.LSF
Latent Variables: Spatial Verbal
Relationships:
VISPERC CUBES LOZENGES = Spatial
PARCOMP SENCOMP WORDMEAN = Verbal
Print Residuals
Number of Decimals = 3
OPTIONS: SC MI
Path Diagram
End of problem
```

You will notice that the SIMPLIS code is similar to the code used for the previous CFA model estimated in LISREL. The biggest difference between this code and the code used previously is that the following syntax is used to read in the raw data: RAW DATA FROM FILE C:\MGGIRLS.LSF. Because raw data are used, the covariance matrix is not included in the code and sample size does not need to be indicated.

You will see the following in the LISREL output related to missing data estimation:

```
     --------------------------------
     EM Algorithm for missing Data:
     --------------------------------

   Number of different missing-value patterns=        46
   Effective sample size:      1250

   Convergence of EM-algorithm in      5 iterations
   -2 Ln(L) =      37795.40985
   Percentage missing values=   15.21
 Note:
   The Covariances and/or Means to be analyzed are estimated
   by the EM procedure and are only used to obtain starting
   values for the FIML procedure
```

There were 46 different missing-value patterns. The LISREL output will include unstandardized estimates, completely standardized estimates, residual matrices (unstandardized, normalized, and standardized), modification indices, expected parameter change values, and abbreviated model fit information. The model fit information provided by LISREL with FIML is below:

```
              Global Goodness of Fit Statistics, FIML case

-2ln(L) for the saturated model =        37795.410
-2ln(L) for the fitted model     =       37803.653
Degrees of Freedom = 8
Full Information ML Chi-Square                  8.243 (P = 0.4101)
Root Mean Square Error of Approximation (RMSEA)  0.00493
90 Percent Confidence Interval for RMSEA        (0.0 ; 0.0338)
P-Value for Test of Close Fit (RMSEA < 0.05)     0.999
```

The abbreviated model fit information indicates good model fit with a non-significant chi-square test of model fit [$\chi^2(8) = 8.243$, $p > .05$] and an acceptable RMSEA estimate [RMSEA = .005 (90% CI: 0.0, 0.34), $p > .05$].

In M*plus*, you will need to read in the raw data file as a .txt, .dat, or .csv file. The **MGGIRLS.TXT** file (which is on the book website) was read into M*plus* and the CFA model was estimated using the following M*plus* code:

```
TITLE:      CFA WITH MISSING DATA

DATA:       FILE is "C:\MGGIRLS.TXT";

VARIABLE:   NAMES are VISPERC CUBES LOZENGES
            PARCOMP SENCOMP WORDMEAN;
            MISSING ARE ALL (-9);

MODEL:      SPATIAL BY VISPERC* CUBES LOZENGES;
            VERBAL  BY PARCOMP* SENCOMP WORDMEAN;
            SPATIAL WITH VERBAL;
            SPATIAL@1;
            VERBAL@1;

OUTPUT: STDYX RESIDUAL MODINDICES(3.84);
```

You will notice that the M*plus* code is very similar to the code used for the previous CFA model analyzed in M*plus*. The biggest difference between this code and the code used previously is that the following syntax is used to read in the raw data: FILE is "C:\MGGIRLS.TXT". Because the raw data are used, the type of summary information (TYPE IS COVARIANCE) and the sample size (NOBSERVATIONS or NOBS) does not need to be indicated in the code. Also, the VARIABLE command in M*plus* is where the missing data flag value must be indicated: MISSING ARE ALL (-9). We recommend using the same missing data flag value for all of the variables. Otherwise, you will have to indicate what values are missing data flag values for the different variables in the data file.

The following output in M*plus* summarizes missing data coverage:

```
SUMMARY OF DATA
      Number of missing data patterns            46

COVARIANCE COVERAGE OF DATA

Minimum covariance coverage value    0.100

          PROPORTION OF DATA PRESENT
              Covariance Coverage
              VISPERC    CUBES    LOZENGES    PARCOMP    SENCOMP
            _____ _____ _____ _____ _____

VISPERC       0.840
CUBES         0.718      0.854
LOZENGES      0.713      0.715      0.844
PARCOMP       0.710      0.722      0.709      0.842
SENCOMP       0.714      0.736      0.718      0.720      0.852
WORDMEAN      0.723      0.733      0.723      0.717      0.729

              Covariance Coverage
                  WORDMEAN
                _____

WORDMEAN          0.855
```

The output indicates that there were 46 different missing data patterns, which was also indicated by LISREL. The minimum covariance coverage value is .100 in M*plus*. If the covariance coverage falls below this default value, M*plus* will not estimate model parameters and will provide a message indicating the low covariance coverage problem. You will see the covariance matrix with covariance coverage values for variances and covariances. The values in the matrix indicate the proportion of data available to calculate the respective element in the matrix.

The M*plus* output will include unstandardized estimates, completely standardized estimates, residual matrices (unstandardized, normalized, and standardized), modification indices (although none were significant with this example), expected parameter change values, and all of the model fit information that is typically provided. The model fit information provided by M*plus* is summarized below:

M*plus* Model Fit of CFA Model with Missing Data

χ^2	*df*	CFI	NNFI/TLI	RMSEA (90% CI)	SRMR
8.242	8	1.00	1.00	.005 (.000, .034)	.012

The chi-square test of model fit and RMSEA from LISREL match the estimates in M*plus*. The CFI, TLI, and SRMR in M*plus* are all suggesting adequate model fit.

When using raw data in M*plus*, the mean structure is freely estimated in addition to the covariance structure. In single group analyses (multiple-group analysis will be discussed in Chapter 9), the mean structure estimates include means of observed continuous covariates, thresholds of observed ordinal covariates, intercepts of continuous observed endogenous variables, and thresholds of ordinal observed endogenous variables. Figure 6.8 illustrates the two-factor model with a mean structure incorporated. To estimate the mean structure, the observed indicator variables are regressed on what is called a *unit predictor* or a *constant* which is equal to a value of 1 for each participant and has no variance. The unit predictor is a pseudo-variable because it is not actually observed within the model or in the data set. However, it is not a latent variable either. That is why it is regularly represented as a triangle containing a value of 1 in SEM illustrations. Hence, you will notice that the triangles containing values of 1 in Figure 6.8 are directly affecting the indicator variables.

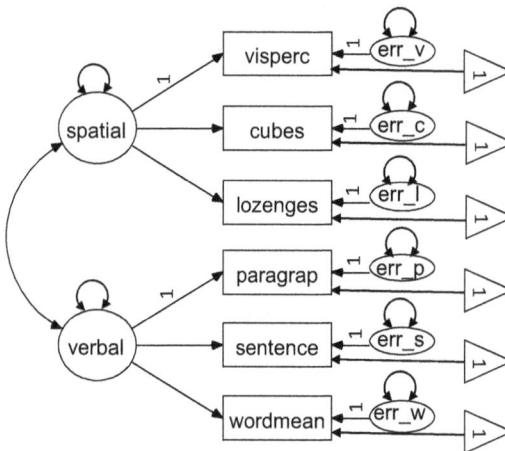

▨ **Figure 6.8:** TWO-FACTOR CORRELATED CFA MODEL WITH A MEAN STRUCTURE

Generally, the estimation of the mean structure estimates will not impact model fit because they are just-identified (e.g., estimating as many indicator intercepts as there are indicators). In this example, you will have estimates of all of the continuous indicator *intercepts* provided in the output. These can be interpreted as in regression models. Below are the unstandardized intercept estimates of the indicators in M*plus* output:

	Estimate	S.E.	Est./S.E.	Two-Tailed P-Value
Intercepts				
VISPERC	0.512	0.209	2.446	0.014
CUBES	0.077	0.139	0.554	0.580
LOZENGES	-0.056	0.230	-0.244	0.808
PARCOMP	0.010	0.102	0.102	0.919
SENCOMP	-0.127	0.143	-0.889	0.374
WORDMEAN	0.060	0.244	0.247	0.805

The intercepts represent the predicted value of the indicators (i.e., *VISPERC*, *CUBES*, *LOZENGES*, *PARCOMP*, *SENCOMP*, and *WORDMEAN*) when their respective factor (*Spatial* or *Verbal*) is equal to zero, holding all else constant. However, the slopes or factor loadings are generally of more interest than the intercepts.

Although the indicator intercepts are not estimated by default in LISREL, they can be requested using the following modified SIMPLIS code:

```
CFA WITH MISSING DATA-INTERCEPT ESTIMATES
Observed Variables:
VISPERC CUBES LOZENGES PARCOMP SENCOMP WORDMEAN
RAW DATA FROM FILE C:\MGGIRLS.LSF
Latent Variables: Spatial Verbal
Relationships:
VISPERC  = CONST Spatial
CUBES    = CONST Spatial
LOZENGES = CONST Spatial
PARCOMP  = CONST Verbal
SENCOMP  = CONST Verbal
WORDMEAN = CONST Verbal
Print Residuals
Number of Decimals = 3
OPTIONS: SC MI
Path Diagram
End of problem
```

You will notice that you have to request a constant or intercept (CONST) for each indicator equation to obtain the intercept estimates. These are provided in the *Measurement Equations* output as follows:

```
Measurement Equations

VISPERC = 0.512 + 4.587*Spatial, Errorvar.= 27.198, R² = 0.436
Standerr             (0.209)          (0.228)          (1.674)
Z-values             2.445            20.107           16.243
P-values             0.014            0.000            0.000

CUBES = 0.0769 + 3.020*Spatial, Errorvar.= 12.376, R² = 0.424
Standerr             (0.139)          (0.152)          (0.745)
Z-values             0.553            19.905           16.622
P-values             0.580            0.000            0.000

LOZENGES = - 0.0560 + 5.998*Spatial, Errorvar.= 23.281, R² = 0.607
Standerr             (0.230)          (0.253)          (2.151)
Z-values             -0.243           23.722           10.824
P-values             0.808            0.000            0.000

PARCOMP = 0.0105 + 3.043*Verbal, Errorvar.= 2.935 , R² = 0.759
Standerr             (0.102)          (0.0887)         (0.238)
Z-values             0.103            34.311           12.336
P-values             0.918            0.000            0.000

SENCOMP = - 0.127 + 4.024*Verbal, Errorvar.= 7.653 , R² = 0.679
Standerr             (0.143)          (0.127)          (0.487)
Z-values             -0.889           31.765           15.700
P-values             0.374            0.000            0.000

WORDMEAN = 0.0604 + 7.191*Verbal, Errorvar.= 18.026, R² = 0.742
Standerr             (0.244)          (0.212)          (1.369)
Z-values             0.247            33.864           13.166
P-values             0.805            0.000            0.000
```

You will notice that the LISREL estimates are the same as those in M*plus*.

CFA CAVEATS

In addition to theory, other important points should be considered when building confirmatory factor models. Observed variable indicators that load on a factor are spuriously related to each other because the factor directly affects the indicator variables. Hence, it is important that the variables are hypothesized to covary because of the factor. A problem can occur if a researcher would like to use a factor to underlie various demographic variables. For instance, one should not include a variable indicating a person's age and a variable indicating a person's racial identity on a "demographics" factor because that would suggest that there is a correlation between the two variables and that the factor has a direct effect on those variables. CFA models were proposed to model measurement error. Demographic variables tend to be fairly reliably measured and could be modeled as stand-alone observed variables. We recommend that the factor represent a true underlying construct and not a general factor on which to load theoretically unrelated indicators.

In a similar vein, it is important that researchers use indicators in a true latent variable system as opposed to what should truly be a formative measurement model. A formative measurement model or an emergent variable system is one in which observed variables (called causal indicators) combine additively and form a linear composite, representing a construct (Bollen & Lennox, 1991). In contrast, a latent variable system, also called a reflective measurement model, is one in which the observed variables (called effect indicators) are linear combinations of the latent construct in addition to measurement error (Bollen & Lennox, 1991). The CFA models discussed in this chapter are reflective measurement models.

An example of a formative measurement model and a reflective measurement model is illustrated in Figure 6.9. As you can see in the top model in Figure 6.9, the observed variables (*education*, *income*, and *occupation*) serve as causal indicators because they directly impact the construct, *socioeconomic status*. Hence, the top model is a formative measurement model. The bottom diagram in Figure 6.9 is the latent variable system or reflective measurement model where the factor (*socioeconomic status*) is directly impacting the three effect indicators (*education*, *income*, and *occupation*). You may notice that the causal direction in the formative measurement model actually makes more theoretical sense than the causal direction in the CFA model presented in Figure 6.9. Specifically, *socioeconomic status* does not directly impact *education*, *income*, and *occupation*. Instead, *education*, *income*, and *occupation* have a direct impact on *socioeconomic status*. Formative measurement models are under-identified models if estimated as illustrated in Figure 6.9. Formative measurement models can be identified if integrated into bigger SEM models where the formative factor directly impacts two endogenous variables (MacCallum & Browne, 1993). In the end, if the indicators make more sense in a formative measurement model, it is not recommended to use them in a latent variable system for the sake of being able to estimate parameters in the CFA model. Again, it is important that the factor represent a true underlying construct that directly impacts the variables serving as effect indicators which is aligned with theory.

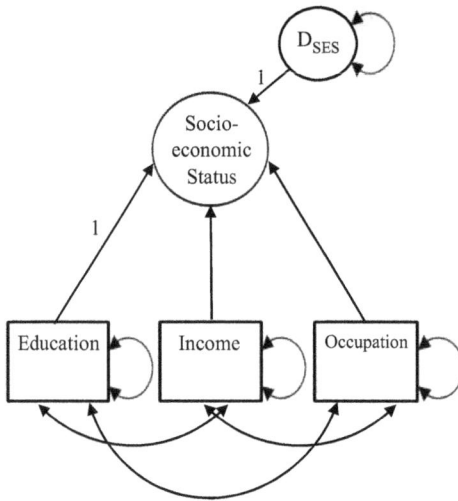

Formative Measurement Model

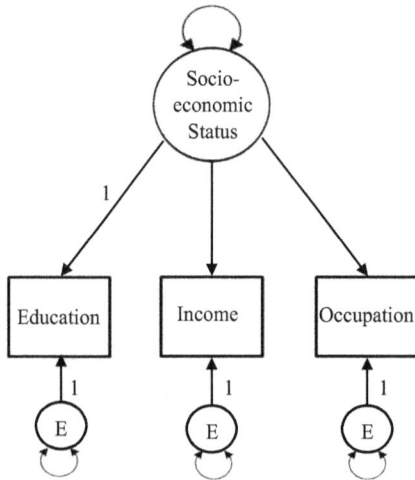

Reflective Measurement Model

▨ **Figure 6.9:** FORMATIVE AND REFLECTIVE MEASUREMENT MODELS

When selecting indicators for a factor, they should provide a fairly comprehensive representation of the construct of interest. For instance, indicators of a *graduate school potential* factor could consist of the graduate record exam (GRE) scores on the *analytical writing*, *verbal*, and *quantitative* subtests. However, there could be other indicators of graduate school success, including *undergraduate GPA, strength of the recommendation letters, strength of the student's statement of purpose*, and *research participation history*. Relatedly, three indicators for a factor results in a just-identified model, because there are six distinct or non-redundant elements in the covariance matrix [3(3 + 1)/2 = 6] and there are six parameters to estimate (three factor loadings and three indicator error variances with the factor variance set equal to 1). With four indicators for a factor, the model is over-identified and can be tested. As the number of indicators per factor increases, there is evidence of more accurate parameter estimates and higher factor reliability (Marsh, Hau, Balla, & Grayson, 1998). Thus, it is strongly recommended to build over-identified CFA models for model testing.

When working with items on scales or instruments, a recommended practice is to include negatively-phrased and positively-phrased items to avoid agreement bias and ensure that participants are in fact reading the questions asked on the survey. It is important that items be coded in the direction that supports the interpretation and/or name of the factor. For instance, if the factor is social support, the items should be positively framed when estimated in order to interpret social support as *positive* social support or the items should be negatively framed when estimated in order to interpret social support as *negative* social support. Essentially, this requires that items be reverse-coded in the proper direction of how the factor should be interpreted (e.g., positively or negatively; higher levels of the latent factor or lower levels of the latent factor, respectively).

When using survey items with ordered response options (e.g., Likert scales) as indicators in a CFA model, they should have at least five categories (e.g., ranging from 1 = *never* to 5 = *all the time*) to be able to be treated as interval-level variables. In this case, ML estimation can be used to estimate the model parameters. When the indicators have fewer than five categories, ML estimation should not be used because it can result in biased parameter estimates (Rhemtulla, Brosseau-Liard, & Savalei, 2012). Instead, robust WLS estimation is recommended when survey items with response option categories less than five are used as indicators. The next section includes an illustration example of a CFA model with missing *ordinal* indicators using robust WLS in LISREL and M*plus*.

CFA WITH MISSING ORDINAL INDICATORS

If you have indicator variables that are ordinal in nature, particularly with fewer than five response categories (as previously mentioned), the robust WLS estimation methods are recommended. We present a CFA example with ordinal indicators with missing data in LISREL and in M*plus*, which requires the raw data. We will use the data described in more detail by Aish and Jöreskog (1990) and is used in LISREL examples with ordinal variables. The data consist of 1719 responses to six variables, three of which are hypothesized indicators of a *Political Efficacy* factor and three of which are hypothesized indicators of a *Political Responsiveness* factor. The six indicators reflect attitudes toward the following abridged statements: no say about government (*NOSAY*); voting is the only way to have any say (*VOTING*); politics and government are complicated (*COMPLEX*); public officials do not care what I think (*NOCARE*); those elected to Parliament lose touch with people (*TOUCH*); and parties are only interested in votes – not opinions (*INTEREST*). The responses were made on a four-point Likert-type scale, ranging from 1 (*agree strongly*) to 4 (*disagree strongly*). Missing data are flagged with the value of –99.00. Again, when using missing data estimation methods in LISREL and M*plus*, the assumption is that the data are missing at random.

In LISREL, the ***EFFICACY.LSF*** was used and is available on the book website. Again, it is important that the user first designate the missing data values in the .LSF file prior to estimating model parameters. It is also important that the variables type be designated as *ordinal* in the .LSF file. This can be done with the .LSF file open and clicking on the **Data** menu and scrolling down to click on the **Define Variables** option. Screenshots of this are presented in Figure 6.10.

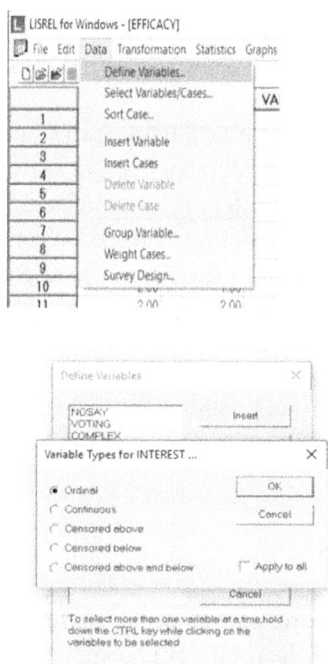

▨ **Figure 6.10:** SCREENSHOTS IN LISREL DESIGNATING VARIABLE TYPE IN EFFICACY.LSF

The CFA model is a two-factor model where the *NOSAY, VOTING,* and *COMPLEX* variables load on the *Political Efficacy* factor and the *NOCARE,*

TOUCH, and *INTEREST* variables load on the *Political Responsiveness* factor. The *Efficacy* and *Responsiveness* factors are hypothesized to be correlated. In LISREL, you will need to request that MCMC imputation be conducted, which will produce 200 polychoric correlation matrices based on the imputed data. The polychoric correlation matrix is necessary to be estimated in order to account for the ordinal nature of the data. The average polychoric correlation will then be used to estimate the CFA model. The SIMPLIS code for this example is below:

```
CFA WITH ORDINAL MISSING DATA
RAW DATA FROM FILE C:\EFFICACY.LSF
Latent Variables: Efficacy Responsive
Relationships:
NOSAY VOTING COMPLEX = Efficacy
NOCARE TOUCH INTEREST = Responsive
Print Residuals
Number of Decimals = 3
OPTIONS: SC MI
LISREL OUTPUT: MI2S IX=76984 NM=200 ME=DWLS
Path Diagram
End of problem
```

In the LISREL OUTPUT command, the **MI2S** option requests the two-stage multiple imputation method in which the model is fitted to the average polychoric correlation matrix for the 200 MCMC imputations (**NM = 200**). The **IX = 76984** sets the initial random seed to be used during imputation, which allows you to reproduce the same output in subsequent analyses when using this same seed number. The **ME = DWLS** option requests that the method of estimation (ME) be set to robust WLS estimation (Diagonally Weighted Least Squares: DWLS).

The two-factor model diagrammed in LISREL with standardized estimates is presented in Figure 6.11. The Satorra-Bentler (2001) scaled chi-square test is C3 in LISREL and is statistically significant. Using the suggested cutoffs provided in Chapter 5, the additional model fit indices suggest acceptable model fit:

LISREL Model Fit of Efficacy CFA Model with Missing Ordinal Data

χ^2	df	CFI	NNFI/TLI	RMSEA (90% CI)	SRMR
63.051*	8	.984	.970	.061 (.047, .076)	.035

Note: $*p < .05.$

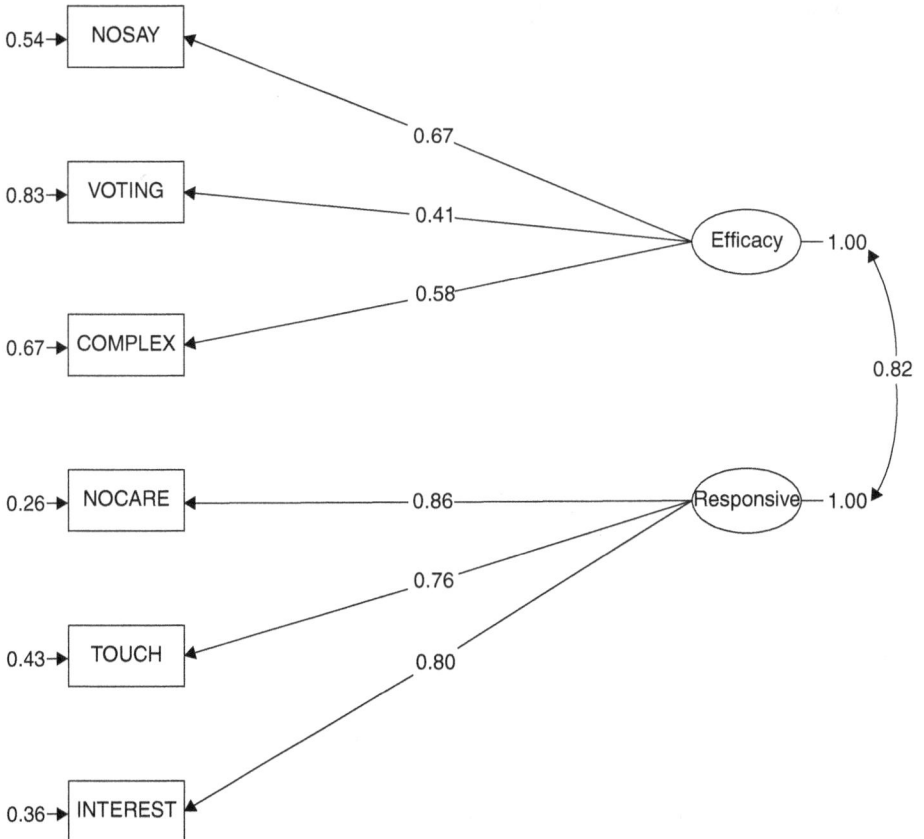

■ **Figure 6.11:** LISREL DIAGRAM OF POLITICAL EFFICACY CFA MODEL

The modification indices could be examined to determine if a parameter could be added to the model that would make theoretical sense and decrease the model chi-square value. Aish and Jöreskog (1990) estimated a CFA model that allowed *NOCARE* to cross-load on the *Political Efficacy* factor. After adding this cross-loading, the Satorra-Bentler chi-square decreased, but it was still significant $[\chi^2(7) = 27.947, p < .05]$.

In M*plus*, the estimation of the CFA model with missing ordinal data can be done using the following code:

```
TITLE:      CFA WITH MISSING ORDINAL DATA
DATA:       FILE is "C:\EFFICACY.TXT";
VARIABLE:   NAMES are NOSAY VOTING COMPLEX
                  NOCARE TOUCH INTEREST;
```

```
                       CATEGORICAL ARE NOSAY VOTING COMPLEX
                                       NOCARE TOUCH INTEREST;
                       MISSING ARE ALL (-99);

MODEL:                 EFFICACY BY NOSAY VOTING COMPLEX;
                       RESPONS BY NOCARE TOUCH INTEREST;
                       EFFICACY WITH RESPONS;

OUTPUT:                STDYX RESIDUAL MODINDICES(3.84);
```

You will notice that a new option in the VARIABLE command is used in this example. The following option indicates that the variables are ordered categorical (ordinal) variables: CATEGORICAL ARE NOSAY VOTING COMPLEX NOCARE TOUCH INTEREST. This option also makes WLSMV the default estimator. The missing data flag is –99 and indicated as follows: MISSING ARE ALL (-99). In M*plus*, the data are not imputed as in LISREL. Instead, the process incorporates FIML estimates in a two-step estimation process (see Asparouhov & Muthén, 2010 for more details concerning estimation with missing ordinal data).

The chi-square test of model fit in M*plus* output with WLSMV looks as follows for this example:

```
Chi-Square Test of Model Fit
      Value                        137.571*
      Degrees of Freedom               8
      P-Value                      0.0000
```

```
* The chi-square value for MLM, MLMV, MLR, ULSMV, WLSM and WLSMV cannot
  be used for chi-square difference testing in the regular way. MLM,
  MLR and WLSM chi-square difference testing is described on the Mplus
  website. MLMV, WLSMV, and ULSMV difference testing is done using the
  DIFFTEST option.
```

As mentioned in Chapter 5, the χ^2 must be modified when using robust estimators, which is being stated in the M*plus* output for this example. As seen above, the chi-square test is significant in M*plus*. The model fit information provided by M*plus* is below:

M*plus* Model Fit of Efficacy CFA Model with Missing Ordinal Data

χ^2	df	CFI	NNFI/TLI	RMSEA (90% CI)	SRMR
137.571*	8	.982	.967	.097 (.083, .112)*	.026

Note: *$p < .05$.

Thus, the chi-square and RMSEA do not support adequate model fit. The modification indices show that the largest drop in chi-square is associated with the cross-loading of *NOCARE* on the *Political Efficacy* factor. After adding this cross-loading, the chi-square decreased, but was still significant [$\chi^2(7) = 61.992$, $p < .05$] and the RMSEA remained unacceptable.

Because the raw data were used in this example, the mean structure estimates are provided in the output by default in M*plus*. The estimates of the indicator thresholds are provided in the M*plus* output because the indicators are ordinal. Specifically, there will be threshold estimates for an ordinal indicator equal to the number of categories (C) minus 1 ($C - 1$). In this example, there were four response options for each indicator; where 1 = *agree strongly*; 2 = *agree*; 3 = *disagree*; and 4 = *disagree strongly*. Thus, there will be three threshold estimates per indicator. The M*plus* output with threshold estimates for this example is below:

	Estimate	S.E.	Est./S.E.	Two-Tailed P-Value
Thresholds				
NOSAY$1	-1.258	0.041	-30.520	0.000
NOSAY$2	-0.221	0.031	-7.167	0.000
NOSAY$3	1.423	0.045	31.642	0.000
VOTING$1	-0.961	0.036	-26.487	0.000
VOTING$2	0.228	0.031	7.406	0.000
VOTING$3	1.669	0.052	31.869	0.000
COMPLEX$1	-0.834	0.035	-24.122	0.000
COMPLEX$2	0.748	0.034	22.176	0.000
COMPLEX$3	1.785	0.057	31.547	0.000
NOCARE$1	-1.042	0.037	-27.848	0.000
NOCARE$2	0.165	0.031	5.362	0.000
NOCARE$3	1.826	0.059	31.152	0.000
TOUCH$1	-0.969	0.037	-26.310	0.000
TOUCH$2	0.532	0.033	16.345	0.000
TOUCH$3	2.149	0.078	27.668	0.000
INTEREST$1	-0.990	0.037	-26.631	0.000
INTEREST$2	0.322	0.032	10.212	0.000
INTEREST$3	2.076	0.073	28.493	0.000

NOSAY$1, *NOSAY$2*, and *NOSAY$3* are the three threshold estimates associated with the *NOSAY* ordinal indicator. With ordinal variables, it is assumed that there is an underlying *latent* continuous distribution of the variable that is normally distributed, which is referred to as a latent response variable (Muthén, 1984). The *NOSAY$1* threshold indicates the value of the *NOSAY* latent response variable where participants would endorse options above the *agree strongly* response option (i.e., *agree*, *disagree*, or *disagree strongly*). The *NOSAY$2* threshold indicates the value of the *NOSAY* latent response variable where participants would endorse options above the *agree* response option (i.e., *disagree* or *disagree strongly*). Finally, the *NOSAY$3* threshold indicates the value of the *NOSAY* latent response variable where participants would endorse the *disagree strongly* option. Figure 6.12 demonstrates the estimated thresholds on the latent response variable distribution.

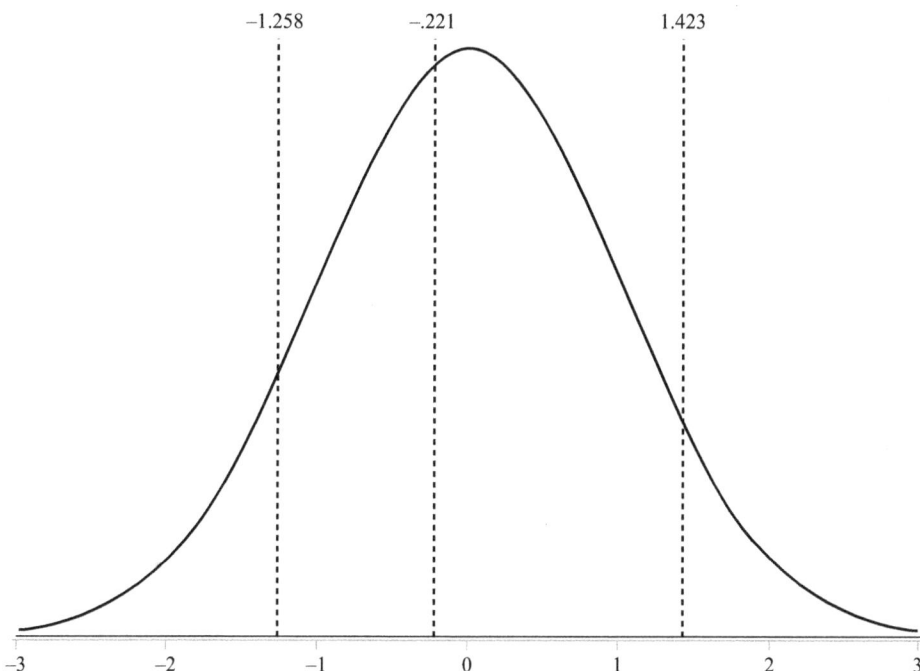

Figure 6.12: NOSAY LATENT RESPONSE VARIABLE WITH ESTIMATED THRESHOLDS

In LISREL, the threshold estimates are not provided by default in the output. However, you can obtain the threshold estimates in the .LSF window. Again, it is important that the variables be designated as ordinal in the .LSF file prior to estimating the thresholds (see Figure 6.10). With the *EFFICACY.LSF* file open, in the **Statistics** menu, scroll down and select **Output Options**. In the *Output Options* dialog box, select **Correlations** from the *Moment Matrix* dropdown list. Click the **OK** button, which will produce the thresholds in the output file. The estimated thresholds from LISREL are below, which match those estimated in M*plus*.

Variable	Mean	St. Dev.	Thresholds		
NOSAY	0.000	1.000	−1.258	−0.221	1.423
VOTING	0.000	1.000	−0.961	0.228	1.669
COMPLEX	0.000	1.000	−0.834	0.748	1.785
NOCARE	0.000	1.000	−1.042	0.165	1.826
TOUCH	0.000	1.000	−0.969	0.532	2.149
INTEREST	0.000	1.000	−0.990	0.322	2.076

CFA MODEL COMPARISONS

Nested and non-nested CFA models can also be compared as described in Chapter 5 using $\Delta\chi^2$ tests and/or information criteria (e.g., AIC, BIC). Remember that models are nested if the set of parameters estimated in the restricted (nested) model is a subset of the parameters estimated in the unrestricted model. Nested CFA models commonly consist of different numbers of correlated factor structures whereas non-nested CFA model comparisons commonly consist of correlated (oblique) versus uncorrelated (orthogonal) factor structures. For example, Galassi, Schanberg, and Ware (1992) examined the factor structure of the Patient Reactions Assessment (PRA) survey, which was developed to measure a patient's perception of the quality of the relationship with their doctor/provider. The PRA is comprised of 15 items thought to be indicators of three factors, including an *Affective* factor (e.g., provider values and respects the patient), an *Information* factor (e.g., provider offers information and explanations to the patient), and a *Communication* factor (e.g., provider can communicate well with the patient). Each factor was comprised of five indicators.

Three factor models were compared, including a three-factor correlated (oblique) model, a three-factor uncorrelated (orthogonal) model, and a one-factor model. These models are illustrated at the factor-structure level for simplicity in Figure 6.13 with respective model fit information and information criteria values. The same 15 observed indicators were used for all three models where the number of distinct or non-redundant elements in the sample covariance matrix is equal to $[15(15 + 1)/2 = 120]$. You will notice that the three-factor correlated model has the fewest degrees of freedom of all the comparison models because it is the most parameterized model (33 total parameters), with 15 factor loadings, 15 indicator error variances, and three covariances to estimate (while setting all three factor variances equal to 1). The three-factor uncorrelated model is estimating three fewer parameters than the three-factor correlated model. Specifically, it does not include the three factor covariances. Thus, the degrees of freedom for the three-factor uncorrelated model are larger by three ($df = 90$) because it is estimating 15 factor loadings and 15 indicator error variances (while setting all three factor variances equal to 1). The three-factor uncorrelated model is nested in the three-factor correlated model. You can arrive at the three-factor uncorrelated model from the three-factor correlated model by setting the three factor covariances equal to zero. The one-factor model is also estimating 15 factor loadings and 15 indicator error variances (while setting the single factor's variance equal to 1). Hence, the degrees of freedom for the one-factor model are the same as the three-factor uncorrelated model ($df = 90$). The one-factor model is nested in the three-factor correlated model. You can arrive at the one-factor model from the three-factor correlated model by setting the factor covariances equal to one and constraining them to equality. The three-factor uncorrelated model and the one-factor model are not nested, which is easily determined since they have the same degrees of freedom.

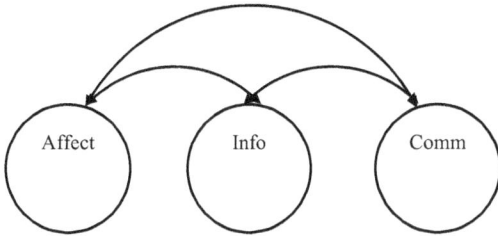

Three-Factor Correlated (Oblique) Model
$[\chi^2(87) = 197.449; NNFI/TLI = 0.942; AIC = 263.449; BIC = 375.439]$

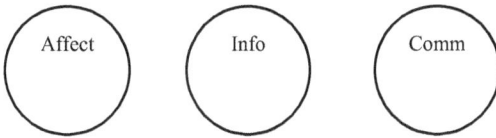

Three-Factor Uncorrelated (Orthogonal) Model
$[\chi^2(90) = 480.884; NNFI/TLI = 0.802; AIC = 546.844; BIC = 658.874]$

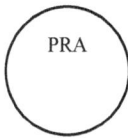

One-Factor PRA Model
$[\chi^2(90) = 225.255; NNFI/TLI = 0.932; AIC = 291.255; BIC = 403.245]$

■ **Figure 6.13:** CFA MODEL COMPARISONS FROM GALASSI, SCHANBERG, AND WARE (1992)

The $\Delta\chi^2$ test between the three-factor correlated and three-factor uncorrelated models resulted in a significant difference $[\Delta\chi^2(3) = 283.435, p < .05]$, indicating that the three factor covariances are contributing significantly to the model. The $\Delta\chi^2$ test between the three-factor correlated and one-factor models resulted in a significant difference $[\Delta\chi^2(3) = 27.806, p < .05]$, also indicating that the three factors and their covariances are contributing significantly to the model. Because the three-factor uncorrelated and one-factor models are not nested, the information criteria (i.e., AIC and BIC) can be used to compare them. The AIC and BIC can actually be used for all of the model comparisons. The AIC and BIC are both smaller for the three-factor correlated model. In the end, the three-factor correlated model would be selected as best representing the relationships in the data.

SUMMARY

EFA is used when a researcher has no prior theoretical basis for knowing the factor structure of a measurement model. The goal of EFA is ultimately data reduction where a few factors explain the variable relations. There are several issues to consider when performing exploratory factor analysis, including adequate sample size, factor and item retention, and factor rotation. A researcher determines which variables go with which factors by selecting those with the highest factor loadings in the pattern matrix, approximating simple structure. Nonetheless, the interpretability of the factor structure is the ultimate benchmark when performing EFA.

After performing EFA, a second sample of data can be obtained and the EFA factor structure can be tested using CFA methods. CFA is used for testing a hypothesized theoretical measurement model where the researcher specifies which variables load on which factors. CFA models should be based on theory and prior research. Compared to EFA, observed variables have fixed loading parameter values of zero on factors they are not hypothesized to measure in CFA. The pattern matrix therefore has fixed zero values to indicate that a variable does not load on that factor.

A CFA example was used to discuss model specification, identification, estimation, and modification. Identification in CFA models includes scaling the factor, which is an important consideration in order to avoid local under-identification of the model. A method not mentioned previously for scaling the factor was suggested by Little, Slegers, and Card (2006) where effects coding is used. Briefly, the loadings for indicators on the same factor are constrained such that their average equals the value of 1. This avoids using a reference indicator approach or fixing the factor variance equal to a value of 1. Scaling methods will not impact the standardized estimates or model fit in single group analyses.

Appropriate estimation in EFA and CFA is determined mainly by the type of indicators used in the analysis. With continuous indicators, ML estimation is appropriate whereas robust WLS estimation is recommended when indicators are responses to Likert-type items with fewer than five response options. CFA examples were provided in LISREL and M*plus* with complete continuous data, missing continuous data, and missing ordinal data. When data are missing, it is important to investigate the cause behind the missing data since missing data estimation methods, such as FIML and multiple imputation, assume that data are missing at random (MAR).

When CFA models do not fit the data well, researchers could use the values in the residual matrix (normalized and standardized), modification indices, expected parameter change values, and the structure coefficient matrix to investigate potential model misspecification. When adding parameters to the CFA model, cross-loadings and error covariances will largely be recommended by the modification indices. Error covariances reflect systematic variance in the pair of indicators remaining after the explanation of variance by the specified factor(s). Again, justification for added parameters, including cross-loadings and error covariances, should be given. Substantial changes would also require new theoretical support for the CFA model tested and subsequent construct validity testing using a different sample.

Model comparisons can also be done by comparing nested and non-nested CFA models. CFA model comparisons are encouraged as a way to investigate the construct validity of surveys or instruments. As described in Chapter 5, model comparisons with nested CFA models can be done using the $\Delta\chi^2$ test and information criteria (e.g., AIC and BIC) whereas model comparisons with non-nested CFA models can be done using information criteria.

When deciding on variables to use as indicators in CFA models, it is important that the variables are hypothesized to covary because of the factor. Indicators should provide a thorough representation of the latent factor, with at least four or more indicators per factor. CFA models are reflective measurement models because the factors are hypothesized to directly affect the responses on the observed indicator variables. If models *theoretically* reflect formative measurement models instead of a reflective measurement model (see Figure 6.9), researchers should not treat causal indicators as effect indicators in a CFA model. There are additional readings that are helpful to learn more about EFA (e.g., Gorsuch, 1983; Comrey & Lee, 1992; Costello & Osborne, 2005) and CFA (e.g., Bollen, 1989; Brown, 2015), which we encourage readers to consult. Reflective measurement models or CFA models are the building blocks for full structural equation models which allow direct and indirect effects to be tested among latent factors, which is discussed in the next chapter.

EXERCISES

You will be using a subset of simulated data described by Marsh, Barnes, Cairns, and Tidman (1984) who examined the self-concept of 251 fifth-graders. These data were simulated to represent 16 items from the Self-Description Questionnaire. Items Y1–Y4 are ratings of abilities in and enjoyment in sports, physical activities, and games. Items Y5–Y8 are ratings about their relative appearance, attractiveness, and how they think others perceive their appearance. Items Y9–Y12 are ratings about their popularity and friendships. Items Y13–Y16 are ratings about their relationship with parents.

There are three alternative models to test for this exercise.

> **Model 1**: Items Y1–Y4 are indicators of a *physical abilities* factor (labeled as *Ability*); items Y5–Y8 are indicators of a *physical appearance* factor (labeled as *Appear*); items Y9–Y12 are indicators of a *peer relationships* factor (labeled as *Peer*); and items Y13–Y16 are indicators of a *parental relationships* factor (labeled as *Parent*). These four factors are correlated.
>
> **Model 2**: Items Y1–Y4 are indicators of a *physical abilities* factor (labeled as *Ability*); items Y5–Y8 are indicators of a *physical appearance* factor (labeled as *Appear*); items Y9–Y12 are indicators of a *peer relationships* factor (labeled as *Peer*); and items Y13–Y16 are indicators of a *parental relationships* factor (labeled as *Parent*). These four factors are **NOT** correlated.
>
> **Model 3**: Items Y1–Y4 are indicators of a *physical abilities* factor (labeled as *Ability*); items Y5–Y8 are indicators of a *physical appearance* factor (labeled as *Appear*); items Y9–Y12 are indicators of a *peer relationships* factor (labeled as *Peer*); and items Y13–Y16 are indicators of a *parental relationships* factor (labeled as *Parent*). *Ability* and *Appear* are correlated. *Peer* and *Parent* are correlated. *Appear* and *Peer* are correlated. *Ability* and *Parent* are correlated.
>
> a. For each model, draw out a complete diagram. For this exercise, you will fix the factor variances equal to 1 and estimate all factor loadings.
> b. How many non-redundant elements are in the sample covariance matrix?
> c. For each model, how many parameters will be estimated?
> d. Determine the degrees of freedom for each model.
> e. Which models are nested and non-nested within other models?
> f. Write programs to estimate each model using ML estimation. For LISREL, *selfconcept.LSF* is available on the book website. For M*plus*, *selfconcept.dat* is available on the book website.
> g. Conduct model comparisons using the $\Delta\chi^2$ test for nested models and the information criteria (AIC and BIC) for nested and non-nested models. Using this information, select the model that best represents the relationships among the variables.

REFERENCES

Aish, A-M., & Jöreskog, K. G. (1990). A panel model for political efficacy and responsiveness: an application of LISREL 7 with weighted least squares. *Quality & Quantity*, 24, 405–426.

Anderson, T. W., & Rubin, H. (1956). Statistical inference in factor analysis. In J. Neyman (Ed.), *Proceedings of the third Berkeley Symposium on Mathematical Statistics and Probability, Volume 5: Contributions to econometrics, industrial research, and psychometry.* Berkeley, CA: University of California Press, pp. 111–150.

Asparouhov, T., & Muthén, B. (2010). Weighted least squares estimation with missing data. Retrieved from www.statmodel.com/download/Gstruc MissingRevision.pdf.

Bartholomew, D. J., Deary, I. J., & Lawn, M. (2009). The origin of factor scores: Spearman, Thomson and Bartlett. *British Journal of Mathematical and Statistical Psychology*, 62(3), 569–582.

Bartlett, M. S. (1937). The statistical conception of mental factors. *British Journal of Psychology*, 28, 97–104.

Bollen, K. A. (1989). *Structural equations with latent variables.* New York: John Wiley & Sons.

Bollen, K. A., & Lennox, R. (1991). Conventional wisdom on measurement: A structural equation perspective. *Psychological Bulletin*, 110(2), 305–314.

Brown, T. A. (2015). *Confirmatory factor analysis for applied research* (2nd ed.). New York: Guilford Press.

Cattell, R. B. (1966). The scree test for the number of factors. *Multivariate Behavioral Research*, 1(2), 245–276.

Cheung, G. W., & Rensvold, R. B. (1999). Testing factorial invariance across groups: A reconceptualization and proposed new method. *Journal of Management*, 25(1), 1–27.

Cochran, W. G. (1968). Errors of measurement in statistics. *Technometrics*, 10, 637–666.

Comrey, A. L. (1973). *A first course in factor analysis.* New York: Academic Press.

Comrey, A. L., & Lee, H. B. (1992). *A first course in factor analysis.* Hillsdale, NJ: Lawrence Erlbaum.

Costello, A. B., & Osborne, J. (2005). Best practices in exploratory factor analysis: Four recommendations for getting the most from your analysis. *Practical Assessment, Research, and Evaluation*, 10(7), 1–9.

Fabrigar, L. R., Wegener, D. T., MacCallum, R. C., & Strahan, E. J. (1999). Evaluating the use of exploratory factor analysis in psychological research. *Psychological Methods*, 4(3), 272–299.

Fuller, W. A. (1987). *Measurement error models.* New York: Wiley.

Galassi, J. P., Schanberg, R., & Ware, W. B. (1992). The Patient Reactions Assessment: A brief measure of the quality of the patient-provider medical relationship. *Psychological Assessment*, 4(3), 346–351.

Gorsuch, R. L. (1983). *Factor analysis* (2nd ed.). Hillsdale, NJ: Lawrence Erlbaum.

Graham, J. M., Guthrie, A. C. & Thompson, B. (2003). Consequences of not interpreting structure coefficients in published CFA research: A reminder. *Structural Equation Modeling*, 10, 142–153.

Green, S. B., Levy, R., Thompson, M. S., Lu, M., & Lo, W-J. (2012). A proposed solution to the problem with using completely random data to assess the number of factors with parallel analysis. *Educational and Psychological Measurement*, 72(3), 357–374.

Guadagnoli, E., & Velicer, W. F. (1988). The relationship of sample size to the stability of component patterns. *Psychological Bulletin*, 103, 265–275.

Hancock, G. R., Stapleton, L. M., & Arnold-Berkovits, I. (2009). The tenuousness of invariance tests within multisample covariance and mean structure models. In T. Teo & M. S. Khine (Eds.), *Structural equation modeling: Concepts and applications in educational research* (pp. 137–174). Rotterdam, Netherlands: Sense Publishers.

Harmon, H. H. (1976). *Modern factor analysis* (3rd ed., rev.). Chicago, IL: University of Chicago Press.

Hendrickson, A. E., & White, O. P. (1964). Promax: A quick method for rotation to oblique simple structure. *British Journal of Statistical Psychology*, 17(1), 65–70.

Holzinger, K. J., & Swineford, F. A. (1939). *A study in factor analysis: The stability of a bi-factor solution.* (Supplementary Educational Monographs, No. 48). Chicago, IL: University of Chicago, Department of Education.

Horn, J. L. (1965). A rationale and test for the number of factors in factor analysis. *Psychometrika*, 30(2), 179–185.

Jennrich, R. I., & Sampson, P. F. (1966). Rotation for simple loadings. *Psychometrika*, 31(3), 313–323.

Jöreskog, K. G. (1969). A general approach to confirmatory maximum likelihood factor analysis. *Psychometrika*, 34, 183–202.

Jöreskog, K. G., & Sörbom, D. (1993). *LISREL 8: Structural equation modeling with the SIMPLIS command language.* Chicago, IL: Scientific Software International.

Kaiser, H. F. (1958). The varimax criterion for analytic rotation in factor analysis. *Psychometrika*, 23(3), 187–200.

Little, T. D., Slegers, D. W., & Card, N. A. (2006). A non-arbitrary method of identifying and scaling latent variables in SEM and MACS models. *Structural Equation Modeling*, 13(1), 59–72.

MacCallum, R. C., & Browne, M. W. (1993). The use of causal indicators in covariance structure models: Some practical issues. *Psychological Bulletin*, 114(3), 533–541.

MacCallum, R. C., Widaman, K. F., Zhang, S., & Hong, S. (1999). Sample size in factor analysis. *Psychological Methods*, 4, 84–99.

Marsh, H. W., Barnes, J., Cairns, L., & Tidman, M. (1984). Self-description questionnaire: Age and sex effects in structure and level of self-concept for preadolescent children. *Journal of Educational Psychology*, 76(5), 940–956.

Marsh, H. W., Hau, K. T., Balla, J. R., & Grayson, D. (1998). Is more ever too much? The number of indicators per factor in confirmatory factor analysis. *Multivariate Behavioral Research*, 33(2), 181–220.

McDonald, R. P. (1985). *Factor analysis and related methods*. Hillsdale, NJ: Lawrence Erlbaum.

Mulaik, S. A. (2005). Looking back on the indeterminacy controversies in factor analysis. In A. Maydeu-Olivares & J. J. McArdle (Eds.), *Contemporary psychometrics* (pp. 173–206). Mahwah, NJ: Lawrence Erlbaum Associates.

Mundfrom, D. J., Shaw, D. G., & Ke, T. L. (2005). Minimum sample size recommendations for conducting factor analyses. *International Journal of Testing*, 5(2), 159–168.

Muthén, B. (1984). A general structural equation model with dichotomous, ordered categorical, and continuous latent variable indicators. *Psychometrika*, 49(1), 115–132.

O'Connor, B. P. (2000). SPSS and SAS programs for determining the number of components using parallel analysis and Velicer's MAP test. *Behavior Research Methods, Instruments, & Computers*, 32(3), 396–402.

Pett, M. A., Lackey, N. R. & Sullivan, J. J. (2003). *Making sense of factor analysis: The use of factor analysis for instrument development in health care research*. Thousand Oaks, CA: Sage Publications.

Preacher, K. J., & MacCallum, R. C. (2003). Repairing Tom Swift's electric factor analysis machine. *Understanding Statistics*, 2(1), 13–43.

Rhemtulla, M., Brosseau-Liard, P. E., & Savalei, V. (2012). When can categorical variables be treated as continuous? A comparison of robust continuous and categorical SEM estimation methods under suboptimal conditions. *Psychological Methods*, 17(3), 354–373.

Satorra, A., & Bentler, P. M. (2001). A scaled chi-square test statistic for moment structure analysis. *Psychometrika*, 66(4), 506–514.

Schumacker, R .E. (2015). *Using R with multivariate statistics*. Thousand Oaks, CA: Sage Publications.

Tabachnick, B. G. & Fidell, L. S. (2007). *Using multivariate statistics* (5th ed.). New York: Pearson Education, Inc.

Thurstone, L. L. (1935). *The vectors of mind*. Chicago, IL: University of Chicago Press.

Whittaker, T. A. (2016). Structural equation modeling. In K. Pituch, *Applied multivariate statistics for the social sciences* (6th ed., pp. 639–746). New York: Routledge.

Whittaker, T. A., & Khojasteh, J. (2013). A comparison of methods to detect invariant reference indicators in structural equation modelling. *International Journal of Quantitative Research in Education*, 1(4), 426–443.

Worthington, R. L., & Whittaker, T. A. (2006). Scale development research: A content analysis and recommendations for best practices. *The Counseling Psychologist*, 34(6), 806–838.

Yoon, M., & Millsap, R. E. (2007). Detecting violations of factor invariance using data-based specification searches: A Monte Carlo study. *Structural Equation Modeling*, 14(3), 435–463.

Chapter 7

FULL SEM

FULL STRUCTURAL EQUATION MODELS

The types of causal relationships specified and estimated among observed variables in the path models described in Chapter 4 can also be specified and estimated among latent variables or constructs. Full structural equation modeling permits the modeling of connections or relationships between latent factors, where factors underlie the observed indicators. With path analysis among observed variables, the variables could be single-score measures (e.g., GPA) or measures with several questions from a survey that are aggregated in some way (e.g., averaged score or summed score). The observed variables are then used to characterize a latent construct (e.g., achievement or motivation). As assumption in path analysis with observed variables is that the variables have no measurement error, particularly the exogenous observed variables in the model as well as the exogenous *or* endogenous *aggregated* variables in the model. When this assumption does not hold, the parameters estimates may be biased (Bollen, 1989). Because researchers are basically concerned with the connections or relationships among latent constructs, full SEM incorporates the latent factors that are hypothesized to cause the observed indicator variables. Further, full SEM offsets the impact of measurement error because the errors of observed indicators are modeled at the measurement or CFA level.

MODEL SPECIFICATION IN FULL SEM

Model specification is the first step in structural equation modeling (also for regression models, path models, and confirmatory factor analysis models). We need theory because a set of observed variables can define a multitude of different latent variables in a measurement model. In addition, many different structural

DOI: 10.4324/9781003044017-7

models can be generated on the basis of different hypothesized relationships among the latent variables.

How does a researcher determine which model is correct? We have already learned that model specification is complicated and we must meet certain data conditions with the observed variables. Basically, structural equation modeling does not determine which model to test; rather, it estimates the parameters in a model once that model has been specified *a priori* by the researcher based on theoretical knowledge. Consequently, theory plays a major role in formulating structural equation models and guides the researcher's decision on which model(s) to specify and test. Once again, we are reminded that model specification is indeed the hardest part of structural equation modeling.

Full SEM models or structural models are extensions of path models with observed variables. To illustrate specification, identification, estimation, and modification of a full SEM model, we will use the model proposed by Duncan and Stoolmiller (1993) examining the impact of self-efficacy and social support on the maintenance of an exercise program across time. The full SEM model is illustrated in Figure 7.1.

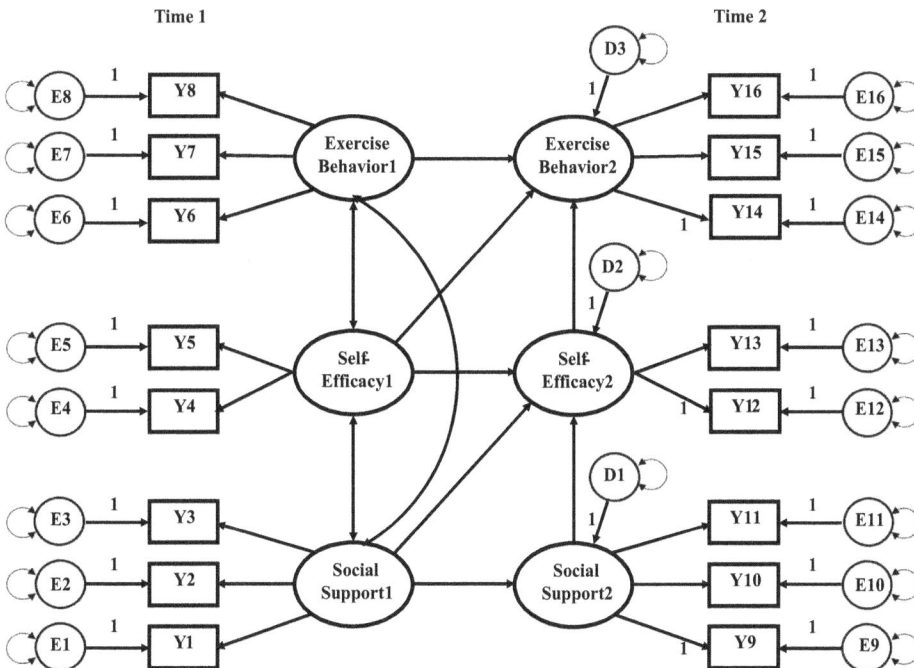

Figure 7.1: FULL SEM MODEL OF EXERCISE BEHAVIOR

This model is a longitudinal model in which the same eight variables are measured at times 1 and 2 and serve as indicators for the same three factors at times 1 and 2. You will notice that relationships can be modeled similarly as with path models using observed variables. Thus, one-headed arrows are direct effects among the factors and can be interpreted as partial regression coefficients. For instance, the *social support* factor at time 1 is theorized to directly affect the *social support* factor at time 2 which, in turn, is theorized to directly impact the *self-efficacy* factor at time 2, which then directly affects the *exercise behavior* factor at time 2. The two-headed arrows between the *exercise behavior*, *self-efficacy*, and *social support* factors at time 1 represent covariances among the factors as opposed to directional relationships.

Factors can be modeled as exogenous or endogenous factors in full SEM or structural models. For instance, the *exercise behavior*, *self-efficacy*, and *social support* factors at time 1 are exogenous and directly impact factors at time 2. *Exercise behavior*, *self-efficacy*, and *social support* factors at time 2 are endogenous factors because they are directly affected by factors at time 1. Again, unexplained variance (represented by errors) will be attached to endogenous observed and latent variables. These errors may be further distinguished to represent errors associated with endogenous *observed* variables (called errors) and errors associated with endogenous latent variables (called *disturbances*). Hence, in Figure 7.1, you will notice E1–E16 directly impact their respective endogenous observed indicator variable whereas D1–D3 directly impact their respective endogenous latent factor. As with errors, disturbances are unobserved. Thus, disturbances are represented in diagrams as circles in which a "D" is enclosed.

Indirect effects among latent factors may also be modeled and tested in full SEM models. In the exercise behavior model, the *self-efficacy* and *social support* factors at time 2 are serving as mediators between the *self-efficacy* and *social support* factors at time 1 and the *exercise behavior* factor at time 2 through three different channels. For example, the *social support* factor at time 1 (SST1) indirectly impacts the *exercise behavior* factor at time 2 (EBT2) through the *social support* (SST2) and *self-efficacy* (SET2) factors at time 2 (SST1 → SST2 → SET2 → EBT2). The *social support* factor at time 1 (SST1) also indirectly impacts the *exercise behavior* factor at time 2 (EBT2) through the *self-efficacy* (SET2) factor at time 2 (SST1 → SET2 → EBT2). Finally, the *self-efficacy* factor at time 1 (SET1) indirectly affects the *exercise behavior* factor at time 2 (EBT2) through the *self-efficacy* (SET2) factor at time 2 (SET1 → SET2 → EBT2). There are two additional indirect effects. Specifically, the *social support* factor at time 1 (SST1) indirectly impacts the *self-efficacy* factor at time 2 (SET2) through the *social support* (SST2) factor at time 2 (SST1 → SST2 → SET2). Further, the *social support* factor at time 2 (SST2) indirectly affects the *exercise behavior* factor at time 2 (EBT2) through the *self-efficacy*

(SET2) factor at time 2 (SST2 → SET2 → EBT2). You can also test the statistical significance of the indirect effects among factors using bootstrapping techniques as described with path models in Chapter 4.

MODEL IDENTIFICATION

In full SEM models, each of the latent constructs represents a CFA or measurement model where the latent factors underlie their respective observed indicator variables. Model identification with full SEM models is determined similarly to that previously described for CFA models, but further discussion about the scaling of factors in the structural model is needed. First, let's compute the number of distinct or non-redundant elements in the sample covariance matrix (**S**) using the following formula (where p is the number of observed variables): $p(p + 1)/2 = 16(16 + 1)/2 = 136$.

Parameters that require estimation in structural equation models include the variances of exogenous variables, direct effects, and covariances. Because the *exercise behavior*, *self-efficacy*, and *social support* factors are unobserved, their scales are not known and must be set for identification purposes. In Chapter 6, we described two ways in which the factors in a CFA or measurement model could be scaled. One way was to use a reference indicator method where the factor loading for one indicator per factor was set equal to 1 and the second way was to set the factor variance equal to 1. With measurement models, either of these factor scaling methods are suitable because the factors are exogenous variables in the model. In full SEM models where factors serve as endogenous variables or are caused by other variables/factors in the model, the only factor scaling option that is suitable is the reference indicator method. Specifically, we do not estimate the variances of endogenous variables (factors in this case). Instead, the disturbance variances of the endogenous factors are estimated. Hence, when setting the scale of a factor that is endogenous in the model, the reference indicator approach must be used.

As illustrated in Figure 7.1, reference indicators must be used for the *exercise behavior*, *self-efficacy*, and *social support* factors at time 2 because they are endogenous. You will notice in Figure 7.1 that loadings have been assigned to equal 1 for the first variable loading on each of the three factors at time 2 (i.e., Y9, Y12, and Y14). All of the other loadings without values of 1 assigned will be estimated. Also, we can assign the variances of the factors at time 1 equal to values of 1.

In the full SEM model in Figure 7.1, the parameters that will be estimated include 16 indicator variable error variances; 3 endogenous factor disturbance variances;

13 factor loadings; 7 direct effects between the factors; and 3 covariances between the three factors at time 1. Hence, a total of 42 parameters will be freely estimated in the hypothesized model. The degrees of freedom (df_{model}) for the full SEM model is: $136 - 42 = 94$.

You will notice that the *self-efficacy* factor at times 1 and 2 only has two indicator variables. Thus, if a researcher tried to estimate the *self-efficacy* measurement model with only two indicators independently of the larger structural model, the model parameters could not be estimated because it would be under-identified. Both the *exercise behavior* and *social support* factors at times 1 and 2 have three indicator variables, resulting in a just-identified model if estimated independently of the larger structural model. As mentioned in Chapter 6, CFA models with a minimum of three indicator variables per factor are needed for identification purposes. Nevertheless, parameters for under-identified CFA models (e.g., the self-efficacy factor) can be estimated when incorporated into larger models, as in this model.

TWO-STEP VERSUS FOUR-STEP SEM MODEL TESTING

When testing full SEM or structural models, poor model fit may be located at the measurement level (relationships among the indicators) and/or at the structural level (relationships among the factors). Thus, if a researcher was going to modify a poor fitting full SEM model, it may be difficult to know where in the model to try and locate potential model misspecification. That is, should a researcher use methods (e.g., modification indices, expected parameter change) to detect potential misspecification at the measurement level (e.g., error covariances among indicator errors) first or at the structural level (e.g., connections between factors) first?

Anderson and Gerbing (1988) proposed a two-step model-building approach that emphasized the analysis of two conceptually distinct models: a measurement model followed by the structural model. The *measurement* model, or confirmatory factor model, specifies the relationships among measured (observed) indicators of the latent variables. The *structural* model specifies relationships among the latent variables as posited by theory. Jöreskog and Sörbom (1993) summarized similar thoughts about two-step modeling by stating:

> The testing of the structural model, i.e., the testing of the initially specified theory, may be meaningless unless it is first established that the measurement model holds. If the chosen indicators for a construct do not measure that construct, the specified theory must be modified before it can be tested. Therefore, the measurement model should be tested before the structural relationships are tested.
>
> *p. 113*

Following the two-step modeling approach suggested by Anderson and Gerbing (1988), the first step is to specify a model where all of the latent factors (and any stand-alone observed variables) covary (or correlate) with all of the other latent factors in the model. This model is the *initial measurement model* or *initial confirmatory factor model*. By covarying all of the latent factors with each other in the initial measurement model, the structural model is a saturated or a just-identified model. That is, no other relationships between the factors could be estimated at the structural level. Recall that just-identified models fit the data perfectly. Since the fit of the structural model (relationships among factors) is perfect, unacceptable fit of the initial measurement model is due to the measurement or CFA model. Specifically, the relationships among the indicators are not being explained well by the measurement or CFA model. Modification indices and/or expected parameter change values could be examined to locate potential parameters that may be added to the model to improve model fit at the measurement level. Parameters that could be added in initial measurement models are commonly cross-loadings or error covariances. If model fit is acceptable after modifying and re-specifying the initial measurement model, a researcher can proceed to the second step of the two-step approach. The initial measurement model may fit the data adequately or it may be re-specified to fit the data adequately. Either way, the well-fitting initial measurement model or the well-fitting re-specified measurement model is called the *final measurement model* or *final confirmatory factor model*.

It must be noted that if you specify an over-identified structural model in step 2 with an underlying measurement model that does not fit the data well in step 1, model fit will not improve. That is, if the hypothesized structural model is not just-identified, releasing some or all of the covariances among the factors at the structural level will decrease the number of parameters to estimate. Decreasing the number of parameters to estimate will not improve fit. On the contrary, decreasing the number of parameters to estimate in a model will generally result in a decline in model fit (although not necessarily statistically speaking).

For the model illustrated in Figure 7.1, the number of parameters to estimate is 42 with 94 *df*. For the initial measurement model, however, there will be 16 error variances (one per indicator); 3 factor variances (one per endogenous factor with reference indicators); 13 factor loadings; and 15 covariances among all of the factors to estimate. Thus, the total number of parameters to estimate in the initial measurement model is 47, with 89 *df* (136 – 47 = 89). There are more parameters to estimate in the initial measurement model than in the originally hypothesized model illustrated in Figure 7.1. Depending upon the complexity of the hypothesized structural model, the initial measurement model could require estimating more parameters than the originally hypothesized structural model. Because models estimating fewer parameters generally fit the data worse than models estimating more parameters, it is essential that the measurement model fit the data well.

The second step of the two-step approach is to specify the structural model with the underlying *final measurement model*. All of the factor covariances should be released and the hypothesized structural relationships among the factors are specified in the model, resulting in the *initial structural model*. The initial structural model is nested in the final measurement model. Again, if the initial structural model is an over-identified model, it will not fit as well as the final measurement model. The hope is that the fit of the initial structural model will not become substantially worse.

If the fit of the initial structural model is adequate, a $\Delta\chi^2$ test may be computed to see if model fit decreased significantly when specifying the structural relationships. With a non-significant $\Delta\chi^2$ test, the fit of the structural model did not significantly decline and no further modification of the model is needed. With a significant $\Delta\chi^2$ test, modification indices could then be examined to determine where potential misspecification is occurring at the structural level among the factors. Once the structural model fits the data adequately, the $\Delta\chi^2$ test may be computed between the *final structural model* and the *final measurement model*. A significant $\Delta\chi^2$ test would signify a significant decline in fit when imposing the structural model on the measurement model. As mentioned in Chapter 5, the sensitivity of the χ^2 to sample size also applies to the $\Delta\chi^2$ test. As a result, some have proposed using differences between fit index values for nested model comparisons as opposed to the $\Delta\chi^2$ test, such as the ΔCFI (e.g., Cheung & Rensvold, 2002; Meade, Johnson, & Braddy, 2008).

Mulaik and Millsap (2000) presented a four-step approach to testing a nested sequence of SEM models:

Step 1: Specify an unrestricted measurement model. Specifically, conduct an exploratory common factor analysis to determine the number of factors (latent variables) that underlie the sample covariance matrix of the observed variables.

Step 2: Specify a confirmatory factor model that tests hypotheses about certain relations among indicator variables and latent variables. Basically, certain factor loadings are fixed to zero in an attempt to have only a single non-zero factor loading for each indicator variable of a latent variable. Sometimes this leads to a lack of measurement model fit because an indicator variable may have a relation with another latent variable. This is the same as step 1 in the two-step modeling approach.

Step 3: Specify relations among the latent variables in a structural model. Certain relations among the latent variables are fixed to zero so that some latent variables are not related to one another. If model fit is poor, the researcher would not proceed to step 4.

Step 4: A researcher plans hypotheses about free parameters in the model. Several approaches are possible: (a) perform simultaneous tests in which free parameters are fixed based on theory or estimates obtained from other research studies; (b) impose fixed parameter values on freed parameters in a nested sequence of models until a misspecified model is achieved (misspecified parameter); or (c) perform a sequence of confidence-interval tests around free parameters using the standard errors of the estimated parameters.

Both the two-step and four-step SEM modeling approaches use the same sample data when fitting models at each of the steps. Thus, both approaches can capitalize on chance occurrences in the data when fitting models at the different stages. There is no extensive comparison of the two approaches in terms of successfully arriving at the correct model. Either method is recommended over a one-step SEM model testing approach that does not distinguish between measurement and structural model fit. For the exercise behavior model example, we will use the two-step SEM modeling approach.

MODEL ESTIMATION

The full SEM model illustrated in Figure 7.1 and investigated by Duncan and Stoolmiller (1993) will be estimated using LISREL and M*plus*. We will be using summary data for this example, which is the covariance matrix. The sample size for this study is fairly small ($n = 84$). The data set for this example is called ***EXBEH. TXT*** and is available on the book website.

First, the initial measurement model will be estimated where all of the factors are specified to covary. In the initial measurement model, the factors at time 2 are not receiving causal inputs from factors at time 1. Hence, the scaling of the factors at time 2 could be done using either the reference indicator approach or by setting the factor variances equal to 1. In this example, however, we will set up the factors at time 2 with reference indicators for an easy transition into the structural model parameterization where they must have a reference indicator. The SIMPLIS code for the initial measurement model is below:

```
INITIAL MEASUREMENT MODEL FOR EXERCISE BEHAVIOR
Observed Variables: Y1-Y16
COVARIANCE MATRIX FROM FILE EXBEH.TXT
```

```
Sample Size: 84
Latent Variables: SST1 SET1 EBT1
SST2 SET2 EBT2
Relationships:
Y1-Y3 = SST1
Y4-Y5 = SET1
Y6-Y8 = EBT1
Y9 = 1*SST2
Y10-Y11 = SST2
Y12 = 1*SET2
Y13 = SET2
Y14 = 1*EBT2
Y15-Y16 = EBT2
Print Residuals
Number of Decimals = 3
OPTIONS: SC MI
Path Diagram
End of Problem
```

With 16 observed indicators, the covariance matrix is quite large, which is why we used the `COVARIANCE MATRIX FROM FILE` command instead of embedding the covariance matrix in the code. **SS** stands for the social support factor, **SE** stands for the self-efficacy factor, **EB** stands for the exercise behavior factor, **T1** stands for time 1, and **T2** stands for time 2. You will notice that Y9, Y12, and Y14 are serving as reference indicators for their respective factor at time 2 (SST2, SET2, and EBT2, respectively):

```
Y9  = 1*SST2
Y12 = 1*SET2
Y14 = 1*EBT2
```

Covariances among the factors will be estimated by default in LISREL. ML estimation was used when estimating model parameters. The path diagram with unstandardized estimates from LISREL is presented in Figure 7.2.

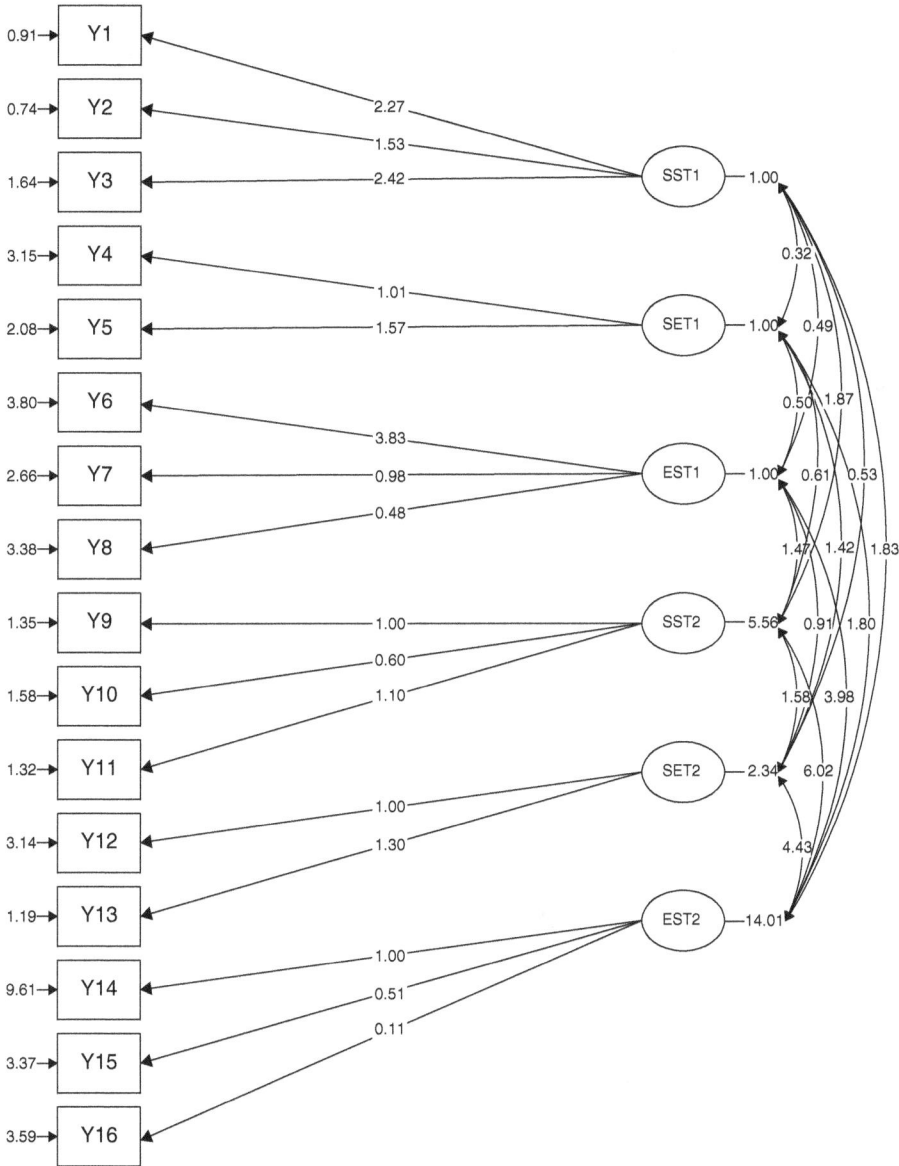

■ **Figure 7.2:** INITIAL MEASUREMENT MODEL FOR EXERCISE BEHAVIOR

The output in LISREL contained the following warning message:

```
W_A_R_N_I_N_G: Matrix above is not positive definite
```

When examining the completely standardized solution in the LISREL output, the correlation between exercise behavior factors at time 1 and time 2 was estimated to be higher than 1.0 (i.e., 1.06). This can sometimes happen in initial measurement models since all factors are allowed to covary, resulting in multicollinearity. Jöreskog (1999, p. 1) stated "that a standardized coefficient of 1.04, 1.40, or even 2.80 does not necessarily imply that something is wrong, although, ..., it might suggest that there is a high degree of multicollinearity in the data." As mentioned in Chapter 5, standardized estimates exceeding values of 1 are called Heywood cases and can occur with multicollinearity in the model, which is occurring in this example with the initial measurement model. Heywood cases could also be the result of small sample size and/or having only two-indicators for *self-efficacy*. It is important that potential misspecification be evaluated when finding standardized estimates larger than anticipated. We will continue with the example because some of the relationships among factors will be relaxed in the structural model, which could remedy this problem with the Heywood case.

The M*plus* code to estimate the initial measurement model is provided below:

```
TITLE:       INITIAL MEASUREMENT MODEL

DATA:        FILE IS EXBEH.TXT;
             TYPE IS COVARIANCE;
             NOBSERVATIONS ARE 84;

VARIABLE:    NAMES ARE Y1 - Y16;

MODEL:       SST1 BY Y1* Y2 Y3;
             SET1 BY Y4* Y5;
             EBT1 BY Y6* Y7 Y8;
             SST2 BY Y9 Y10 Y11;
             SET2 BY Y12 Y13;
             EBT2 BY Y14 Y15 Y16;
             SST1@1; SET1@1; EBT1@1;

OUTPUT:      STDYX MODINDICES(3.84);
```

You will notice that the variances of the three factors at time 1 are set equal to 1:

```
SST1@1; SET1@1; EBT1@1;
```

The factor variances were set equal to 1 because their default reference indicators (immediately following the BY statements) were estimated, as indicated with the asterisks:

```
SST1 BY Y1* Y2 Y3;
SET1 BY Y4* Y5;
EBT1 BY Y6* Y7 Y8;
```

The reference indicator approach was used for the three factors at time 2 since they will be endogenous factors in the structural model. Covariances among the factors will also be estimated by default in M*plus*. However, you could write the following code in the M*plus* input file to covary all factors:

```
SST1 WITH SET1 EBT1 SST2 SET2 EBT2;
SET1 WITH EBT1 SST2 SET2 EBT2;
EBT1 WITH SST2 SET2 EBT2;
SST2 WITH SET2 EBT2;
SET2 WITH EBT2;
```

Similar to the warning message in LISREL, the following warning message was displayed in the M*plus* output:

```
THE MODEL ESTIMATION TERMINATED NORMALLY

WARNING:    THE  LATENT  VARIABLE  COVARIANCE  MATRIX
(PSI) IS NOT POSITIVE DEFINITE.  THIS COULD INDICATE
A NEGATIVE VARIANCE/RESIDUAL VARIANCE FOR A LATENT
VARIABLE,  A  CORRELATION  GREATER  OR  EQUAL  TO  ONE
BETWEEN TWO LATENT VARIABLES, OR A LINEAR DEPENDENCY
AMONG  MORE  THAN  TWO  LATENT  VARIABLES.  CHECK  THE
TECH4 OUTPUT FOR MORE INFORMATION. PROBLEM INVOLVING
VARIABLE EBT2.
```

Even though the model parameters were estimated [THE MODEL ESTIMATION TERMINATED NORMALLY], M*plus* indicated that there is a problem involving the *exercise behavior* factor at time 2. Again, the correlation between the *exercise behavior* factors at time 1 and 2 was estimated to be greater than 1 (i.e., 1.06). The next subsection continues the model testing and model modification of the exercise behavior model.

MODEL TESTING AND MODEL MODIFICATION

The model fit for the initial measurement model from LISREL and M*plus* output are presented in Tables 7.1 and 7.2, respectively. Both LISREL and M*plus* indicated that the initial measurement model fits poorly [LISREL: $\chi^2(89) = 180.38$, $p < .05$; CFI = .882; NNFI/TLI = .841; RMSEA = .11 (90% CI: .09, .13) with $p < .05$; SRMR = .09] and [M*plus*: $\chi^2(89) = 182.55$, $p < .05$; CFI = .881; NNFI/TLI = .839; RMSEA = .11 (90% CI: .09, .13) with $p < .05$; SRMR = .09].

Table 7.1: Model Fit for the Initial Measurement Model in LISREL Output

```
                                 Log-likelihood Values

                          Estimated Model      Saturated Model
                          ---------------      ---------------
Number of free parameters(t)           47                  136
-2ln(L)                          3010.081             2829.699
AIC (Akaike, 1974)*              3104.081             3101.699
BIC (Schwarz, 1978)*             3217.767             3430.661

*LISREL uses AIC= 2t - 2ln(L) and BIC = tln(N)- 2ln(L)


                       Goodness-of-Fit Statistics

Degrees of Freedom for (C1)-(C2)              89
Maximum Likelihood Ratio Chi-Square (C1)      180.382 (P = 0.0000)
Browne's (1984) ADF Chi-Square (C2_NT)        164.463 (P = 0.0000)

Estimated Non-centrality Parameter (NCP)      91.382
90 Percent Confidence Interval for NCP        (56.865 ; 133.676)

Minimum Fit Function Value                    2.173
Population Discrepancy Function Value (F0)     1.101
90 Percent Confidence Interval for F0         (0.685 ; 1.611)
Root Mean Square Error of Approximation (RMSEA)  0.111
90 Percent Confidence Interval for RMSEA      (0.0877 ; 0.135)
P-Value for Test of Close Fit (RMSEA < 0.05)  0.000

Expected Cross-Validation Index (ECVI)        3.306
90 Percent Confidence Interval for ECVI       (2.890 ; 3.815)
ECVI for Saturated Model                      3.277
ECVI for Independence Model                   11.148

Chi-Square for Independence Model (120 df)    893.257

Normed Fit Index (NFI)                        0.798
Non-Normed Fit Index (NNFI)                   0.841
Parsimony Normed Fit Index (PNFI)             0.592
Comparative Fit Index (CFI)                   0.882
Incremental Fit Index (IFI)                   0.886
Relative Fit Index (RFI)                      0.728

Critical N (CN)                               57.571

Root Mean Square Residual (RMR)               0.484
Standardized RMR                              0.0897
Goodness of Fit Index (GFI)                   0.801
Adjusted Goodness of Fit Index (AGFI)         0.697
Parsimony Goodness of Fit Index (PGFI)        0.525
```

Table 7.2: Model Fit for the Initial Measurement Model in M*plus* Output

```
MODEL FIT INFORMATION

Number of Free Parameters                                    47

Loglikelihood
     H0 Value                                        -2733.133
     H1 Value                                        -2641.855

Information Criteria
     Akaike (AIC)                                     5560.265
     Bayesian (BIC)                                   5674.513
     Sample-Size Adjusted BIC                         5526.251
       (n* = (n + 2) / 24)

Chi-Square Test of Model Fit
     Value                                             182.555
     Degrees of Freedom                                     89
     P-Value                                            0.0000

RMSEA (Root Mean Square Error Of Approximation)
     Estimate                                            0.112
     90 Percent C.I.                          0.089    0.135
     Probability RMSEA <= .05                            0.000

CFI/TLI
     CFI                                                 0.881
     TLI                                                 0.839

Chi-Square Test of Model Fit for the Baseline Model
     Value                                             904.019
     Degrees of Freedom                                    120
     P-Value                                            0.0000

SRMR (Standardized Root Mean Square Residual)
     Value                                               0.090
```

The modification indices in LISREL and M*plus* output for the initial measurement model are presented in Tables 7.3 and 7.4, respectively. The modification indices suggested that the error covariance between Y14 and Y6 would result in the largest drop in the chi-square model fit statistic of approximately 25 points. The error covariance between Y14 and Y6 is theoretically plausible since the two indicators are the same item measured at time 1 and time 2. Specifically, the response a person makes to a survey item or question at one measurement occasion is expected to be comparable to the response made to the same survey item or question at a later measurement occasion.

Table 7.3: Modification Indices in LISREL Output for the Initial Measurement Model

```
The Modification Indices Suggest to Add an Error Covariance
Between              and          Decrease in Chi-Square    New Estimate
          Y9              Y1              11.8                  0.66
          Y10             Y2              10.9                  0.47
          Y11             Y3              12.4                  0.85
          Y12             Y4               8.8                  1.22
          Y12             Y5              20.3                 -2.11
          Y13             Y4               9.8                 -1.43
          Y13             Y5              21.0                  2.56
          Y14             Y6              25.2                  8.77
          Y14             Y7               9.0                 -2.11
          Y14             Y8              11.1                 -2.18
          Y16             Y8              10.1                  1.22
```

Table 7.4: Modification Indices in M*plus* Output for the Initial Measurement Model

```
MODEL MODIFICATION INDICES

NOTE:  Modification indices for direct effects of observed dependent
variables regressed on covariates may not be included.  To include
these, request MODINDICES (ALL).

Minimum M.I. value for printing the modification index    3.840

                      M.I.       E.P.C.      Std E.P.C.    StdYX E.P.C.

BY Statements

SST1    BY Y4         4.988      0.538         0.538         0.265
SST1    BY Y5         4.987     -0.839        -0.839        -0.396
SST1    BY Y12        5.069      0.547         0.547         0.235
SST1    BY Y13        5.069     -0.712        -0.712        -0.316
SET1    BY Y6         5.689     -2.056        -2.056        -0.481
SET1    BY Y12        4.392     -1.313        -1.313        -0.564
SET1    BY Y13        4.393      1.709         1.709         0.758
EBT1    BY Y12        4.481      0.746         0.746         0.321
EBT1    BY Y13        4.481     -0.972        -0.972        -0.431
SST2    BY Y1         4.818     -0.288        -0.675        -0.276
SST2    BY Y12        8.441      0.320         0.750         0.322
SST2    BY Y13        8.442     -0.416        -0.976        -0.433
SET2    BY Y6         4.275     -1.425        -2.165        -0.507
EBT2    BY Y8         5.052      0.338         1.259         0.667

WITH Statements

Y7      WITH Y5       4.054      0.605         0.605         0.260
Y8      WITH Y7       5.900      0.804         0.804         0.271
Y9      WITH Y1      11.901      0.653         0.653         0.596
Y9      WITH Y3       6.376     -0.570        -0.570        -0.387
Y10     WITH Y1       5.655     -0.414        -0.414        -0.349
```

Table 7.4: (*cont.*)

Y10	WITH Y2	11.059	0.462	0.462	0.431
Y11	WITH Y1	6.992	-0.527	-0.527	-0.486
Y11	WITH Y3	12.597	0.841	0.841	0.578
Y12	WITH Y4	8.887	1.204	1.204	0.387
Y12	WITH Y5	20.527	-2.088	-2.088	-0.827
Y13	WITH Y4	9.909	-1.416	-1.416	-0.740
Y13	WITH Y5	21.248	2.529	2.529	1.627
Y14	WITH Y6	25.479	8.663	8.663	1.451
Y14	WITH Y7	9.114	-2.085	-2.085	-0.418
Y14	WITH Y8	11.184	-2.151	-2.151	-0.382
Y15	WITH Y6	6.092	-2.239	-2.239	-0.633
Y16	WITH Y6	5.341	-1.289	-1.289	-0.353
Y16	WITH Y8	10.175	1.206	1.206	0.350
Y16	WITH Y15	6.602	0.970	0.970	0.282

The error covariance between Y14 and Y6 was added to the model in LISREL with the following statement:

```
Set Error Covariance Between Y14 and Y6 to be Free
```

and in M*plus* with the following statement:

```
Y14 WITH Y6;
```

After estimating the re-specified model, the fit was still unacceptable, although improved [LISREL: $\chi^2(88) = 156.624$, $p < .05$; CFI = .911; NNFI/TLI = .879; RMSEA = .097 (90% CI: .07, .12) with $p < .05$; SRMR = .077] and [M*plus*: $\chi^2(88) = 158.511$, $p < .05$; CFI = .910; NNFI/TLI = .877; RMSEA = .098 (90% CI: .07, .12) with $p < .05$; SRMR = .077]. It must be noted that the same warning messages that appeared after estimating the initial measurement model still appeared with the estimation of this re-specified model. However, the correlation between the *exercise behavior* factor at times 1 and 2 was not estimated to be greater than 1. The estimated correlation between the two factors was close to a value of 1 (.913). Further, no other standardized estimates were greater than 1 and no variances were negative.

The modification indices were examined to see which parameters could be added to improve fit by decreasing the chi-square test statistic significantly and would be theoretically plausible. This process was consecutively repeated until the model fit of the measurement model was adequate, which resulted in including error covariances between the following variables: Y13 and Y5; Y11 and Y3; Y10 and Y2; and Y16 and Y8. The error covariances that were added were all between the same item measured at times 1 and 2. The final measurement model fit the data adequately [LISREL: $\chi^2(84) = 92.482$, $p > .05$; CFI = .989; NNFI/ TLI = .984; RMSEA = .035 (90% CI: .00, .07) with $p > .05$; SRMR = .074] and [M*plus*: $\chi^2(84) = 93.596$, $p > .05$; CFI = .983; NNFI/TLI = .983; RMSEA = .037 (90% CI: .00, .07) with $p > .05$; SRMR = .073]. It is important to note that the warning message in LISREL and in M*plus* no longer appeared in the output once the second error covariance (between Y13 and Y5) was added.

Now that we have the final measurement model, the next step is to specify the structural model for estimation. The SIMPLIS code is provided below:

```
INITIAL STRUCTURAL MODEL
Observed Variables: Y1-Y16
COVARIANCE MATRIX FROM FILE EXBEH.TXT
Sample Size: 84
Latent Variables: SST1 SET1 EBT1
SST2 SET2 EBT2
Relationships:
Y1-Y3 = SST1
Y4-Y5 = SET1
Y6-Y8 = EBT1
Y9 = 1*SST2
Y10-Y11 = SST2
Y12 = 1*SET2
Y13 = SET2
Y14 = 1* EBT2
Y15-Y16 = EBT2
Set Error Covariance Between Y14 and Y6 to be Free
Set Error Covariance Between Y13 and Y5 to be Free
Set Error Covariance Between Y11 and Y3 to be Free
Set Error Covariance Between Y10 and Y2 to be Free
Set Error Covariance Between Y16 and Y8 to be Free
EBT2 = EBT1 SET1 SET2
SET2 = SET1 SST1 SST2
SST2 = SST1
Print Residuals
Number of Decimals = 3
OPTIONS: SC MI
Path Diagram
End of Problem
```

You will notice the five error covariances included in the code based on the modification indices. There are also three new equations included, which are specifying the relationships among the factors at the structural level. For example, the following equation specifies that the *self-efficacy* factor at time 1 (SET1), the *self-efficacy* factor at time 2 (SET2), and the *exercise behavior* factor at time 1 (EBT1) all directly impact the *exercise behavior* factor at time 2:

```
EBT2 = EBT1 SET1 SET2
```

The covariances among the three exogenous factors at time 1 will be estimated by default.

The M*plus* code to estimate the initial structural model is provided below:

```
TITLE:       INITIAL STRUCTURAL MODEL
DATA:        FILE IS EXBEH.TXT;
             TYPE IS COVARIANCE;
             NOBSERVATIONS ARE 84;
VARIABLE:    NAMES ARE Y1 - Y16;
MODEL:       SST1 BY Y1* Y2 Y3;
             SET1 BY Y4* Y5;
             EBT1 BY Y6* Y7 Y8;
             SST2 BY Y9 Y10 Y11;
             SET2 BY Y12 Y13;
             EBT2 BY Y14 Y15 Y16;
             SST1@1; SET1@1; EBT1@1;
             Y14 WITH Y6;
             Y13 WITH Y5;
             Y11 WITH Y3;
             Y10 WITH Y2;
             Y16 WITH Y8;
             EBT2 ON EBT1 SET1 SET2;
             SET2 ON SET1 SST1 SST2;
             SST2 ON SST1;
OUTPUT:      STDYX MODINDICES(3.84);
```

Again, you will notice the five error covariances that were added during the model modification phase of the measurement model. Also, three new structural equations are included. For instance, the following statement specifies that the *self-efficacy* factor at time 1 (SET1), the *social support* factor at time 1 (SST1), and the *social support* factor at time 2 (SST2) all directly affect the *self-efficacy* factor at time 2 (SET2):

```
SET2 ON SET1 SST1 SST2;
```

In other words, SET2 is regressed **ON** SET1, SST1, and SST2. The covariances among the three exogenous factors at time 1 will be estimated by default, but the following statements could still be included in the M*plus* code to model those covariances:

```
EBT1 WITH SET1 SST1;
SET1 WITH SST1;
```

The model fit for the initial structural model from LISREL and M*plus* output are presented in Tables 7.5 and 7.6, respectively. Both LISREL and M*plus* indicated that the initial structural model fits the data adequately [LISREL: $\chi^2(89) = 105.116$, $p > .05$; CFI = .979; NNFI/TLI = .972; RMSEA = .047 (90% CI: .00, .08) with $p > .05$; SRMR = .10] and [M*plus*: $\chi^2(89) = 106.383$, $p > .05$; CFI = .978; NNFI/TLI = .970; RMSEA = .048 (90% CI: .00, .08) with $p > .05$; SRMR = .10].

Table 7.5: Model Fit of the Initial Structural Model in LISREL Output

	Log-likelihood Values	
	Estimated Model	Saturated Model
Number of free parameters(t)	47	136
$-2\ln(L)$	2934.815	2829.699
AIC (Akaike, 1974)*	3028.815	3101.699
BIC (Schwarz, 1978)*	3142.501	3430.661

*LISREL uses AIC= 2t − 2ln(L) and BIC = tln(N)− 2ln(L)

Goodness-of-Fit Statistics

Degrees of Freedom for (C1)-(C2)	89
Maximum Likelihood Ratio Chi-Square (C1)	105.116 (P = 0.1168)
Browne's (1984) ADF Chi-Square (C2_NT)	96.337 (P = 0.2792)
Estimated Non-centrality Parameter (NCP)	16.116
90 Percent Confidence Interval for NCP	(0.0 ; 45.906)
Minimum Fit Function Value	1.266
Population Discrepancy Function Value (F0)	0.194
90 Percent Confidence Interval for F0	(0.0 ; 0.553)
Root Mean Square Error of Approximation (RMSEA)	0.0467
90 Percent Confidence Interval for RMSEA	(0.0 ; 0.0788)
P-Value for Test of Close Fit (RMSEA < 0.05)	0.540
Expected Cross-Validation Index (ECVI)	2.399
90 Percent Confidence Interval for ECVI	(2.205 ; 2.758)
ECVI for Saturated Model	3.277
ECVI for Independence Model	11.148
Chi-Square for Independence Model (120 df)	893.257

Table 7.5: (*cont.*)

Normed Fit Index (NFI)	0.882
Non-Normed Fit Index (NNFI)	0.972
Parsimony Normed Fit Index (PNFI)	0.654
Comparative Fit Index (CFI)	0.979
Incremental Fit Index (IFI)	0.980
Relative Fit Index (RFI)	0.841
Critical N (CN)	98.077
Root Mean Square Residual (RMR)	0.910
Standardized RMR	0.100
Goodness of Fit Index (GFI)	0.873
Adjusted Goodness of Fit Index (AGFI)	0.806
Parsimony Goodness of Fit Index (PGFI)	0.571

Table 7.6: Model Fit of the Initial Structural Model in M*plus* Output

```
MODEL FIT INFORMATION

Number of Free Parameters                          47

Loglikelihood
    H0 Value                               -2695.046
    H1 Value                               -2641.855

Information Criteria
    Akaike (AIC)                            5484.092
    Bayesian (BIC)                          5598.341
    Sample-Size Adjusted BIC                5450.078
      (n* = (n + 2) / 24)

Chi-Square Test of Model Fit
    Value                                    106.383
    Degrees of Freedom                            89
    P-Value                                   0.1010

RMSEA (Root Mean Square Error Of Approximation)
    Estimate                                   0.048
    90 Percent C.I.                      0.000 0.080
    Probability RMSEA <= .05                   0.514

CFI/TLI
    CFI                                        0.978
    TLI                                        0.970

Chi-Square Test of Model Fit for the Baseline Model
    Value                                    904.019
    Degrees of Freedom                           120
    P-Value                                   0.0000

SRMR (Standardized Root Mean Square Residual)
    Value                                      0.097
```

Because the initial structural model is nested in the final measurement model, a $\Delta\chi^2$ test can be conducted between the two models. Using the chi-square statistics provided by LISREL and M*plus*, the $\Delta\chi^2$ test between the final measurement model and the initial structural model was statistically significant [LISREL: $\Delta\chi^2 (5) = 12.634, p < .05$] and [M*plus*: $\Delta\chi^2 (5) = 12.787, p < .05$]. Hence, specifying our structural model on the final measurement model resulted in a significant drop in fit, which is not a desirable result. Nonetheless, the final structural model does fit the data adequately. Thus, the initial structural model is also the final structural model. The model fit information from LISREL is summarized in Table 7.7.

Table 7.7: Model Fit Summary Table for Full SEM Model of Exercise Behavior

Model	χ^2	df	CFI	NNFI/ TLI	RMSEA (90% CI) p-value	SRMR
Initial measurement model	180.38*	89	.88	.84	.11 (.09, .13) p < .05	.09
Final measurement model	92.48	84	.99	.98	.03 (.00, .07) p > .05	.07
Initial/final structural model	105.12	89	.98	.97	.05 (.00, .08) p > .05	.10

Note: * $p < .05$. Values were taken from LISREL output.

The parameter estimates at the measurement level, including unstandardized factor loadings and respective standard errors, standardized factor loadings, and *R* square values from M*plus* output are presented in Table 7.8. When reference indicators are used, their loadings are fixed to a value of 1. Thus, unstandardized loadings for reference indicators and their respective standard errors will not be estimated. You will notice that the unstandardized loadings (and corresponding standard errors) for Y9, Y12, and Y14, which served as reference indicators, are also presented in Table 7.8. In order to obtain the unstandardized loadings for the reference indicators, you will need to select an alternate indicator to serve as the reference indicator for the respective factor. In order to obtain unstandardized loading estimates for Y9, Y12, and Y14, we selected Y10, Y13, and Y15 as reference indicators for the *social support* factor at time 2, the *self-efficacy* factor at time 2, and the *exercise behavior* factor at time 2, respectively, and re-estimated the initial (and final) structural model. This can be done in LISREL with the following SIMPLIS code:

```
STRUCTURAL MODEL-REFERENCE INDICATORS
Observed Variables: Y1-Y16
COVARIANCE MATRIX FROM FILE EXBEH.TXT
Sample Size: 84
```

```
Latent Variables: SST1 SET1 EBT1
SST2 SET2 EBT2
Relationships:
Y1-Y3 = SST1
Y4-Y5 = SET1
Y6-Y8 = EBT1
Y9 = SST2
Y10 = 1*SST2
Y11 = SST2
Y12 = SET2
Y13 = 1*SET2
Y14 = EBT2
Y15 = 1*EBT2
Y16 = EBT2
Set Error Covariance Between Y14 and Y6 to be Free
Set Error Covariance Between Y13 and Y5 to be Free
Set Error Covariance Between Y11 and Y3 to be Free
Set Error Covariance Between Y10 and Y2 to be Free
Set Error Covariance Between Y16 and Y8 to be Free
EBT2 = EBT1 SET1 SET2
SET2 = SET1 SST1 SST2
SST2 = SST1
Print Residuals
Number of Decimals = 3
OPTIONS: SC MI AD=OFF
Path Diagram
End of Problem
```

It is important to note that the **AD = OFF** option was used because the admissibility test of the regularity of the theta-delta matrix (i.e., matrix of exogenous variables) detected inadmissible solutions (i.e., a negative residual variance) and did not converge after 50 iterations. LISREL provided the following warning messages:

```
W_A_R_N_I_N_G: THETA-DELTA is not positive definite
Number of Iterations = 50
```

After turning off the admissibility test, the model converged in 110 iterations. The following M*plus* code will set alternate loadings equal to 1:

```
TITLE:      STRUCTURAL MODEL-REFERENCE
            INDICATORS
DATA:       FILE IS EXBEH.TXT;
            TYPE IS COVARIANCE;
            NOBSERVATIONS ARE 84;
VARIABLE:   NAMES ARE Y1 - Y16;
MODEL:      SST1 BY Y1* Y2 Y3;
            SET1 BY Y4* Y5;
            EBT1 BY Y6* Y7 Y8;
            SST2 BY Y9* Y10@1 Y11;
            SET2 BY Y12* Y13@1;
            EBT2 BY Y14* Y15@1 Y16;
            SST1@1; SET1@1; EBT1@1;
            Y14 WITH Y6;
            Y13 WITH Y5;
            Y11 WITH Y3;
            Y10 WITH Y2;
            Y16 WITH Y8;
            EBT2 ON EBT1 SET1 SET2;
            SET2 ON SET1 SST1 SST2;
            SST2 ON SST1;
            EBT1 WITH SET1 SST1;
            SET1 WITH SST1;
OUTPUT:     STDYX MODINDICES(3.84);
```

Table 7.8: Unstandardized Estimates, Standard Errors, Standardized Estimates, and
R Square Values for the Exercise Behavior Model

Item	Unstandardized Estimates	S.E.	Standardized Estimates	*R* Square
Social support time 1				
Y1	2.34*	.20	.96	.92
Y2	1.50*	.15	.85	.73
Y3	2.31*	.24	.86	.73
Self-efficacy time 1				
Y4	1.46*	.26	.72	.52
Y5	0.96*	.24	.45	.20
Exercise behavior time 1				
Y6	3.08*	.43	.73	.54
Y7	1.19*	.21	.63	.40
Y8	.74*	.22	.39	.15

Table 7.8: (*cont.*)

Item	Unstandardized Estimates	S.E.	Standardized Estimates	R Square
Social support time 2				
Y9	1.85*	.22	.95	.90
Y10	.54*	.06	.72	.52
Y11	.98*	.08	.87	.75
Self-efficacy time 2				
Y12	1.41*	.24	.89	.79
Y13	.71*	.12	.65	.43
Exercise behavior time 2				
Y14	1.04*	.24	.52	.27
Y15	.96*	.22	.91	.83
Y16	.28*	.10	.35	.12
Error covariances				
Y6 ↔ Y14	8.40*	1.87	.75	–
Y5 ↔ Y13	2.16*	.47	.68	–
Y3 ↔ Y11	1.02*	.29	.52	–
Y2 ↔ Y10	.52*	.16	.44	–
Y16 ↔ Y8	1.06*	.37	.34	–

Note: * $p < .05$.

As seen in Table 7.8, all of the unstandardized factor loadings and error covariances are statistically significant. R square values for the indicator variables are also presented in Table 7.8, indicating the proportion of variance in the indicator that is explained by the model. For instance, 92% of the variance in Y1 is explained by the model, whereas only 12% of the variance in Y16 is explained by the model.

The parameter estimates at the structural level, including unstandardized estimates with corresponding standard errors, standardized estimates, and R square values from M*plus* output are presented in Table 7.9. As seen in Table 7.9, the unstandardized covariances and unstandardized direct effects among the factors at the structural level are statistically significant, with the exception of the direct effect from the *social support* factor at time 1 to the *self-efficacy* factor at time 2 (–.44). Using the standardized estimates, for instance, as *exercise behavior* at time 1 (EBT1) increases by one standard deviation, *exercise behavior* at time 2 (EBT2) is estimated to increase by about .84 standard deviation units, holding all else constant.

Table 7.9: Unstandardized Estimates, Standard Errors, Standardized Estimates, and R Square Values at the Structural Level for the Exercise Behavior Model

Item	Unstandardized Estimates	S.E.	Standardized Estimates	R Square
SST1 ↔ SET1	.46*	.10	.46	–
SST1 ↔ EBT1	.46*	.10	.46	–
SET1 ↔ EBT1	.54*	.13	.54	–
SST1 → SST2	1.88*	.25	.76	–
SST1 → SET2	−.44	.34	−.21	–
SET1 → SET2	1.32*	.32	.64	–
SET1 → EBT2	−1.25*	.56	−.52	–
EBT1 → EBT2	2.01*	.55	.84	–
SST2 → SET2	.37*	.13	.45	–
SET2 → EBT2	.76*	.24	.65	–
SST2				.58
SET2				.59
EBT2				.90

Note: * $p < .05$.

You will also notice that R square values are associated with the three endogenous factors at time 2. For example, the model explains roughly 58% of the variance in *social support* at time 2. The model explains approximately 59% of the variance in *self-efficacy* at time 2. Finally, the model explains about 90% of the variance in *exercise behavior* at time 2.

In addition to presenting results in tables as illustrated in Tables 7.8 and 7.9, you could also portray the standardized estimates as in the diagram shown in Figure 7.3. You will notice that the five error covariances added during the measurement model modification (i.e., Y14 ↔ Y6; Y13 ↔ Y5; Y11 ↔ Y3, Y10 ↔ Y2; and Y16 ↔ Y8) are not included in the diagram for simplicity.

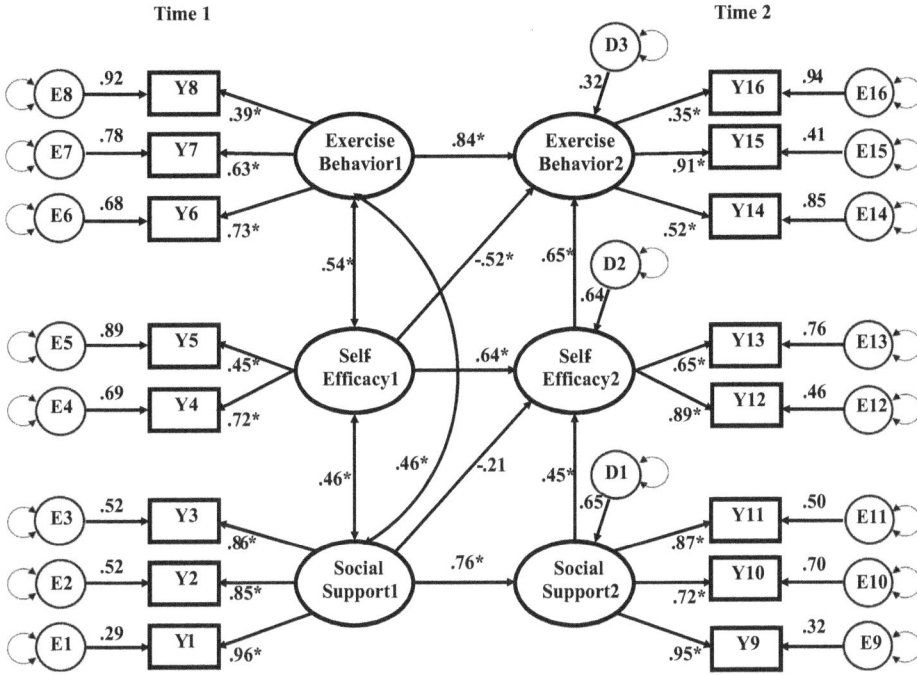

Figure 7.3: STANDARDIZED ESTIMATES FOR THE FULL SEM MODEL OF EXERCISE BEHAVIOR

You could also simply present the standardized estimates for the structural model like the diagram presented in Figure 7.4 in addition to tables similar to Tables 7.8 and 7.9.

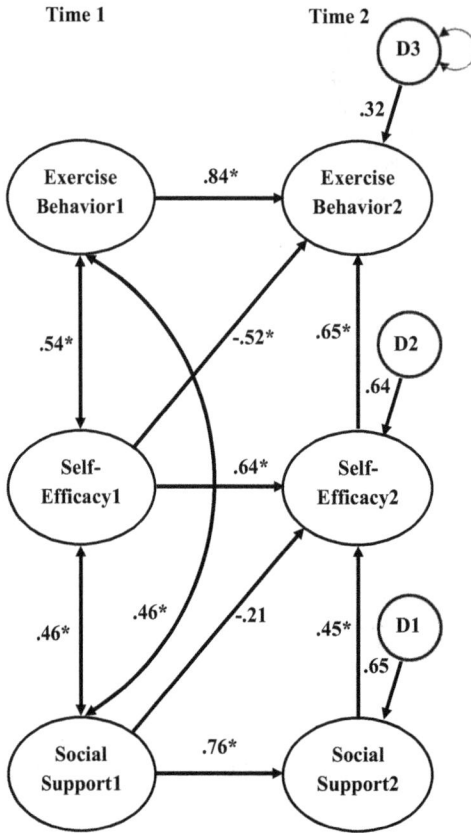

■ **Figure 7.4:** STANDARDIZED ESTIMATES FOR THE STRUCTURAL MODEL OF EXERCISE BEHAVIOR

You will notice in the diagrams that the paths from errors to their respective indicators and the paths from disturbances to their respective latent factors have values attached. There are different ways that you can present information about error and disturbance variances. In Figures 7.3 and 7.4, the square root of the standardized residual variances for the indicators are presented for the error paths and the square root of the standardized residual variances for the endogenous factors are presented for the disturbance paths. Remember that the standardized residual variances are equal to $1 - R^2$ and the square root of this value represents the direct effect of the error or disturbance on its respective observed or latent variable, respectively. The residual variances could also be included in the diagram as shown in Figure 7.5. If unstandardized estimates are illustrated in a diagram, those unstandardized estimates (direct effects, covariances, and error/disturbance variances) would simply replace the standardized estimates in Figure 7.5. Another option when presenting the results is to report the R square value for the endogenous variables (latent or observed) in the diagram.

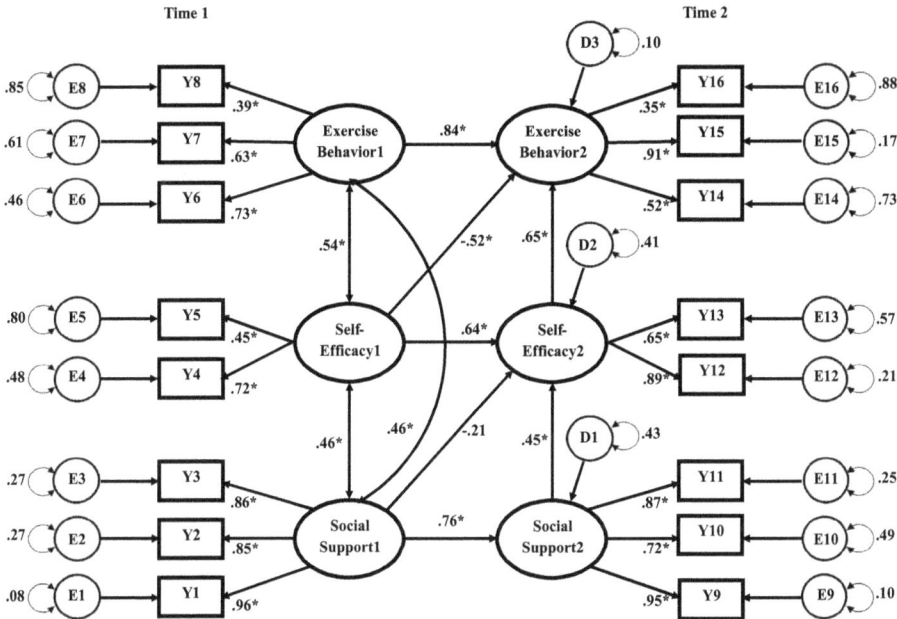

Figure 7.5: STANDARDIZED ESTIMATES FOR THE FULL SEM MODEL OF EXERCISE BEHAVIOR WITH RESIDUAL VARIANCES

INDIRECT EFFECTS IN FULL SEM MODELS

Indirect effects among factors can also be tested similarly to that described with observed variable path analysis models in Chapter 4. There are five specific indirect effects among the factors in the Exercise Behavior model: 1) SST1→ SET2 → EBT2; 2) SST1 → SST2 → SET2 → EBT2; 3) SET1 → SET2 → EBT2; 4) SST1 → SST2 → SET2; and 5) SST2 → SET2 → EBT2. In this exercise behavior example, we only have the summary data (i.e., the covariance matrix). Thus, we will not be able to test indirect effects among the factors using bootstrapping techniques with only the covariance matrix. If you did have raw data, the following M*plus* code could then be used to obtain bias-corrected bootstrapped confidence intervals for the indirect effects among factors in the exercise behavior model as follows:

```
TITLE:     STRUCTURAL MODEL WITH BC-
           BOOTSTRAPPED INDIRECT EFFECTS

DATA:      FILE IS EXBEH.DAT;

VARIABLE: NAMES ARE Y1 - Y16;

ANALYSIS: BOOTSTRAP = 5000;
```

```
MODEL:      SST1 BY Y1* Y2 Y3;
            SET1 BY Y4* Y5;
            EBT1 BY Y6* Y7 Y8;
            SST2 BY Y9 Y10 Y11;
            SET2 BY Y12 Y13;
            EBT2 BY Y14 Y15 Y16;
            SST1@1; SET1@1; EBT1@1;
            Y14 WITH Y6;
            Y13 WITH Y5;
            Y11 WITH Y3;
            Y10 WITH Y2;
            Y16 WITH Y8;
            EBT2 ON EBT1 SET1 SET2;
            SET2 ON SET1 SST1 SST2;
            SST2 ON SST1;

MODEL INDIRECT: EBT2 IND SET1;
                EBT2 IND SST1;
                EBT2 IND SST2;
                SET2 IND SST1;

OUTPUT:     STDYX CINTERVAL(BCBOOTSTRAP);
```

Again, the raw data should be delimited (e.g., space, tab, comma) without variable names on the top row. The ANALYSIS command with the **Bootstrap = 5000** option specifies that the number of bootstrapped draws (with replacement) should be equal to 5000. Thus, the program will take 5000 bootstrapped draws with replacement, where the sample size in each of the k data sets drawn is equal to the original number of observations (i.e., $n = 84$). The **CINTERVAL** option in the OUTPUT command requests confidence intervals be constructed around parameter estimates and the **(BCBOOTSTRAP)** option with the **CINTERVAL** option requests that bias-corrected bootstrapped confidence intervals be constructed around parameter estimates. If you want percentile bootstrapped confidence intervals instead, the following option in the OUTPUT command should be used: **INTERVAL(BOOTSTRAP)**.

Again, with the covariance matrix, the indirect effects can only be tested by dividing the product of the indirect effect by the standard error obtained using the delta method in LISREL and M*plus*. As discussed in Chapter 4, bootstrapped estimates are recommended because of the non-normal distribution associated with indirect effects. Since researchers will regularly have the raw data with which to work, bias-corrected bootstrapped confidence intervals for the testing of indirect effects among factors is recommended. Following the example in Chapter 4, you could then present the results similarly to those presented in Table 4.9 with latent factors instead of observed variables.

The cautions about interpreting indirect effects discussed in Chapter 4 still hold with full SEM models. Specifically, the temporal ordering of the variables involved in the mediating relationships is important for stronger claims to be made about the results when testing indirect effects. It should also be mentioned that not every specific indirect effect in a model has to be tested if temporal ordering and/or theory does not support the testing of a specific indirect effect.

SUMMARY

This chapter introduced full structural equation modeling, which is an extension of observed variable path analysis to a path analysis among latent factors. We provided an example of a structural equation model for exercise behavior in order to discuss model specification, identification, estimation, testing, and modification. We followed a two-step SEM modeling approach where the measurement model is examined first, followed by the structural model. It is important to ensure that adequate measurement of the latent factors is supported prior to testing the hypothesized relationships among those latent factors. If the measurement model does not fit the data adequately, model modifications can be used to help improve model fit if theory supports the addition of parameters to the model. If the measurement model does not achieve adequate fit to the data, testing structural relationships among latent factors is a moot point.

There is also a four-step SEM model testing approach that has been recommended with full SEM models. Both the two-step and four-step approaches allow the comparison of nested measurement and structural models. The $\Delta\chi^2$ test can be used when comparing nested models to determine if a significant gain or loss in fit has occurred. Researchers can also use the difference between the nested models' fit indices (e.g., ΔCFI) as an alternative to the $\Delta\chi^2$ test. Similar to observed variable path models, indirect effects can be tested in full SEM models using bootstrapping methods as described in Chapter 4.

EXERCISES

1. Figure 7.6 illustrates the political democracy structural equation model presented in Bollen (1989). A brief description of the factors and variables are provided below:

   D1960 (Democracy in 1960 Factor)
   Y1 = freedom of the press in 1960
   Y2 = freedom of political opposition in 1960
   Y3 = fairness of elections in 1960
   Y4 = effectiveness of elected legislature in 1960

D1965 (Democracy in 1965 Factor)
Y5 = freedom of the press in 1965
Y6 = freedom of political opposition in 1965
Y7 = fairness of elections in 1965
Y8 = effectiveness of elected legislature in 1965

I1960 (Industrialization in 1960 Factor)
X1 = Gross national product (GNP) per capita in 1960
X2 = energy consumption per capita in 1960
X3 = percentage of labor force in industry in 1960

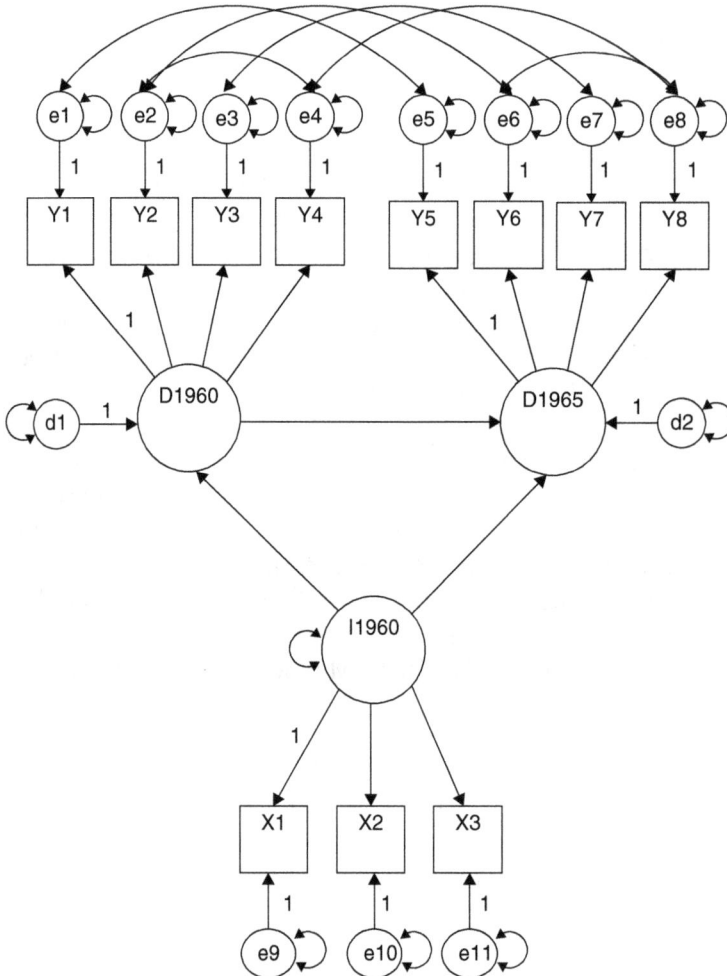

■ **Figure 7.6:** BOLLEN'S (1989) POLITICAL DEMOCRACY MODEL

Answer the following questions based on the political democracy SEM model.

a. How many distinct or non-redundant elements are in the sample covariance matrix?

b. How many parameters are to be estimated in the political democracy SEM model presented in Figure 7.6? What are the degrees of freedom (df) for the model in Figure 7.6?

c. In two-step SEM, the initial CFA or measurement model is recommended to be estimated prior to the structural model. In the initial CFA/measurement model, all factors are intercorrelated with each other so that the structural model is just-identified. Draw the initial CFA/measurement model that would be tested prior to the structural model presented in Figure 7.6.

d. How many parameters are to be estimated in the initial CFA/measurement model? What are the degrees of freedom (df) for the initial CFA/measurement model?

e. What is the difference between the df for the model presented in Figure 7.6 and the df for the initial CFA/measurement model? Explain why the difference in the two model's df is equal to the value you calculated.

f. How will the structural model fit compared to the initial CFA/measurement model? Explain your answer.

g. The covariance matrix for the political democracy data is presented below in the following variable order: Y1–Y8 X1–X3. The data are based on 75 countries. Write a SIMPLIS or M*plus* program to estimate the initial CFA/measurement model that you drew for 1.c. and the structural model illustrated in Figure 7.6 to confirm your answers to 1.e and 1.f.

```
6.89
6.25 15.58
5.84 5.84 10.76
6.09 9.51 6.69 11.22
5.06 5.60 4.94 5.70 6.83
5.75 9.39 4.73 7.44 4.98 11.38
5.81 7.54 7.01 7.49 5.82 6.75 10.80
5.67 7.76 5.64 8.01 5.34 8.25 7.59 10.53
0.73 0.62 0.79 1.15 1.08 0.85 0.94 1.10 0.54
1.27 1.49 1.55 2.24 2.06 1.81 2.00 2.23 0.99 2.28
0.91 1.17 1.04 1.84 1.58 1.57 1.63 1.69 0.82 1.81 1.98
```

2. The following SIMPLIS and M*plus* programs were used to estimate a structural model based on an example (example 5.12) provided on the M*plus* website.

SIMPLIS Code:

```
STRUCTURAL MODEL CH7-Q2.a.
RAW DATA FROM FILE Chapter7Q2.LSF
Latent Variables: F1-F4
Relationships:
Y1        = 1*F1
Y2-Y3     = F1
Y4        = 1*F2
Y5-Y6     = F2
Y7        = 1*F3
Y8-Y9     = F3
Y10       = 1*F4
Y11-Y12 = F4
F4 = F3
F3 = F1 F2
Set Covariance Between F1 and F2 to be free
OPTIONS: SC EF ND=3
Path Diagram
End of Problem
```

M*plus* Code:

```
TITLE: STRUCTURAL MODEL CH7-Q2.a.

DATA:   FILE IS Chapter7Q2.TXT;

VARIABLE: NAMES ARE Y1-Y12;

MODEL: F1 BY Y1-Y3;
       F2 BY Y4-Y6;
       F3 BY Y7-Y9;
       F4 BY Y10-Y12;
       F4 ON F3;
       F3 ON F1 F2;
       F1 WITH F2;

OUTPUT: STDYX;
```

a. Based on the LISREL and M*plus* code provided above, draw the hypothesized model that is to be estimated.
b. Draw the initial CFA/measurement model that would be tested prior to the structural model you drew in 2.a.
c. Determine the *df* for the initial CFA/measurement model and the *df* for the structural model based on the code provided above.

d. How will the structural model fit compared to the initial CFA/measurement model? Explain your answer.

e. Use the *Chapter7Q2.LSF* file or the *Chapter7Q2.txt* file on the book website and go through the two-step SEM process by first estimating the initial CFA/measurement model followed by the structural model. If the initial CFA/measurement model does not fit the data well, use modification indices to detect potential model misspecifications and add parameters to help improve model fit prior to estimating the initial structural model.

f. There are two indirect effects among the factors in the structural model. Specifically, F1 and F2 indirectly affect F4 through F3. Test these two indirect effects in LISREL using the **EF** option in the Options command. In M*plus*, test the two indirect effects using bias-corrected bootstrapped confidence intervals with 5000 bootstrapped draws.

REFERENCES

Anderson, J. C., & Gerbing, D. W. (1988). Structural equation modeling in practice: A review and recommended two-step approach. *Psychological Bulletin*, 103, 411–423.

Bollen, K. A. (1989). *Structural equations with latent variables*. New York: Wiley.

Cheung, G. W., & Rensvold, R. B. (2002). Evaluating goodness-of-fit indexes for testing measurement invariance. *Structural Equation Modeling*, 9(2), 233–255.

Duncan, T.E., & Stoolmiller, M. (1993). Modeling social and psychological determinants of exercise behaviors via structural equation systems. *Research Quarterly for Exercise and Sport*, 64(1), 1–16.

Jöreskog, K. G. (1999). *How large can a standardized coefficient be?* Retrieved from www.statmodel.com/download/Joreskog.pdf

Jöreskog, K. G., & Sörbom, D. (1993). *LISREL 8: Structural equation modeling with the SIMPLIS command language*. Hillsdale, NJ: Lawrence Erlbaum.

Meade, A. W., Johnson, E. C., & Braddy, P. W. (2008). Power and sensitivity of alternative fit indices in tests of measurement invariance. *Journal of Applied Psychology*, 93(3), 568–592.

Mulaik, S. A., & Millsap, R. E. (2000). Doing the four-step right. *Structural Equation Modeling*, 7, 36–73.

Chapter 8

EXTENSIONS OF CFA MODELS

SECOND-ORDER FACTOR MODEL

Model Specification

The first model we introduce is a *second-order* or *higher-order* factor model. A second-order factor model is specified where a higher-order factor explains the relationships among first-order factors. Theory plays an important role in justifying a higher-order factor. A quality of life (QOL) second-order factor model is illustrated in Figure 8.1 using data from the AIDS Time-Oriented Health Outcome Study (ATHOS) and illustrated by Chen, West, and Sousa (2006) based on Stewart and Ware's (1992) study. Twelve items related to quality of life in terms of health-related outcomes (which is a subset of 17) were used as indicators in the second-order model. The 12 variables are *diffeas*, *sloract*, *confused*, *forget*, *diffconc*, *tired*, *enrgtic*, *worout*, *peppy*, *afraid*, *frust*, and *hlthwry*. *Cognition*, *Vitality*, and *Disease Worry* (DISWORR) are three first-order factors that are hypothesized to be caused by a second-order factor, namely *Quality of Life* (QOL). The data provided by Chen et al. (2006) were based on 403 subjects receiving care from community-based providers in San Francisco for HIV-related illnesses.

DOI: 10.4324/9781003044017-8

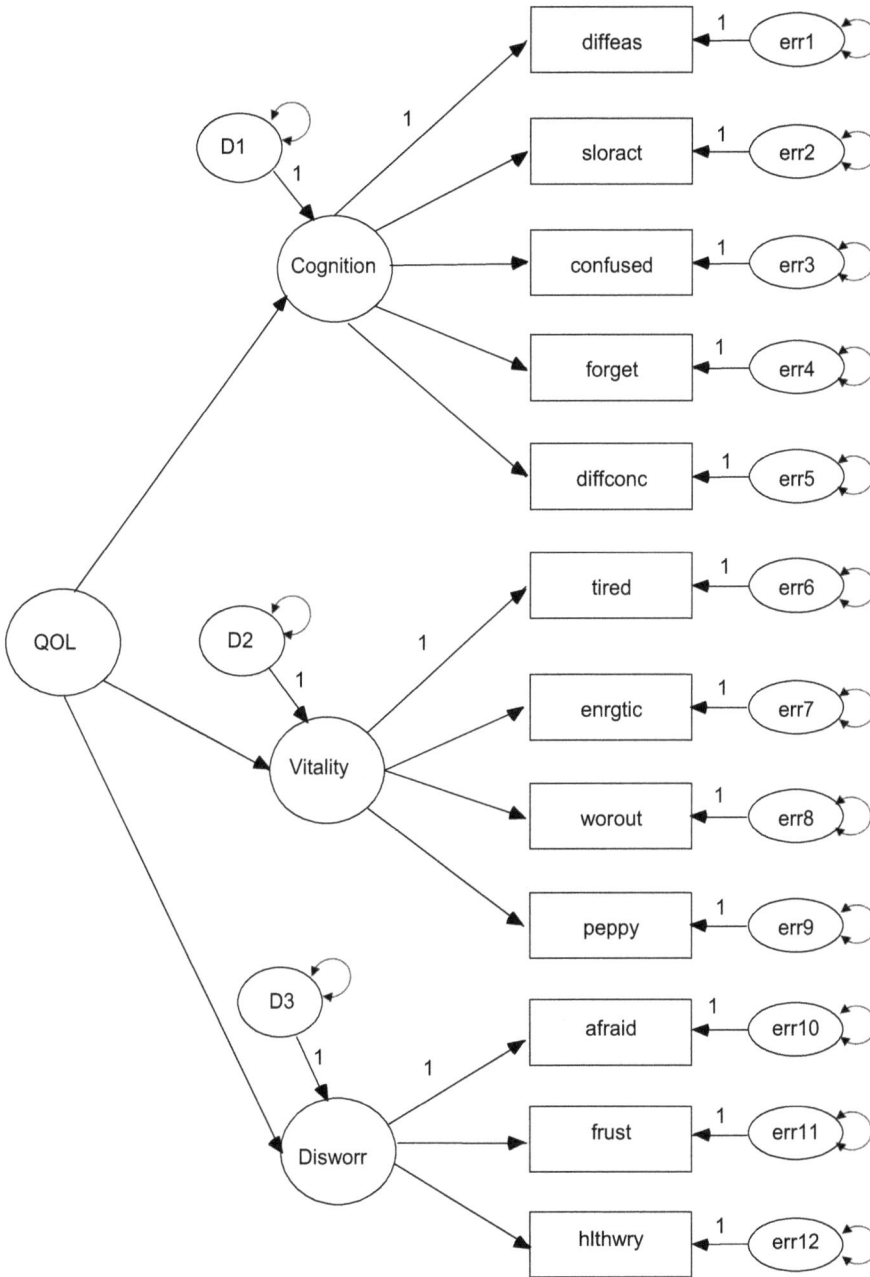

■ **Figure 8.1:** QOL SECOND-ORDER FACTOR MODEL

You will notice in Figure 8.1 that the first-order factors are serving as indicators for the second-order QOL factor. Thus, the QOL factor has direct effects on the first-order factors. Just as the first-order factor loadings represent the direct effects from the first-order factors to their respective indicators, the second-order factor loadings represent the direct effects from the second-order factor to the first-order factors. Theoretically, the second-order model suggests that the QOL factor is the reason that the first-order factors are related or correlated.

Model Identification

Recalling from earlier chapters, the number of distinct or non-redundant values in the sample covariance matrix (**S**) is $p(p + 1)/2$. With the 12 indicator variables in the second-order factor model, there are 78 distinct values (variances and covariances): $p(p + 1)/2 = 12(13)/2 = 78$. You will notice that the three-first order factors (*Cognition*, *Vitality*, and *Disease Worry*) have observed indicators (i.e., *diffeas*, *tired*, and *afraid*, respectively) with loadings set to equal 1. Remember from Chapter 7 that endogenous latent factors must use a reference indicator approach in order to set their scale for identification purposes. Because the first-order factors are directly affected by the second-order factor, they are endogenous and will have disturbance variances estimated. To set the scale of the second-order factor (*QOL*), you can either set its variance equal to 1 or use the reference indicator approach (e.g., set the direct effect from *QOL* to *Cognition* equal to 1).

The parameters to be estimated in the model are 9 first-order factor loadings; 3 second-order factor loadings (while setting the QOL factor variance equal to 1); 12 error variances (1 for each observed indicator variable); and 3 disturbance variances (one for each first-order factor), equaling 27 total parameters. The degrees of freedom for the model is computed as $78 - 27 = 51$. With 51 degrees of freedom and the factors properly scaled, the model is over-identified.

An important consideration with second-order factor models is the number of first-order factors included in the model. Second-order factor models are comprised of a measurement model and a structural model. Specifically, the indicators are related because of their respective first-order factors whereas the first-order factors are related because of the second-order factor. With two first-order factors, the structural level model is an under-identified model. Hence, the overall or global degrees of freedom for a second-order model with only two first-order factors may be equal to 1 or greater, but the model would not be able to be estimated with the under-identified structural model.

With three first-order factors, the structural level model is a just-identified model and parameters will be able to be estimated. However, the higher-order structural model cannot be tested since it is just-identified. With four or more first-order factors, the structural model is an over-identified model and the higher-order structural model can be tested. One way to calculate degrees of freedom at the structural level is to consider the first-order factors as pseudo-variables. This can be done to determine the number of distinct or non-redundant elements in the pseudo-covariance matrix of the first-order factors. The number of parameters at the structural level can then be counted. Subtracting the number of parameters from the number of distinct elements in the pseudo-covariance matrix of first-order factors will provide degrees of freedom associated with the structural model. For instance, with three first-order factors, there would be $[p(p + 1)/2 = 6]$ 6 distinct elements in the pseudo-covariance matrix of first-order factors, represented with η below (3 factor variances and 3 factor covariances):

$$
\begin{array}{ccc}
\eta_{\text{Cognition}} & \eta_{\text{Vitality}} & \eta_{\text{Disworr}}
\end{array}
$$

$$
\begin{bmatrix}
\sigma^2_{\text{Cognition}} & \sigma_{\text{Cog\_Vit}} & \sigma_{\text{Cog\_Disw}} \\
\sigma_{\text{Vit\_Cog}} & \sigma^2_{\text{Vitality}} & \sigma_{\text{Vit\_Disw}} \\
\sigma_{\text{Disw\_Cog}} & \sigma_{\text{Disw\_Vit}} & \sigma^2_{\text{Disworr}}
\end{bmatrix}.
$$

The structural level model of the second-order factor model is presented in Figure 8.2. At the structural level, there are 3 disturbance variances and 3 second-order factor loadings needing to be estimated (while setting the QOL factor variance equal to 1), equaling 6 parameters to estimate. Thus, the higher-order structural model is just-identified with three first-order factors.

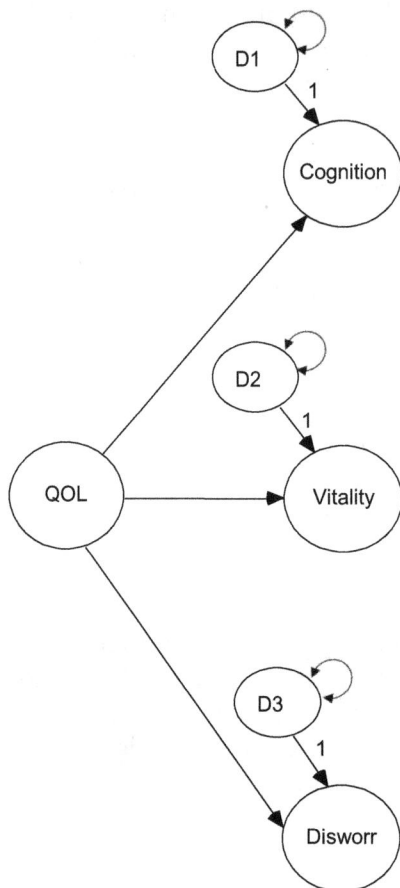

■ **Figure 8.2:** STRUCTURAL MODEL OF THE QOL SECOND-ORDER FACTOR MODEL

With four first-order factors, we would have $[p(p + 1)/2 = 10]$ 10 non-redundant elements in the pseudo-covariance matrix of first-order factors:

$$
\begin{array}{cccc}
\eta_1 & \eta_2 & \eta_3 & \eta_4
\end{array}
$$
$$
\begin{bmatrix}
\sigma_1^2 & \sigma_{12} & \sigma_{13} & \sigma_{14} \\
\sigma_{21} & \sigma_2^2 & \sigma_{23} & \sigma_{24} \\
\sigma_{31} & \sigma_{32} & \sigma_3^2 & \sigma_{34} \\
\sigma_{41} & \sigma_{42} & \sigma_{43} & \sigma_4^2
\end{bmatrix}.
$$

A second-order structural model with four first-order factors is illustrated in Figure 8.3. In this model, there are 4 disturbance variances and 4 second-order factor loadings to be estimated while setting the higher-order factor (η) variance equal to 1. This results in an over-identified structural model with 2 degrees of freedom $(10 - 8 = 2)$.

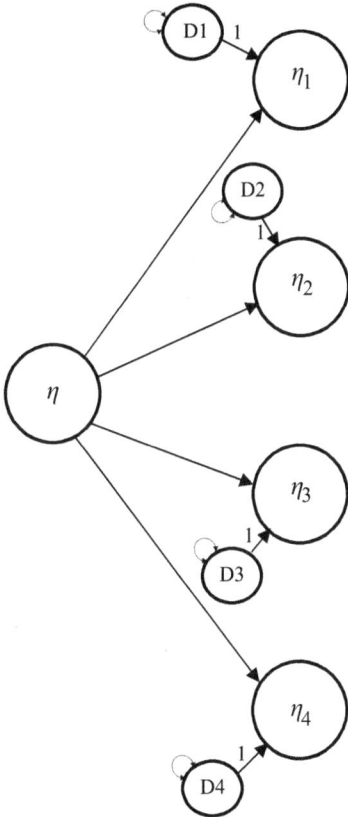

■ **Figure 8.3:** STRUCTURAL MODEL WITH FOUR FIRST-ORDER FACTORS

Model Estimation

The covariance matrix available in Chen et al. (2006) is available in **QOL.txt** on the book website. ML estimation is used as the estimator in this example. It must be noted that there are variables in the covariance matrix that will not be used in the second-order model. The SIMPLIS code to estimate the second-order model is provided below:

```
QOL SECOND-ORDER FACTOR MODEL
OBSERVED VARIABLES: SF DIFFEAS SLORACT CONFUSED
FORGET DIFFCONC TIRED ENRGTIC WOROUT PEPPY ATPEACE FEELBLUE
HAPPY NERVOUS DOWN AFRAID FRUST HLTHWRY
COVARIANCE MATRIX FROM FILE QOL.TXT
SAMPLE SIZE: 403
```

```
LATENT VARIABLES: COGNITN VITALITY DISWORR QOL
DIFFEAS = 1*COGNITN
SLORACT - DIFFCONC = COGNITN
TIRED = 1*VITALITY
ENRGTIC - PEPPY = VITALITY
AFRAID = 1*DISWORR
FRUST - HLTHWRY = DISWORR
COGNITN - DISWORR = QOL
OPTIONS: SC MI
PATH DIAGRAM
END OF PROBLEM
```

You will notice that we designated which items should serve as reference indicators for the first-order factors. Hence, *DIFFEAS*, *TIRED*, and *AFRAID* will serve as reference indicators. You will also notice that you will define the second-order *QOL* factor in the **Latent Variables** command and the first-order factors (*COGNITN – DISWORR*) will serve as indicators for the *QOL* factor:

```
COGNITN - DISWORR = QOL
```

The M*plus* code to estimate the second-order factor model is provided below:

```
TITLE:      QOL SECOND ORDER MODEL

DATA:       FILE IS QOL.TXT;
            TYPE = COVARIANCE;
            NOBS = 403;

VARIABLE:   NAMES ARE SF DIFFEAS SLORACT CONFUSED FORGET DIFFCONC
                      TIRED ENRGTIC WOROUT PEPPY ATPEACE FEELBLUE
                      HAPPY NERVOUS DOWN AFRAID FRUST HLTHWRY;
            USEVARIABLES ARE DIFFEAS - PEPPY AFRAID - HLTHWRY;

MODEL:      COGNITN  BY DIFFEAS - DIFFCONC;
            VITALITY BY TIRED - PEPPY;
            DISWORR  BY AFRAID - HLTHWRY;
            QOL BY COGNITN - DISWORR*;
            QOL@1;

OUTPUT:     SAMP STDYX MODINDICES(3.84);
```

Because we are not using all of the variables in the covariance matrix, the **USEVARIABLES** option in the VARIABLE command was used. To avoid the default of using the reference indicator method for the second-order *QOL* factor, we estimated all of the second-order loadings by including an asterisk:

```
QOL BY COGNITN - DISWORR*;
```

and set the *QOL* factor variance equal to 1:

```
QOL@1;
```

Model Testing and Model Modification

The model fit information in the output for the second-order factor model can be examined to assess its fit to the data. The model fit indices indicated that the hypothesized second-order factor model, with the exception of the chi-square test of model fit, is exhibiting adequate fit to the data [LISREL: $\chi^2(51) = 93.602$, $p < .05$; CFI = .985; NNFI/TLI = .981; RMSEA = .046 (90% CI: .031, .060) with $p > .05$; SRMR = .033] and [M*plus*: $\chi^2(51) = 93.835$, $p < .05$; CFI = .985; NNFI/TLI = .981; RMSEA = .046 (90% CI: .031, .060) with $p > .05$; SRMR = .033]. The chi-square test of model fit does not support adequate model fit whereas the remaining indices (i.e., CFI, NNFI/TLI, RMSEA, and SRMR) support acceptable model fit, which could be due to the large sample size ($n = 403$). The modification indices could be examined to detect potential theoretically meaningful re-specifications in the model. For instance, the largest drop in chi-square (approximately 20 points) was associated with the cross-loading of the *frust* indicator on the *Vitality* factor. Nonetheless, we decided not to modify the model given the majority of support for adequate model fit.

Model Interpretation

The measurement equations with unstandardized *first-order* factor loadings and corresponding standard errors in LISREL are provided below:

```
Measurement Equations

DIFFEAS = 1.000*COGNITN, Errorvar.= 0.181, R² = 0.728
Standerr        (0.0174)
Z-values        10.387
P-values         0.000

 SLORACT = 0.759*COGNITN, Errorvar.= 0.354  , R² = 0.442
Standerr  (0.0515)                (0.0270)
Z-values  14.734                  13.083
P-values   0.000                   0.000

CONFUSED = 1.053*COGNITN, Errorvar.= 0.224  , R² = 0.706
Standerr  (0.0509)                (0.0208)
Z-values  20.703                  10.783
P-values   0.000                   0.000
```

```
    FORGET = 0.953*COGNITN, Errorvar.= 0.270 , R² = 0.620
Standerr  (0.0508)                       (0.0227)
Z-values  18.739                         11.893
P-values   0.000                          0.000

  DIFFCONC = 1.052*COGNITN, Errorvar.= 0.226 , R² = 0.704
Standerr  (0.0509)                        (0.0209)
Z-values  20.655                          10.818
P-values   0.000                           0.000

     TIRED = 1.000*VITALITY, Errorvar.= 0.165 , R² = 0.711
Standerr  (0.0170)
Z-values   9.741
P-values   0.000

    ENRGTIC = 0.934*VITALITY, Errorvar.= 0.419 , R² = 0.458
Standerr  (0.0639)                        (0.0329)
Z-values  14.627                          12.706
P-values   0.000                           0.000

    WOROUT = 1.039*VITALITY, Errorvar.= 0.158 , R² = 0.735
Standerr  (0.0522)                        (0.0172)
Z-values  19.897                           9.203
P-values   0.000                           0.000

     PEPPY = 0.970*VITALITY, Errorvar.= 0.259 , R² = 0.596
Standerr  (0.0557)                        (0.0224)
Z-values  17.408                          11.569
P-values   0.000                           0.000

    AFRAID = 1.000*DISWORR, Errorvar.= 0.254, R² = 0.729
Standerr  (0.0294)
Z-values   8.640
P-values   0.000

     FRUST = 0.925*DISWORR, Errorvar.= 0.271  , R² = 0.684
Standerr  (0.0497)                       (0.0279)
Z-values  18.611                          9.708
P-values   0.000                          0.000

   HLTHWRY = 0.947*DISWORR, Errorvar.= 0.328  , R² = 0.652
Standerr  (0.0522)                       (0.0317)
Z-values  18.132                         10.344
P-values   0.000                          0.000
```

There are 12 measurement equations, one for each of the 12 indicator variables. With the exception of the first-order loadings for the three reference indicators (*DIFFEAS*, *TIRED*, and *AFRAID*), which are fixed to equal 1, all of the first-order loadings are statistically significant. As described in Chapter 7, in order to obtain the unstandardized estimates of the loadings (and respective standard errors) for the reference indicators, you would rerun the model with alternative reference indicators for the three first-order factors.

The structural equations with unstandardized *second-order* loadings and respective standard errors in LISREL are presented below:

```
Structural Equations

  COGNITN = 0.468*QOL, Errorvar.= 0.266  , R² = 0.452
Standerr   (0.0404)               (0.0328)
Z-values   11.610                  8.120
P-values    0.000                  0.000

VITALITY = 0.530*QOL, Errorvar.= 0.125  , R² = 0.693
Standerr   (0.0393)               (0.0281)
Z-values   13.482                  4.433
P-values    0.000                  0.000

  DISWORR = 0.631*QOL, Errorvar.= 0.284  , R² = 0.584
Standerr   (0.0498)               (0.0467)
Z-values   12.676                  6.072
P-values    0.000                  0.000
```

All of the second-order factor loadings are statistically significant. With the exception of minor differences in the standard error estimates, the unstandardized estimates are the same in the M*plus* output provided below:

```
MODEL RESULTS
                                                    Two-Tailed
                    Estimate    S.E.    Est./S.E.    P-Value
  COGNITN  BY
    DIFFEAS          1.000     0.000     999.000     999.000
    SLORACT          0.759     0.051      14.753       0.000
    CONFUSED         1.053     0.051      20.729       0.000
    FORGET           0.953     0.051      18.762       0.000
    DIFFCONC         1.052     0.051      20.681       0.000

  VITALITY BY
    TIRED            1.000     0.000     999.000     999.000
    ENRGTIC          0.934     0.064      14.645       0.000
    WOROUT           1.039     0.052      19.922       0.000
    PEPPY            0.970     0.056      17.430       0.000

DISWORR  BY
    AFRAID           1.000     0.000     999.000     999.000
    FRUST            0.925     0.050      18.634       0.000
    HLTHWRY          0.947     0.052      18.155       0.000
```

QOL BY				
COGNITN	0.468	0.040	11.610	0.000
VITALITY	0.530	0.039	13.482	0.000
DISWORR	0.631	0.050	12.676	0.000
Variances				
QOL	1.000	0.000	999.000	999.000
Residual Variances				
DIFFEAS	0.181	0.017	10.400	0.000
SLORACT	0.353	0.027	13.100	0.000
CONFUSED	0.224	0.021	10.796	0.000
FORGET	0.269	0.023	11.908	0.000
DIFFCONC	0.225	0.021	10.832	0.000
TIRED	0.165	0.017	9.753	0.000
ENRGTIC	0.418	0.033	12.722	0.000
WOROUT	0.158	0.017	9.214	0.000
PEPPY	0.258	0.022	11.583	0.000
AFRAID	0.253	0.029	8.651	0.000
FRUST	0.270	0.028	9.720	0.000
HLTHWRY	0.327	0.032	10.357	0.000
COGNITN	0.266	0.033	8.120	0.000
VITALITY	0.124	0.028	4.433	0.000
DISWORR	0.283	0.047	6.072	0.000

The measurement equations in LISREL provide R^2 values associated with each indicator variable, indicating the proportion of variance in the indicator variable explained by the model. For instance, about 62% of the variance in the *FORGET* indicator is explained by the model. More interesting is the proportion of variance in the first-order factors explained by the second-order *QOL* factor, which is provided by the structural equations in LISREL. For example, approximately 45% of the variance in the *Cognition* factor is explained by the second-order QOL factor. Approximately 69% of the variance in the *Vitality* factor and approximately 58% of the variance in the *Disease Worry* factor is explained by the *QOL* factor.

The R^2 values associated with the indicator variables and the first-order factors in M*plus* are provided with the standardized model results:

STANDARDIZED MODEL RESULTS

STDYX Standardization

	Estimate	S.E.	Est./S.E.	Two-Tailed P-Value
COGNITN BY				
DIFFEAS	0.853	0.017	49.989	0.000
SLORACT	0.665	0.030	21.849	0.000
CONFUSED	0.840	0.018	46.640	0.000
FORGET	0.787	0.022	35.954	0.000
DIFFCONC	0.839	0.018	46.332	0.000

```
VITALITY BY
  TIRED               0.843       0.019       43.775       0.000
  ENRGTIC             0.677       0.030       22.297       0.000
  WOROUT              0.857       0.018       46.607       0.000
  PEPPY               0.772       0.024       32.178       0.000

DISWORR BY
  AFRAID              0.854       0.020       43.037       0.000
  FRUST               0.827       0.021       38.790       0.000
  HLTHWRY             0.807       0.022       35.946       0.000

QOL BY
  COGNITN             0.672       0.041       16.260       0.000
  VITALITY            0.832       0.040       20.737       0.000
  DISWORR             0.764       0.041       18.733       0.000

Variances
  QOL                 1.000       0.000      999.000      999.000

Residual Variances
  DIFFEAS             0.272       0.029        9.320       0.000
  SLORACT             0.558       0.040       13.801       0.000
  CONFUSED            0.294       0.030        9.701       0.000
  FORGET              0.380       0.034       11.013       0.000
  DIFFCONC            0.296       0.030        9.736       0.000
  TIRED               0.289       0.032        8.907       0.000
  ENRGTIC             0.542       0.041       13.169       0.000
  WOROUT              0.265       0.032        8.420       0.000
  PEPPY               0.404       0.037       10.907       0.000
  AFRAID              0.271       0.034        7.997       0.000
  FRUST               0.316       0.035        8.978       0.000
  HLTHWRY             0.348       0.036        9.611       0.000
  COGNITN             0.548       0.056        9.869       0.000
  VITALITY            0.307       0.067        4.600       0.000
  DISWORR             0.416       0.062        6.669       0.000

R-SQUARE

  Observed                                              Two-Tailed
  Variable          Estimate       S.E.     Est./S.E.    P-Value

  DIFFEAS             0.728       0.029       24.995       0.000
  SLORACT             0.442       0.040       10.925       0.000
  CONFUSED            0.706       0.030       23.320       0.000
  FORGET              0.620       0.034       17.977       0.000
  DIFFCONC            0.704       0.030       23.166       0.000
  TIRED               0.711       0.032       21.888       0.000
  ENRGTIC             0.458       0.041       11.148       0.000
  WOROUT              0.735       0.032       23.303       0.000
  PEPPY               0.596       0.037       16.089       0.000
  AFRAID              0.729       0.034       21.518       0.000
  FRUST               0.684       0.035       19.395       0.000
  HLTHWRY             0.652       0.036       17.973       0.000
```

Latent Variable	Estimate	S.E.	Est./S.E.	Two-Tailed P-Value
COGNITN	0.452	0.056	8.130	0.000
VITALITY	0.693	0.067	10.368	0.000
DISWORR	0.584	0.062	9.367	0.000

The standardized estimates in M*plus* are identical to the estimates provided in the completely standardized solution in LISREL given below:

```
Completely Standardized Solution

          LAMBDA-Y
               COGNITN    VITALITY    DISWORR
               --------   --------    --------
  DIFFEAS       0.853        --          --
  SLORACT       0.665        --          --
 CONFUSED       0.840        --          --
   FORGET       0.787        --          --
 DIFFCONC       0.839        --          --
    TIRED        --        0.843         --
  ENRGTIC        --        0.677         --
   WOROUT        --        0.857         --
    PEPPY        --        0.772         --
   AFRAID        --          --        0.854
    FRUST        --          --        0.827
  HLTHWRY        --          --        0.807

          GAMMA
               QOL
               --------
  COGNITN       0.672
 VITALITY       0.832
  DISWORR       0.764

          Correlation Matrix of ETA and KSI
               COGNITN    VITALITY    DISWORR       QOL
               --------   --------    --------   --------
  COGNITN       1.000
 VITALITY       0.559       1.000
  DISWORR       0.514       0.636       1.000
      QOL       0.672       0.832       0.764      1.000

          PSI
          Note: This matrix is diagonal.
               COGNITN    VITALITY    DISWORR
               --------   --------    --------
                0.548       0.307       0.416
```

```
THETA-EPS
    DIFFEAS      SLORACT      CONFUSED      FORGET      DIFFCONC       TIRED
   --------     --------     --------     --------     --------     --------
      0.272        0.558        0.294        0.380        0.296        0.289

THETA-EPS
    ENRGTIC       WOROUT        PEPPY       AFRAID        FRUST       HLTHWRY
   --------     --------     --------     --------     --------     --------
      0.542        0.265        0.404        0.271        0.316        0.348
```

BIFACTOR MODEL

Model Specification

The second model we introduce is the *bifactor model* (Holzinger & Swineford, 1937). A *bifactor factor* model is specified where all of the variables serve as indicators for a general factor as well as for their respective group (e.g., subscale) factor. According to Reise (2012, p. 668), "the general factor represents the conceptually broad 'target' construct an instrument was designed to measure, and the group factors represent more conceptually narrow subdomain constructs." Another important specification in the bifactor model is that all of the factors are orthogonal or uncorrelated. Hence, theory is an important consideration when justifying a bifactor model. Using the same 12 items from the AIDS Time-Oriented Health Outcome Study (ATHOS) that were used for the second-order factor model, *QOL* was specified as the general factor and the *Cognition*, *Vitality*, and *Disease Worry* factors were specified as the group factors. This bifactor model is illustrated in Figure 8.4.

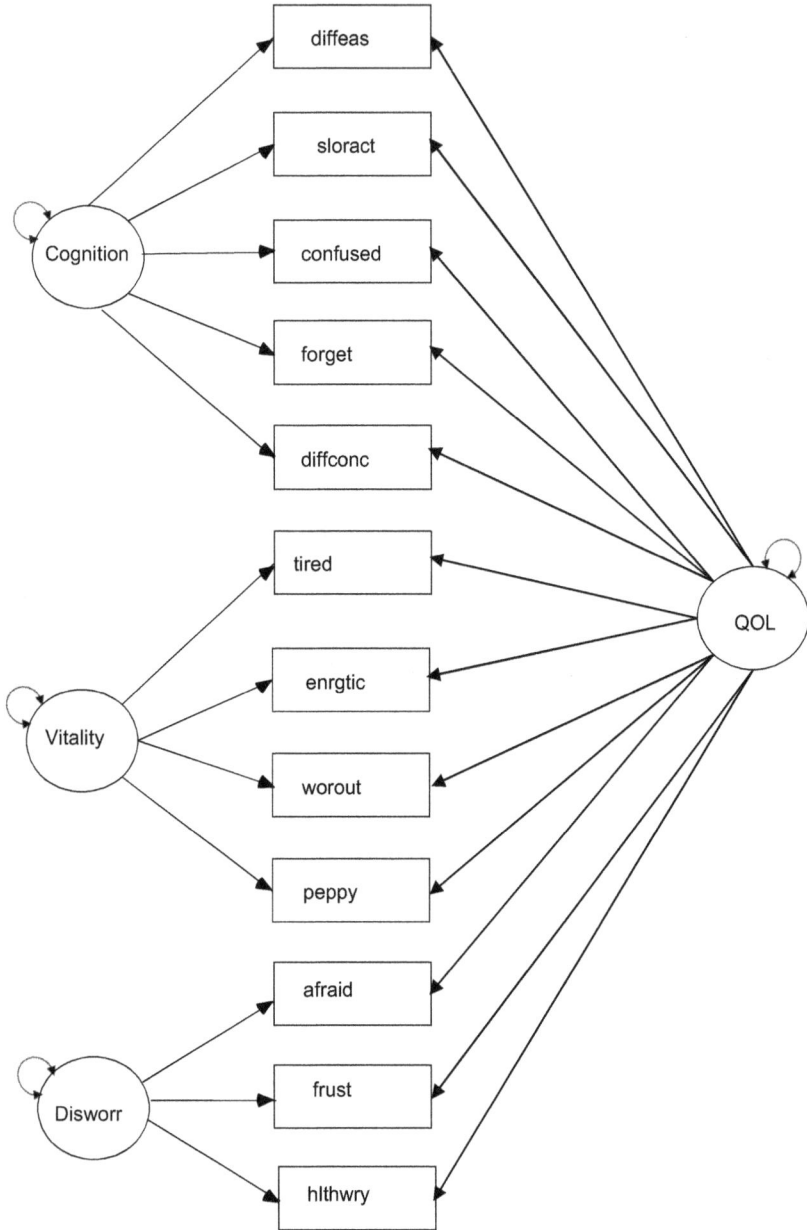

■ **Figure 8.4:** QOL BIFACTOR MODEL

You will notice in Figure 8.4 that all 12 items are serving as indicators for the general QOL factor while also serving as indicators for their respective target factor (i.e., *Cognition, Vitality, Disease Worry*). Theoretically, the bifactor model suggests that the QOL factor is the reason that all of the 12 items are related in addition to being related to other items loading on their respective target factor. Although they are not illustrated for simplicity in Figure 8.4, error variances for all 12 indicators are still estimated in the bifactor model.

Model Identification

Again, the number of non-redundant elements in the sample covariance matrix is $p(p + 1)/2 = 12(13)/2 = 78$. In the bifactor model, either factor scaling method could be used since the factors are not endogenous variables in the model. Thus, you can use a reference indicator for the general *QOL* factor and each of the three target factors or you can estimate all of the factor loadings and set the variance of the general *QOL* factor and each of the three target factors equal to 1. For this example, we will set the variances of the factors equal to 1 and estimate all of the factor loadings.

The parameters to be estimated in the model are 12 *QOL* factor loadings; 12 target factor loadings (5 for *Cognition*; 4 for *Vitality*; and 3 for *Disease Worry*); and 12 error variances (one for each observed indicator variable), equaling 36 total parameters. The degrees of freedom for the model is computed as $78 - 36 = 42$. With 42 degrees of freedom and the factors properly scaled, the model is over-identified.

Again, the factors do not covary in traditional bifactor models. That is not to say that a covariance between factors could not be estimated in a bifactor model. But convergence and/or identification problems could be encountered if factors are allowed to covary in a bifactor model. Yet again, theory concerning the relationships among factors is an important consideration in a bifactor model.

Model Estimation

The same covariance matrix that was used for the second-order factor model in the **QOL.txt** file is used for the bifactor model. Again, ML estimation is used in this example. The SIMPLIS code to estimate the bifactor model is provided below:

```
QOL BIFACTOR MODEL
OBSERVED VARIABLES: SF DIFFEAS SLORACT CONFUSED
FORGET DIFFCONC TIRED ENRGTIC WOROUT PEPPY ATPEACE FEELBLUE
HAPPY NERVOUS DOWN AFRAID FRUST HLTHWRY
COVARIANCE MATRIX FROM FILE QOL.TXT
SAMPLE SIZE: 403
LATENT VARIABLES: COGNITN VITALITY DISWORR QOL
DIFFEAS - DIFFCONC = COGNITN
```

```
TIRED - PEPPY = VITALITY
AFRAID - HLTHWRY = DISWORR
DIFFEAS - PEPPY AFRAID - HLTHWRY = QOL
SET COVARIANCES OF COGNITN - QOL TO 0
OPTIONS: SC MI
PATH DIAGRAM
END OF PROBLEM
```

You will notice that the covariances between all of the factors have been set equal to 0 to avoid the default estimation of factor covariances with the following statement: SET COVARIANCES OF COGNITN - QOL TO 0. LISREL sets the factor variances equal to 1 by default. The M*plus* code to estimate the bifactor model is provided below:

```
TITLE:       BIFACTOR MODEL OF QOL

DATA:        FILE IS QOL.TXT;
             TYPE = COVARIANCE;
             NOBS = 403;

VARIABLE:    NAMES ARE SF DIFFEAS SLORACT CONFUSED FORGET DIFFCONC
                       TIRED ENRGTIC WOROUT PEPPY ATPEACE FEELBLUE
                       HAPPY NERVOUS DOWN AFRAID FRUST HLTHWRY;
             USEVARIABLES ARE DIFFEAS - PEPPY AFRAID - HLTHWRY;

ANALYSIS:    MODEL = NOCOVARIANCES;

MODEL:       COGNITN BY DIFFEAS - DIFFCONC*;
             VITALITY BY TIRED - PEPPY*;
             DISWORR  BY AFRAID - HLTHWRY*;
             QOL BY DIFFEAS - HLTHWRY*;
             COGNITN@1 VITALITY@1 DISWORR@1 QOL@1;

OUTPUT:      SAMP STDYX MODINDICES(3.84);
```

Similar to LISREL, the default in M*plus* is to estimate factor covariances. To prevent the factor covariances from being estimated, the **MODEL = NOCOVARIANCES** option in the ANAYSIS command was used. You will also notice that asterisks are placed after the last indicator for each factor specification in order to prevent the default of using a reference indicator. Specifically:

```
COGNITN BY DIFFEAS - DIFFCONC*;
VITALITY BY TIRED - PEPPY*;
DISWORR BY AFRAID - HLTHWRY*;
QOL BY DIFFEAS - HLTHWRY*;
```

Likewise, all of the factor variances were set to equal 1:

```
COGNITN@1 VITALITY@1 DISWORR@1 QOL@1;
```

Model Testing and Model Modification

The model fit information in the output for the bifactor model was examined to assess its fit to the data. The model fit indices indicated that the hypothesized bifactor model, with the exception of the chi-square test of model fit, fits the data acceptably [LISREL: $\chi^2(42) = 61.211$, $p < .05$; CFI = .993; NNFI/TLI = .989; RMSEA = .034 (90% CI: .012, .051) with $p > .05$; SRMR = .020] and [M*plus*: $\chi^2(42) = 61.363$, $p < .05$; CFI = .993; NNFI/TLI = .989; RMSEA = .034 (90% CI: .012, .051) with $p > .05$; SRMR = .020]. Again, the chi-square test of model fit does not support adequate model fit whereas the remaining indices (i.e., CFI, NNFI/TLI, RMSEA, and SRMR) support acceptable model fit, which could be due to the large sample size ($n = 403$).

Model Interpretation

The measurement equations with unstandardized loadings on the target factors and the general *QOL* factor (with corresponding standard errors) in LISREL are provided below:

```
Measurement Equations

  DIFFEAS = 0.535*COGNITN + 0.444*QOL, Errorvar.= 0.184 , R² = 0.725
Standerr   (0.0356)         (0.0432)              (0.0180)
Z-values   15.036           10.263               10.212
P-values   0.000            0.000                0.000

  SLORACT = 0.353*COGNITN + 0.395*QOL, Errorvar.= 0.354 , R² = 0.442
Standerr   (0.0398)         (0.0428)              (0.0268)
Z-values   8.866            9.228                13.186
P-values   0.000            0.000                0.000

 CONFUSED = 0.549*COGNITN + 0.488*QOL, Errorvar.= 0.223 , R² = 0.708
Standerr   (0.0383)         (0.0460)              (0.0208)
Z-values   14.325           10.620               10.740
P-values   0.000            0.000                0.000

   FORGET = 0.567*COGNITN + 0.372*QOL, Errorvar.= 0.251 , R² = 0.647
Standerr   (0.0386)         (0.0460)              (0.0234)
Z-values   14.685           8.091                10.737
P-values   0.000            0.000                0.000

 DIFFCONC = 0.537*COGNITN + 0.496*QOL, Errorvar.= 0.228 , R² = 0.701
Standerr   (0.0384)         (0.0458)              (0.0208)
Z-values   13.976           10.827               10.943
P-values   0.000            0.000                0.000
```

```
     TIRED = 0.425*VITALITY + 0.506*QOL, Errorvar.= 0.134 , R² = 0.765
Standerr   (0.0492)             (0.0416)            (0.0267)
Z-values   8.642                12.162              5.030
P-values   0.000                0.000               0.000

   ENRGTIC = 0.281*VITALITY + 0.521*QOL, Errorvar.= 0.422 , R² = 0.454
Standerr   (0.0629)             (0.0493)            (0.0328)
Z-values   4.472                10.571              12.871
P-values   0.000                0.000               0.000

    WOROUT = 0.328*VITALITY + 0.567*QOL, Errorvar.= 0.166 , R² = 0.721
Standerr   (0.0532)             (0.0413)            (0.0181)
Z-values   6.180                13.723              9.186
P-values   0.000                0.000               0.000

     PEPPY = 0.267*VITALITY + 0.554*QOL, Errorvar.= 0.262 , R² = 0.591
Standerr   (0.0564)             (0.0435)            (0.0221)
Z-values   4.736                12.751              11.876
P-values   0.000                0.000               0.000

    AFRAID = 0.594*DISWORR + 0.601*QOL, Errorvar.= 0.222 , R² = 0.763
Standerr   (0.0535)             (0.0520)            (0.0385)
Z-values   11.103               11.569              5.753
P-values   0.000                0.000               0.000

     FRUST = 0.386*DISWORR + 0.654*QOL, Errorvar.= 0.277 , R² = 0.676
Standerr   (0.0506)             (0.0481)            (0.0263)
Z-values   7.630                13.594              10.535
P-values   0.000                0.000               0.000

   HLTHWRY = 0.554*DISWORR + 0.565*QOL, Errorvar.= 0.313 , R² = 0.667
Standerr   (0.0548)             (0.0527)            (0.0378)
Z-values   10.109               10.717              8.294
P-values   0.000                0.000               0.000
```

There is a measurement equation for each of the 12 indicator variables with factor loadings for their respective target factor and for the general *QOL* factor. All of the target factor loadings and general factor loadings are statistically significant. The unstandardized estimates in M*plus* are similar and are provided below:

```
MODEL RESULTS

                                                      Two-Tailed
                   Estimate    S.E.    Est./S.E.      P-Value
 COGNIT BY
   DIFFEAS          0.535      0.036     15.054        0.000
   SLORACT          0.353      0.040      8.877        0.000
   CONFUSED         0.548      0.038     14.342        0.000
   FORGET           0.566      0.039     14.704        0.000
   DIFFCONC         0.537      0.038     13.993        0.000
```

VITALITY BY				
TIRED	0.424	0.049	8.654	0.000
ENRGTIC	0.281	0.063	4.478	0.000
WOROUT	0.328	0.053	6.188	0.000
PEPPY	0.267	0.056	4.742	0.000
DISWORR BY				
AFRAID	0.593	0.053	11.116	0.000
FRUST	0.386	0.051	7.640	0.000
HLTHWRY	0.554	0.055	10.121	0.000
QOL BY				
DIFFEAS	0.443	0.043	10.276	0.000
SLORACT	0.394	0.043	9.239	0.000
CONFUSED	0.488	0.046	10.634	0.000
FORGET	0.372	0.046	8.101	0.000
DIFFCONC	0.496	0.046	10.840	0.000
TIRED	0.505	0.042	12.177	0.000
ENRGTIC	0.521	0.049	10.584	0.000
WOROUT	0.567	0.041	13.740	0.000
PEPPY	0.554	0.043	12.767	0.000
AFRAID	0.600	0.052	11.583	0.000
FRUST	0.654	0.048	13.611	0.000
HLTHWRY	0.564	0.053	10.731	0.000
Variances				
COGNITN	1.000	0.000	999.000	999.000
VITALITY	1.000	0.000	999.000	999.000
DISWORR	1.000	0.000	999.000	999.000
QOL	1.000	0.000	999.000	999.000
Residual Variances				
DIFFEAS	0.183	0.018	10.225	0.000
SLORACT	0.353	0.027	13.203	0.000
CONFUSED	0.222	0.021	10.753	0.000
FORGET	0.250	0.023	10.750	0.000
DIFFCONC	0.228	0.021	10.957	0.000
TIRED	0.134	0.027	5.037	0.000
ENRGTIC	0.421	0.033	12.887	0.000
WOROUT	0.166	0.018	9.198	0.000
PEPPY	0.262	0.022	11.891	0.000
AFRAID	0.221	0.038	5.760	0.000
FRUST	0.277	0.026	10.548	0.000
HLTHWRY	0.313	0.038	8.304	0.000

The R^2 values associated with each indicator variable are provided in the measurement equations in LISREL and indicate the proportion of variance in the indicator variable explained by the bifactor model. The largest R^2 value is associated with the *TIRED* indicator, with approximately 76.5% of its variance explained by the model. The smallest R^2 value is associated with the *SLORACT* indicator, with approximately 44% of its variance explained by the model.

Again, the R^2 values associated with the indicator variables in M*plus* are provided with the standardized model results:

STANDARDIZED MODEL RESULTS

STDYX Standardization

	Estimate	S.E.	Est./S.E.	Two-Tailed P-Value
COGNITN BY				
DIFFEAS	0.655	0.037	17.619	0.000
SLORACT	0.444	0.046	9.679	0.000
CONFUSED	0.629	0.038	16.504	0.000
FORGET	0.672	0.037	18.281	0.000
DIFFCONC	0.615	0.039	15.961	0.000
VITALITY BY				
TIRED	0.562	0.063	8.894	0.000
ENRGTIC	0.320	0.070	4.557	0.000
WOROUT	0.426	0.068	6.260	0.000
PEPPY	0.334	0.070	4.800	0.000
DISWORR BY				
AFRAID	0.614	0.052	11.767	0.000
FRUST	0.418	0.053	7.825	0.000
HLTHWRY	0.572	0.053	10.823	0.000
QOL BY				
DIFFEAS	0.543	0.044	12.422	0.000
SLORACT	0.495	0.046	10.783	0.000
CONFUSED	0.559	0.043	13.022	0.000
FORGET	0.441	0.048	9.129	0.000
DIFFCONC	0.568	0.042	13.377	0.000
TIRED	0.670	0.043	15.550	0.000
ENRGTIC	0.593	0.046	12.827	0.000
WOROUT	0.735	0.040	18.522	0.000
PEPPY	0.693	0.042	16.659	0.000
AFRAID	0.621	0.043	14.603	0.000
FRUST	0.708	0.038	18.510	0.000
HLTHWRY	0.583	0.044	13.125	0.000
Variances				
COGNITN	1.000	0.000	999.000	999.000
VITALITY	1.000	0.000	999.000	999.000
DISWORR	1.000	0.000	999.000	999.000
QOL	1.000	0.000	999.000	999.000

Residual Variances

DIFFEAS	0.275	0.030	9.207	0.000
SLORACT	0.558	0.040	13.915	0.000
CONFUSED	0.292	0.030	9.663	0.000
FORGET	0.353	0.036	9.945	0.000
DIFFCONC	0.299	0.030	9.841	0.000
TIRED	0.235	0.048	4.877	0.000
ENRGTIC	0.546	0.041	13.405	0.000
WOROUT	0.279	0.033	8.447	0.000
PEPPY	0.409	0.037	11.198	0.000
AFRAID	0.237	0.043	5.526	0.000
FRUST	0.324	0.034	9.653	0.000
HLTHWRY	0.333	0.042	7.868	0.000

R-SQUARE

Observed Variable	Estimate	S.E.	Est./S.E.	Two-Tailed P-Value
DIFFEAS	0.725	0.030	24.253	0.000
SLORACT	0.442	0.040	11.037	0.000
CONFUSED	0.708	0.030	23.404	0.000
FORGET	0.647	0.036	18.218	0.000
DIFFCONC	0.701	0.030	23.075	0.000
TIRED	0.765	0.048	15.866	0.000
ENRGTIC	0.454	0.041	11.147	0.000
WOROUT	0.721	0.033	21.834	0.000
PEPPY	0.591	0.037	16.178	0.000
AFRAID	0.763	0.043	17.800	0.000
FRUST	0.676	0.034	20.104	0.000
HLTHWRY	0.667	0.042	15.730	0.000

The standardized estimates in M*plus* are similar to the estimates provided in the completely standardized solution in LISREL given below:

Completely Standardized Solution

LAMBDA-X

	COGNITN	VITALITY	DISWORR	QOL
	--------	--------	--------	--------
DIFFEAS	0.655	- -	- -	0.543
SLORACT	0.444	- -	- -	0.495
CONFUSED	0.629	- -	- -	0.559
FORGET	0.672	- -	- -	0.442
DIFFCONC	0.615	- -	- -	0.568
TIRED	- -	0.562	- -	0.670
ENRGTIC	- -	0.320	- -	0.593
WOROUT	- -	0.425	- -	0.735

PEPPY	- -	0.334	- -	0.693
AFRAID	- -	- -	0.614	0.621
FRUST	- -	- -	0.418	0.708
HLTHWRY	- -	- -	0.572	0.583

PHI

Note: This matrix is diagonal.

COGNITN	VITALITY	DISWORR	QOL
--------	--------	--------	--------
1.000	1.000	1.000	1.000

THETA-DELTA

DIFFEAS	SLORACT	CONFUSED	FORGET	DIFFCONC	TIRED
--------	--------	--------	--------	--------	--------
0.275	0.558	0.292	0.353	0.299	0.235

THETA-DELTA

ENRGTIC	WOROUT	PEPPY	AFRAID	FRUST	HLTHWRY
--------	--------	--------	--------	--------	--------
0.546	0.279	0.409	0.237	0.324	0.333

MODEL COMPARISONS BETWEEN THE SECOND-ORDER AND BIFACTOR MODELS

If both the second-order model and the bifactor model are theoretically reasonable, they can be compared using information criteria as described with other model comparisons. Table 8.1 provides the model fit information from M*plus* for both the second-order and bifactor models, including the information criteria. As seen in Table 8.1, the AIC is smaller for the bifactor model whereas the BIC is smaller for the second-order model. The information criteria in LISREL also demonstrated this discrepancy between the two information criteria. The sample size adjusted BIC (nBIC) criterion in M*plus* is also provided in Table 8.1, which agrees with the AIC in the selection of the bifactor model. Hence, a researcher would select the bifactor model when using the information criteria comparisons. Table 8.2 presents the unstandardized factor loading estimates with corresponding standard errors and Table 8.3 presents the standardized factor loading estimates for the second-order and bifactor models. Alternate reference indicators were used when estimating the unstandardized loadings (and corresponding standard errors) for the *DIFFEAS*, *TIRED*, and *AFRAID* indicators in the second-order factor model for Table 8.2 results.

Table 8.1: Model Fit Summary Table for the Second-order and Bifactor Models

Model	χ^2	df	CFI	NNFI/ TLI	RMSEA (90% CI) p-value	SRMR	AIC	BIC	nBIC
Second-order model	93.84*	51	.99	.98	.05 (.03, .06) p > .05	.03	9393.84	9501.81	9416.13
Bifactor model	61.36*	42	.99	.99	.03 (.01, .05) p > .05	.02	9379.36	9523.33	9409.09

Note: * p < .05. Values were taken from M*plus* output.

Table 8.2: Unstandardized Factor Loadings (and Corresponding Standard Errors) for the Second-order and Bifactor Models for Quality of Life

	Second-Order QOL Factor Model				*QOL Bifactor Model*			
Item/Factor	Cognition	Vitality	Disease Worry	QOL	Cognition	Vitality	Disease Worry	QOL
diffeas	.95 (.05)*				.54 (.04)*			.44 (.04)*
sloract	.76 (.05)*				.35 (.04)*			.40 (.04)*
confused	1.05 (.05)*				.55 (.04)*			.49 (.05)*
forget	.95 (.05)*				.57 (.04)*			.37 (.05)*
diffconc	1.05 (.05)*				.54 (.04)*			.50 (.05)*
tired		.96 (.05)*				.43 (.04)*		.51 (.04)*
enrgtic		.93 (.06)*				.28 (.06)*		.52 (.05)*
worout		1.04 (.05)*				.33 (.05)*		.57 (.04)*
peppy		.97 (.06)*				.27 (.06)*		.55 (.04)*
afraid			1.06 (.06)*				.59 (.05)*	.60 (.05)*
frust			.93 (.05)*				.39 (.05)*	.65 (.05)*
hlthwry			.95 (.05)*				.55 (.05)*	.57 (.05)*
Cognition				.47 (.04)*				--
Vitality				.53 (.04)*				--
Disease Worry				.63 (.05)*				--

Note: *p < .05. In order to estimate loadings and standard errors for diffeas, tired, and afraid items that served as reference indicators in second-order model, the confused, worout, and hlthwry items were subsequently used as reference indicators and the model was re-estimated.

Table 8.3: Standardized Factor Loadings for the Second-order and Bifactor Models for Quality of Life

Item/Factor	Second-Order QOL Factor Model				QOL Bifactor Model			
	Cognition	Vitality	Disease Worry	QOL	Cognition	Vitality	Disease Worry	QOL
diffeas	.85				.66			.54
sloract	.67				.44			.50
confused	.84				.63			.56
forget	.79				.67			.44
diffconc	.84				.62			.57
tired		.84				.56		.67
enrgtic		.68				.32		.59
worout		.86				.43		.74
peppy		.77				.33		.69
afraid			.85				.61	.62
frust			.83				.42	.71
hlthwry			.81				.57	.58
Cognition				.67				--
Vitality				.83				--
Disease Worry				.76				--

A benefit of using the bifactor model is that different omega indices can be calculated that quantify the proportion of total variance in the responses to all of the items due to all of the factors combined and due to each of the factors individually (Reise et al., 2013). These omega indices do assume that the factors are uncorrelated in the bifactor model. Once again, theoretical interpretation of the bifactor model structure is crucial and has serious implications in scientific research and practice.

It must be noted that the bifactor model has been shown to suffer from *overfitting*, which may be the reason that it tends to fit the data more adequately than other models to which it is compared (e.g., second-order factor models). For instance, Bonifay and Cai (2017) found that the bifactor model demonstrated better fit to data sets with random patterns than competing models. Likewise, Reise, Kim, Mansolf, and Widaman (2016) demonstrated that the bifactor model is better at modeling unlikely item response patterns in the data than comparison models.

SUMMARY

The second-order factor model and the bifactor model permit alternative specifications of the factors underlying responses to items on surveys and instruments. The specifications of the second-order factor model and the bifactor model imply very different relationships among the factors as well as between the factors and item responses. Although it was not included in the model comparisons subsection, confirmatory factor analysis models with all factors intercorrelating can also be a model that may be compared to the second-order factor model and the bifactor model. For instance, a quality of life CFA model where the three factors (i.e., *Cognition*, *Vitality*, and *Disease Worry*) were all correlated could be included as a comparison model. It is important to note, however, that the three-factor correlated quality of life CFA model would fit that data exactly the same as the higher-order factor model. The reason for this is because the structural-level of both the second-order and three-factor correlated models is just-identified. Thus, model comparisons with only three first-order factors will limit comparisons to correlated CFA models. With four or more first-order factors, model comparisons between second-order factor models and correlated CFA models will result in different degrees of freedom between the two models, allowing for more meaningful model comparisons. In fact, the second-order factor model will be nested in the correlated CFA model with four or more first-order factors.

Our introduction to both second-order and bifactor models in this chapter discussed issues relating to specification, identification, estimation, testing and modification, and model interpretation. Once again, we must mention the importance of theoretical justifications when examining the second-order factor model and/or the bifactor model. It is also important to caution users again about the potential overfitting problems associated with the bifactor model. We hope that our discussion in this chapter has provided you with a basic overview about the potential uses for the second-order and bifactor models, which should support various theoretical perspectives.

EXERCISES

The data in ***Ch8Ex. LSF*** and ***Ch8Ex.dat*** contain raw data for 12 variables (Y1–Y12) and are provided for example 5.6 on the M*plus* website. It is hypothesized that items Y1–Y3 load on F1 (Factor 1); items Y4–Y6 load on F2 (Factor 2); items Y7–Y9 load on F3 (Factor 3); and items Y10–Y12 load on F4 (Factor 4).

Using this data set, answer the following questions:

1. How many distinct or non-redundant observations are in the sample covariance matrix?
2. A researcher wants to compare the following three models using these data:
   a. A four-factor CFA model with all factors (F1–F4) correlating.
   b. A second-order factor model where F1–F4 are first-order factors and F5 (Factor 5) is the higher-order factor on which F1–F4 load.

c. A bifactor model where F1–F4 are the group factors and F5 (Factor 5) is the general factor.

For each of the three models, calculate the number of parameters to be estimated and degrees of freedom for each of the models.

3. Which of the three models is the most parameterized model?
4. Which of the three models is the least parameterized model?
5. Write three programs (either in LISREL or M*plus*) to estimate and compare the three models described above.

REFERENCES

Bonifay, W., & Cai, L. (2017). On the complexity of item response theory models. *Multivariate Behavioral Research*, 52(4), 465–484.

Chen, F. F., West, S. G., & Sousa, K. H. (2006). A comparison of bifactor and second-order models of quality of life. *Multivariate Behavioral Research*, 41(2), 189–255.

Holzinger, K. J., & Swineford, F. (1937). The bi-factor method. *Psychometrika*, 2(1), 41–54.

Reise, S. P. (2012). The rediscovery of bifactor measurement models. *Multivariate Behavioral Research*, 47(5), 667–696.

Reise, S. P., Bonifay, W. E., & Haviland, M. G. (2013). Scoring and modeling psychological measures in the presence of multidimensionality. *Journal of Personality Assessment*, 95(2), 129–140.

Reise, S. P., Kim, D. S., Mansolf, M., & Widaman, K. (2016). Is the bifactor model a better model or is it just better at modeling implausible responses? Application of iteratively reweighted least squares to the Rosenberg self-esteem scale. *Multivariate Behavioral Research*, 51(6), 818–838.

Stewart, A. L., & Ware, J. E. (1992). *Measuring functioning and well-being: The medical outcome study approach*. Durham, NC: Duke University Press.

Chapter 9

MULTIPLE GROUP (SAMPLE) MODELS

BRIEF SUMMARY OF MULTIPLE GROUP MODELING

A popular use of multiple group modeling is to test for measurement invariance. *Measurement invariance* implies that two or more groups have the same measurement or CFA model. That is, the latent construct is measured similarly across the two groups being compared, which provides support for construct validity. Once a researcher establishes that the latent variable constructs are measured the same across groups, a comparison of the groups in terms of structural model parameters can then be conducted. Byrne and Sunita (2006) as well as Vandenberg and Lance (2000) provide a step-by-step approach for examining measurement invariance in SEM. A unified approach to multi-group modeling is explained in Marcoulides and Schumacker (2001).

When testing for parameter equality across groups, it is possible for a measurement model or specific SEM model parameters to be different for each group. Differences across groups can actually lead to meaningful discussions of and understandings about the differences found across groups. Also, a misconception in multiple group modeling is that all paths in the model have to be the same. Multiple group modeling permits the testing of group differences in the specified model without requiring all parameters in the model to be equal (Jöreskog & Sörbom, 1993).

When testing for the equality of model parameters across groups, it is generally referred to as invariance or equivalence testing. Multiple group SEM procedures commonly go through a sequential process, beginning with a baseline multiple group model without equality constraints which is compared to models with

DOI: 10.4324/9781003044017-9

increasingly restrictive equality constraints for model parameters across groups. The $\Delta\chi^2$ test is frequently used when comparing the multiple group models sequentially being estimated because the models with equality constraints are nested within the baseline multiple group model. When the $\Delta\chi^2$ test is significant, the equality constraints are too restrictive, suggesting that one or more of the model parameters are not equal or non-invariant across groups. Identifying non-invariant model parameters can then be done using modification indices. Multiple group path analysis is first illustrated using an example, followed by multiple group CFA.

MULTIPLE GROUP PATH ANALYSIS MODEL

The multiple group approach can be used to test a theoretical path model for differences in parameter estimates across samples of data. A theoretical observed variable path analysis model is diagrammed in Figure 9.1, which is adapted from Howard and Maxwell's (1982) study investigating the relationship between course grades and course evaluations in a research methods course taught in the Psychology department. This model was introduced in the Chapter 1 Exercises. The model hypothesizes that a student's motivation to take the course (*Motivation*) directly impacts student progress in the class (*Progress*), the student's grade in the class (*Grade*), and the student's satisfaction with the class (*Satisfaction*). The model also hypothesizes that student progress and the student's grade in the class directly affect satisfaction with the class. The data set contains simulated responses to the four variables (*Progress*, *Grade*, *Satisfaction*, and *Motivation*). The data set also includes a grouping variable (*Group*) indicating whether or not the student is a Psychology major ($N = 200$) or a non-major ($N = 150$), where 1 = major and 2 = non-major. The data were split into two files for use in LISREL, one for the majors and one for the non-majors (*MAJORS.LSF* and *NONMAJORS.LSF*, respectively). The whole data set is contained in the *MGPATH.DAT* file for use in M*plus*. All of the data files are on the book website.

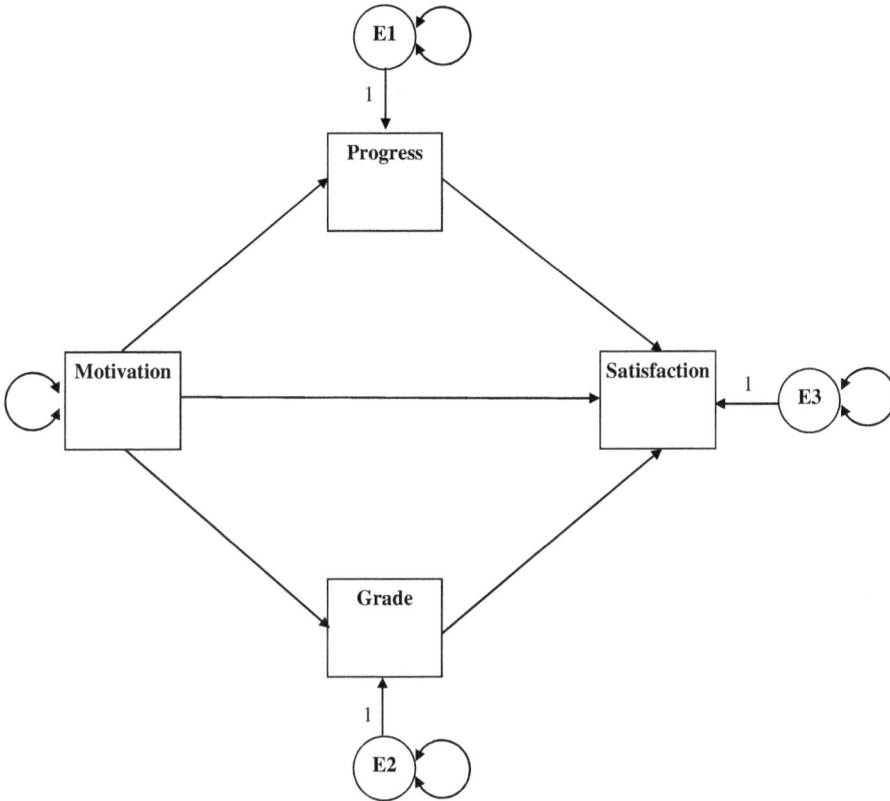

■ **Figure 9.1:**   OBSERVED VARIABLE PATH MODEL FOR MULTIPLE GROUP ANALYSIS

For this example, we will compare the parameter estimates in the observed variable path model illustrated in Figure 9.1 across the Psychology majors and non-majors to determine whether they differ significantly. Specifically, we will compare the 5 direct effects across the two groups to determine if the relationships between the variables are equal or invariant. Prior to conducting multiple group analysis, it is important to first make sure that the models fit the data adequately in each group separately. The reason this is important is because equality constraints will be imposed on the model parameters across groups to test for invariance, which will reduce the number of parameters to be estimated. Thus, model fit will not improve with multiple group models if the models do not fit the data adequately in each group separately in the beginning.

Model Identification in Separate Groups

For this path model, there are $p(p + 1)/2 = 4(5)/2 = 10$ distinct or non-redundant elements in the sample covariance matrix for each group. The parameters to be estimated include 1 exogenous variable's variance (*Motivation*); 3 error variances (for *Progress*, *Satisfaction*, and *Grade*); and 5 direct effects. Thus, 9 parameters are to be estimated. This is an over-identified model with 1 degree of freedom (10 – 9 =1).

Model Estimation in Separate Groups

The SIMPLIS code to fit the data in the Psychology major and non-major groups separately is provided below, respectively:

```
PATH MODEL FOR MAJORS
RAW DATA FROM FILE MAJORS.LSF
RELATIONSHIPS:
SATISFAC = PROGRESS MOTIVAT GRADE
PROGRESS = MOTIVAT
GRADE = MOTIVAT
OPTIONS: SC MI
PATH DIAGRAM
END OF PROBLEM

PATH MODEL FOR NONMAJORS
RAW DATA FROM FILE NONMAJORS.LSF
RELATIONSHIPS:
SATISFAC = PROGRESS MOTIVAT GRADE
PROGRESS = MOTIVAT
GRADE = MOTIVAT
OPTIONS: SC MI
PATH DIAGRAM
END OF PROBLEM
```

The M*plus* code for testing the models in each group is provided below:

```
TITLE:      PATH MODEL FOR MAJORS

DATA:       FILE IS MGPATH.DAT;

VARIABLE:   NAMES ARE PROGRESS GRADE SATISFAC MOTIVAT GROUP;
            USEVAR ARE PROGRESS GRADE SATISFAC MOTIVAT;
            USEOBS ARE GROUP EQ 1;

MODEL:      SATISFAC ON PROGRESS MOTIVAT GRADE;
            PROGRESS ON MOTIVAT;
            GRADE ON MOTIVAT;

OUTPUT:     STDYX MODINDICES(3.84);
```

```
TITLE:       PATH MODEL FOR NONMAJORS

DATA:        FILE IS MGPATH.DAT;

VARIABLE:    NAMES ARE PROGRESS GRADE SATISFAC MOTIVAT GROUP;
             USEVAR ARE PROGRESS GRADE SATISFAC MOTIVAT;
             USEOBS ARE GROUP EQ 2;

MODEL:       SATISFAC ON PROGRESS MOTIVAT GRADE;
             PROGRESS ON MOTIVAT;
             GRADE ON MOTIVAT;

OUTPUT:      STDYX MODINDICES(3.84);
```

You will notice in the M*plus* code that the **USEVAR** (short for **USEVARIABLES**) option in the VARIABLE command is used to avoid an error message that we are not using the GROUP variable in the model because it is the group identifying variable. You will also notice in the M*plus* code that the **USEOBS** (short for **USEOBSERVATIONS**) option in the VARIABLE command is used to select the specific groups. Specifically, USEOBS ARE GROUP EQ 1 selects cases to include in the analysis if the *Group* variable equals 1, which selects data for the Psychology majors group to be analyzed. ML estimation is used as the estimator in this example.

Model Testing in Separate Groups

The model fit the data well in each group separately. Table 9.1 provides model fit information from LISREL and M*plus* for the model in each of the groups. If the models did not fit the data well, researchers can attempt to improve model fit in each group separately with use of the modification indices, expected parameter change, and the residuals. It is important to note that you can still proceed with multiple group modeling procedures even if different parameters are estimated in different groups. For instance, suppose the model did not fit well in the non-majors group and the modification indices suggested adding a direct effect from *Progress* to *Grade*. This direct effect was originally hypothesized by Howard and Maxwell (1982) and would result in a just-identified model. Thus, if adding this direct effect for the non-majors group, the model would fit the data perfectly. Also suppose that the model fit the data well in the majors group and no model modifications were suggested to improve model fit. You could still continue to test the equality of the same model parameters across the groups, which would be the original 5 direct effects. But, you would not be able to test for equality of the direct effect from *Progress* to *Grade* that was only added to the non-majors' model.

Table 9.1: Model Fit Information for the Path Model for Motivation in Each Student Group

Software/Group	χ^2	df	CFI	NNFI/ TLI	RMSEA (90% CI) p-value	SRMR
LISREL						
Psychology majors	.304	1	1.00	1.00	.00 (.00, .15) p > .05	.005
Non-majors	.027	1	1.00	1.01	.00 (.00, .11) p > .05	.003
M*plus*						
Psychology majors	.306	1	1.00	1.00	.00 (.00, .15) p > .05	.010
Non-majors	.027	1	1.00	1.00	.00 (.00, .11) p > .05	.012

Note: N = 200 for Psychology majors; *N* = 150 for Non-majors.

Model Identification of the Baseline Multiple Group Model – With No Equality Constraints

Since the models fit the data adequately in each of the groups separately, the next step is to estimate a baseline multiple group model with no parameter equality constraints imposed. This model will then serve as the *baseline model* which is compared to the model with parameters constrained to be equal across groups. The constrained multiple group model is nested in the baseline model. Thus, a $\Delta\chi^2$ can be calculated to determine whether adding the equality constraints across groups (which reduces the number of parameters to estimate in the model) decreases model fit significantly. If not, this supports the conclusion that the model parameters are equal or invariant across groups. If the $\Delta\chi^2$ is statistically significant, this suggests that one or more of the parameters constrained to equality across groups is unequal or non-invariant and should be freely estimated in each of the groups. A representation of the baseline multiple group path model is presented in Figure 9.2.

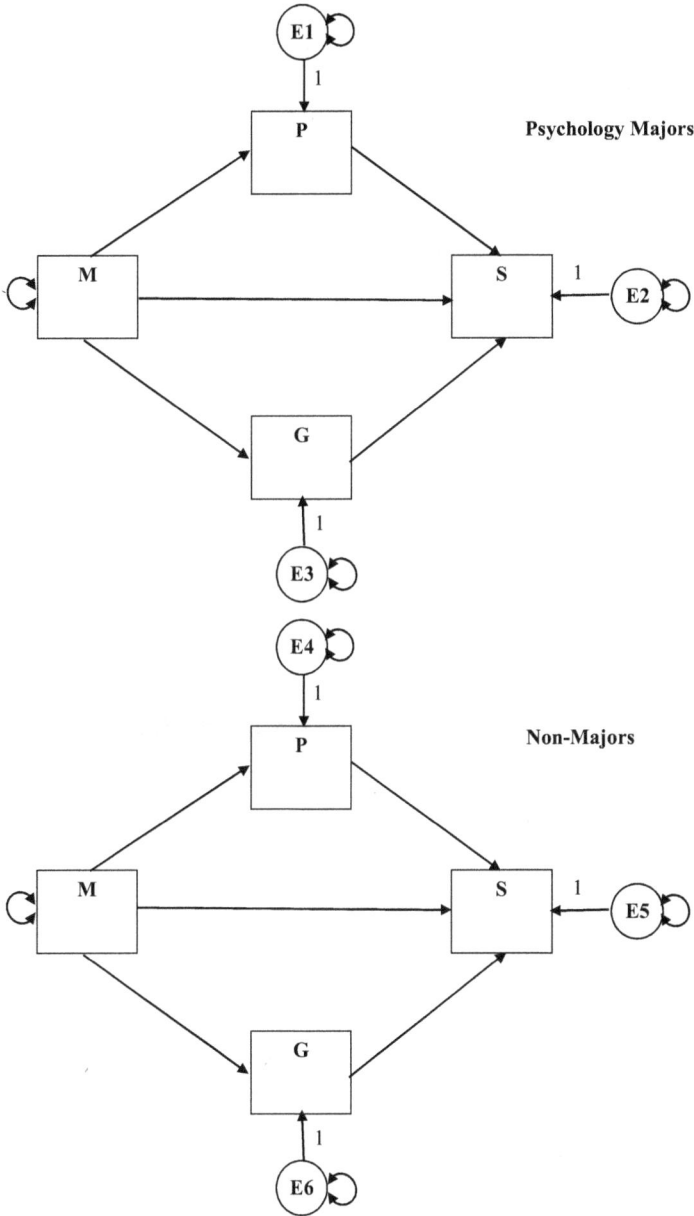

Figure 9.2: BASELINE MULTIPLE GROUP PATH MODEL OF MOTIVATION

Model identification with multiple group models is similar to determining model identification as described for previous models with a slight change. Specifically, multiple group models estimate model parameters separately for each group, but simultaneously. For the baseline multiple group path model, we first count the number of distinct or non-redundant elements in the sample covariance matrix separately for each group and then sum them for a total number of distinct elements. As previously calculated, there are $p(p + 1)/2 = 4(5)/2 = 10$ distinct or nonredundant elements in the sample covariance matrix for each group. Summing up the distinct elements for each group is 20 (10 + 10 = 20).

The next step is to calculate the number of parameters to be estimated in the model for each group separately and then sum them for a total number of parameters. As previously determined, the parameters to be estimated in the model for each group include 1 exogenous variable's variance (*Motivation*); 3 error variances (for *Progress*, *Satisfaction*, and *Grade*); and 5 direct effects. Thus, 9 parameters are to be estimated in the model for each group. Summing up the number of parameters for each group is 18 (9 + 9 = 18).

To determine the degrees of freedom for the baseline model, the total number of parameters to estimate (18) is subtracted from the total number of distinct elements (20). Thus, the degrees of freedom for the baseline multiple group model is 2 (*20 – 18 = 2*). You may notice that the *df* for the baseline model is also the sum of the *df* for the path model in each group (1 + 1). The multiple group baseline model is an over-identified model with 2 *df*.

Model Estimation of the Baseline Multiple Group Model – With No Equality Constraints

The SIMPLIS code to estimate the baseline multiple group path model is provided below:

```
GROUP 1: BASELINE MULTIPLE GROUP PATH MODEL
RAW DATA FROM FILE MAJORS.LSF
RELATIONSHIPS:
SATISFAC = PROGRESS MOTIVAT GRADE
PROGRESS = MOTIVAT
GRADE = MOTIVAT
GROUP 2: BASELINE MULTIPLE GROUP PATH MODEL
RAW DATA FROM FILE NONMAJORS.LSF
RELATIONSHIPS:
SATISFAC = PROGRESS MOTIVAT GRADE
PROGRESS = MOTIVAT
GRADE = MOTIVAT
SET THE ERROR VARIANCE OF SATISFAC FREE
SET THE ERROR VARIANCE OF PROGRESS FREE
SET THE ERROR VARIANCE OF GRADE FREE
OPTIONS: SC
END OF PROBLEM
```

In LISREL, you specify a multiple group model with **GROUP 1** and **GROUP 2** headings. Different data sets are read in for each group in the code provided. For both groups, the same model is estimated under the RELATIONSHIPS heading because no equality constraints are imposed in the baseline model. You will notice for GROUP 2 that the error variances for *Satisfaction*, *Progress*, and *Grade* are set free or freely estimated. The default with multiple group analysis in LISREL is that the error variances are assumed to be equal across groups. Thus, unless otherwise specified, the error variances for the endogenous variables (*Satisfaction*, *Progress*, and *Grade*) will be constrained to equality across groups. The three statements setting the endogenous variables' error variances to be free (e.g., SET THE ERROR VARIANCE OF GRADE FREE) will release the default equality constraints.

The M*plus* code for testing the baseline multiple group path model is provided below:

```
TITLE:      BASELINE MULTIPLE GROUP PATH MODEL

DATA:       FILE IS MGPATH.DAT;

VARIABLE:   NAMES ARE PROGRESS GRADE SATISFAC MOTIVAT GROUP;
            USEVAR ARE PROGRESS GRADE SATISFAC MOTIVAT;
            GROUPING IS GROUP (1=MAJORS 2=NONMAJORS);

MODEL:      SATISFAC ON PROGRESS MOTIVAT GRADE;
            PROGRESS ON MOTIVAT;
            GRADE ON MOTIVAT;

OUTPUT:     STDYX;
```

For the baseline multiple group path model code in M*plus*, you will notice that the **GROUPING IS GROUP** option in the VARIABLE command is used. The **GROUPING IS** option indicates that a multiple group model is to be estimated in M*plus* and **GROUP** identifies the variable used as the grouping identifier. You will notice that the groups are also defined when using this option **(1=MAJORS 2= NONMAJORS)**.

Model Testing of the Baseline Multiple Group Model – With No Equality Constraints

The baseline model fit the data well [LISREL: $\chi^2(2) = .331$, $p > .05$; CFI = 1.00; NNFI/TLI = 1.01; RMSEA = .00 (90% CI: .00, .08) with $p > .05$; Group 1 SRMR = .005; Group 2 SRMR = .003] and [M*plus*: $\chi^2(2) = .332$, $p > .05$; CFI = 1.00; NNFI/TLI = 1.00; RMSEA = .00 (90% CI: .00, .08) with $p > .05$; SRMR = .011]. LISREL does not provide an overall SRMR, but provides one for each group whereas M*plus* calculates an overall SRMR.

Both LISREL and M*plus* indicate the chi-square contribution from each of the groups to the baseline chi-square test statistic. For instance, the "Group Goodness of Fit Statistics" for Groups 1 and 2 in LISREL output is provided below, respectively:

```
             Group Goodness of Fit Statistics

Contribution to Chi-Square                        0.305
Percentage Contribution to Chi-Square             91.956

Root Mean Square Residual (RMR)                   0.0212
Standardized RMR                                  0.00491
Goodness of Fit Index (GFI)                       0.999

             Group Goodness of Fit Statistics

Contribution to Chi-Square                        0.0267
Percentage Contribution to Chi-Square             8.044

Root Mean Square Residual (RMR)                   0.00775
Standardized RMR                                  0.00336
Goodness of Fit Index (GFI)                       1.00
```

M*plus* provides the following information about the chi-square contribution from each group in the Model Fit Information section:

```
Chi-Square Contribution From Each Group

    MAJORS                      0.306
    NONMAJORS                   0.027
```

You will notice that the sum of each group's chi-square value equals the baseline model's chi-square value.

Model Identification of the Multiple Group Model – With Equality Constraints

After estimating the baseline multiple group model, the next step is to constrain the model parameters to equality across the groups. A representation of these model constraints is presented in Figure 9.3. As shown in Figure 9.3, the 5 direct effects are marked with the same letters across the two groups to indicate which parameters are constrained to be equal. For instance, the direct effect from *Motivation* to *Progress* (M → P) is parameter "**a**" and is constrained to be equal for both Psychology Majors and for Non-Majors. It must be noted that it is the unstandardized parameter estimates that are constrained to be equal across groups and not the standardized estimates. The reason for this is that it is important to account for the variances associated with variables when testing for equality across groups. That is, standardized data assume the same variances for all variables *and* for all groups, which is highly unlikely. Thus, all of the 5 unstandardized direct effects (**a–e**) will be constrained to be equal to test whether the relationships among variables in the model are the same for Psychology Majors and Non-Majors.

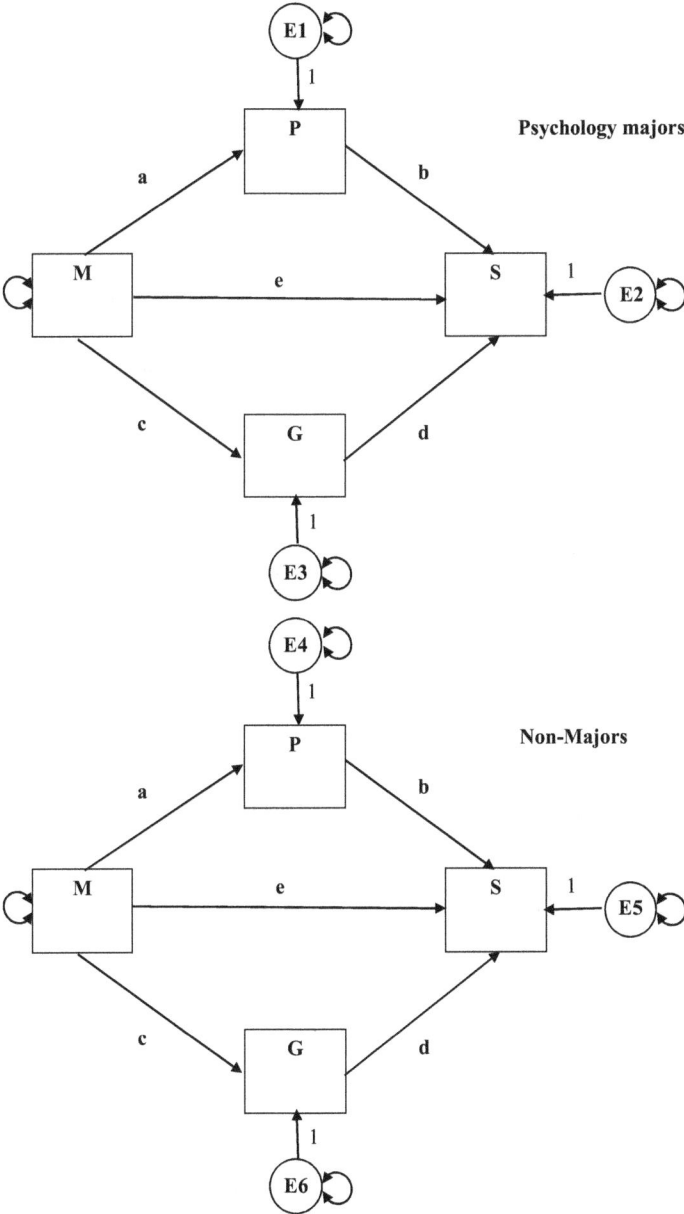

Figure 9.3: MULTIPLE GROUP PATH MODEL OF MOTIVATION WITH EQUALITY CONSTRAINTS

The number of distinct or non-redundant elements in the sample covariance matrix for each group was summed previously for the baseline model and is used to calculate *df* for the constrained multiple group path model. Recall that the total number of distinct elements is 20 (10 + 10 = 20). The next step is to calculate the number of parameters to be estimated in the constrained multiple group path model. Using Figure 9.3, the variance of *Motivation* in each group will be freely estimated (2 parameters); the error variances for *Progress*, *Satisfaction*, and *Grade* in each group will be freely estimated (6 parameters); and the 5 direct effects (**a–e**) that are constrained to equality across groups will be estimated (5 parameters). Thus, there are 13 parameters to estimate in the constrained multiple group path model. The *df* for the constrained multiple group path model are 7 (20 – 13 = 7). You will notice that the difference in *df* between the constrained multiple group (*df* = 7) and baseline (*df* =2) models is equal to the number of constrained parameters, which are the 5 direct effects (7 – 2 = 5)

Model Estimation of the Multiple Group Model – With Equality Constraints

The SIMPLIS code to estimate the constrained multiple group path model is provided below:

```
GROUP 1: CONSTRAINED MULTIPLE GROUP PATH MODEL
RAW DATA FROM FILE MAJORS.LSF
RELATIONSHIPS:
SATISFAC = PROGRESS MOTIVAT GRADE
PROGRESS = MOTIVAT
GRADE = MOTIVAT
GROUP 2: CONSTRAINED MULTIPLE GROUP PATH MODEL
RAW DATA FROM FILE NONMAJORS.LSF
RELATIONSHIPS:
SET THE ERROR VARIANCE OF SATISFAC FREE
SET THE ERROR VARIANCE OF PROGRESS FREE
SET THE ERROR VARIANCE OF GRADE FREE
OPTIONS: SC MI
END OF PROBLEM
```

You will notice that the relationships among observed variables are specified for Group 1, but not for Group 2. By specifying the direct effects among the observed variables in Group 1 only, LISREL will constrain those effects to equality across groups. You will notice that we are not constraining error variances to equality across groups. Thus, the three statements for error variances of the endogenous variables are set to be free and not constrained to equality across groups.

The M*plus* code for testing the constrained multiple group path model is provided below:

```
TITLE:       CONSTRAINED MULTIPLE GROUP PATH MODEL

DATA:        FILE IS MGPATH.DAT;

VARIABLE:    NAMES ARE PROGRESS GRADE SATISFAC MOTIVAT GROUP;
             USEVAR ARE PROGRESS GRADE SATISFAC MOTIVAT;
             GROUPING IS GROUP (1=MAJORS 2=NONMAJORS);

MODEL:       SATISFAC ON PROGRESS (1)
                         MOTIVAT  (2)
                         GRADE    (3);
               PROGRESS ON MOTIVAT (4);
                  GRADE ON MOTIVAT (5);

OUTPUT:      STDYX MODINDICES(3.84);
```

In the M*plus* code, you will notice numbers 1–5 in parenthesis after statements in the MODEL command. The numbers in parenthesis in the multiple group model impose equality constraints on that particular parameter across groups. Thus, the 5 direct effects will be constrained to equality across groups using the M*plus* code above. There can only be one equality constraint or number in parenthesis per line in the code. Otherwise, you will receive an error message about the format of the equality constraints. The numbers used in the parenthesis are arbitrary, meaning that you could have used 101–105 instead of 1–5, respectively. It must also be noted that although the numbers themselves are arbitrary, you do not want to put the same number for each equality constraint because that would constrain all direct effects to be the same within and across each group.

Model Testing and Modification of the Multiple Group Model – With Equality Constraints

The constrained multiple group path model did not fit the data well [LISREL: $\chi^2(7) = 160.945$, $p < .05$; CFI = .863; NNFI/TLI = .766; RMSEA = .356 (90% CI: .309, .404) with $p < .05$; Group 1 SRMR = .157; Group 2 SRMR = .343] and [M*plus*: $\chi^2(7) = 161.983$, $p < .05$; CFI = .905; NNFI/TLI = .836; RMSEA = .356 (90% CI: .309, .404) with $p < .05$; SRMR = 1.467]. This suggests that one or more of the unstandardized parameter estimates are not equal across groups. As mentioned previously, a $\Delta\chi^2$ test can be calculated between the baseline and constrained models to determine if the equality constraints resulted in a significant loss in fit. The $\Delta\chi^2$ test indicated a significant difference between baseline and constrained multiple group models [LISREL: $\Delta\chi^2 = 160.614$, $\Delta df = 5$, $p < .05$] and [M*plus*: $\Delta\chi^2 = 161.651$, $\Delta df = 5$, $p < .05$].

In order to determine which equality constraints should be freed or released, the modification indices can be used. The modification indices indicate how much the chi-square would drop if a parameter was freely estimated in the model. This applies to constrained parameters as well because the modification indices will suggest which constrained parameter should be estimated freely in each group instead of constrained to equality. In LISREL and M*plus*, modification indices are provided for both groups. The modification indices can also suggest adding parameters that are not constrained. Hence, it is important that you examine the modification indices only for *constrained parameters*.

The modification indices in LISREL that involve equality constraints across Group 1 and Group 2 are provided below, respectively:

```
GROUP 1: CONSTRAINED MULTIPLE GROUP MODEL

Modification Indices and Expected Change

        The Modification Indices Suggest to Add the
Path to       from            Decrease in Chi-Square     New Estimate
    PROGRESS      PROGRESS             85.8              0.76 IN GROUP 1
    PROGRESS         GRADE              8.1              0.06 IN GROUP 1
    PROGRESS      SATISFAC             38.7              0.11 IN GROUP 1
    SATISFAC      PROGRESS             13.5              1.53 IN GROUP 1
    SATISFAC         GRADE             10.6              0.83 IN GROUP 1
    SATISFAC      SATISFAC             22.5              0.12 IN GROUP 1

        The Modification Indices Suggest to Add the
Path to       from            Decrease in Chi-Square     New Estimate
    PROGRESS       MOTIVAT             85.8              1.33 IN GROUP 1
    SATISFAC       MOTIVAT             36.1              1.53 IN GROUP 1

GROUP 2: CONSTRAINED MULTIPLE GROUP PATH MODEL

Modification Indices and Expected Change

        The Modification Indices Suggest to Add the
Path to       from            Decrease in Chi-Square     New Estimate
    PROGRESS      PROGRESS             85.8             -0.76 IN GROUP 2
    PROGRESS         GRADE             29.5              0.17 IN GROUP 2
    PROGRESS      SATISFAC             65.7              0.16 IN GROUP 2
    SATISFAC      PROGRESS             12.1              1.19 IN GROUP 2
    SATISFAC         GRADE             12.0              0.69 IN GROUP 2
    SATISFAC      SATISFAC             22.5             -0.12 IN GROUP 2
```

```
        The Modification Indices Suggest to Add the
Path to      from             Decrease in Chi-Square    New Estimate
   PROGRESS      MOTIVAT            94.1                 0.50 IN GROUP 2
   SATISFAC      MOTIVAT            41.1                 0.89 IN GROUP 2
```

The modification indices in M*plus* that involve equality constraints across Group 1 and Group 2 are provided below:

	M.I.	E.P.C.	Std E.P.C.	StdYX E.P.C.
Group MAJORS				
ON Statements				
PROGRESS ON PROGRESS	86.393	0.762	0.762	0.762
PROGRESS ON GRADE	8.209	0.062	0.062	0.124
PROGRESS ON SATISFAC	38.985	0.109	0.109	0.379
PROGRESS ON MOTIVAT	86.394	0.236	0.236	0.183
SATISFAC ON PROGRESS	13.622	0.165	0.165	0.048
SATISFAC ON GRADE	10.678	0.056	0.056	0.032
SATISFAC ON SATISFAC	22.627	0.115	0.115	0.115
SATISFAC ON MOTIVAT	36.353	0.263	0.263	0.059
Group NONMAJORS				
ON Statements				
PROGRESS ON PROGRESS	86.395	-0.762	-0.762	-0.762
PROGRESS ON GRADE	29.695	-0.169	-0.169	-0.304
PROGRESS ON SATISFAC	66.131	-0.164	-0.164	-0.521
PROGRESS ON MOTIVAT	86.395	-0.598	-0.598	-0.416
SATISFAC ON PROGRESS	13.622	-0.126	-0.126	-0.040
SATISFAC ON GRADE	10.678	-0.079	-0.079	-0.045
SATISFAC ON SATISFAC	22.627	-0.115	-0.115	-0.115
SATISFAC ON MOTIVAT	36.353	-0.380	-0.380	-0.083

Modification indices can be associated with all fixed and constrained parameters, which may not make sense and could lead to models that do not converge. Again, it is important that you examine the modification indices associated with parameter equality constraints with multiple group models. In this case, the five direct effects are of interest because they were constrained to equality. Both LISREL and M*plus* suggest that the largest chi-square decrease (by about 86 points) would result when releasing the equality constraint of the direct effect from *Motivation* to *Progress*.

Based on the modification indices, the direct effect from *Motivation* to *Progress* was allowed to be freely estimated in each group and the model was re-estimated. The SIMPLIS code with this path freely estimated in both groups is provided below:

```
GROUP 1: PARTIAL INVARIANCE MULTIPLE GROUP MODEL
RAW DATA FROM FILE MAJORS.LSF
RELATIONSHIPS:
SATISFAC = PROGRESS MOTIVAT GRADE
PROGRESS = MOTIVAT
GRADE = MOTIVAT
GROUP 2: PARTIAL INVARIANCE MULTIPLE GROUP MODEL
RAW DATA FROM FILE NONMAJORS.LSF
RELATIONSHIPS:
PROGRESS = MOTIVAT
SET THE ERROR VARIANCE OF SATISFAC FREE
SET THE ERROR VARIANCE OF PROGRESS FREE
SET THE ERROR VARIANCE OF GRADE FREE
OPTIONS: SC MI
END OF PROBLEM
```

You will notice that the relationship between *Motivation* and *Progress* (PROGRESS = MOTIVAT) is specified in Group 2. This will estimate this direct effect freely across the two groups and release the equality constraint.

The M*plus* code to freely estimate the direct effect from *Motivation* to *Progress* is provided below:

```
TITLE:      PARTIAL INVARIANCE MULTIPLE GROUP PATH MODEL

DATA:       FILE IS MGPATH.DAT;

VARIABLE:   NAMES ARE PROGRESS GRADE SATISFAC MOTIVAT GROUP;
            USEVAR ARE PROGRESS GRADE SATISFAC MOTIVAT;
            GROUPING IS GROUP(1=MAJORS 2=NONMAJORS);

MODEL:      SATISFAC ON PROGRESS (1)
                         MOTIVAT  (2)
                         GRADE    (3);
            PROGRESS ON MOTIVAT;
            GRADE    ON MOTIVAT (5);

OUTPUT:     STDYX MODINDICES(3.84);
```

In M*plus*, you will delete the number 4 in parenthesis [(4)] that was used to constrain the direct effect from *Motivation* to *Progress* across groups.

Because we released a parameter equality constraint, we are estimating one additional model parameter for the re-specified *partial invariance* multiple group model. Thus, the *df* for the partial invariance multiple group model decreased by 1, making *df* for this model equal to 6. We are calling this model a partial invariance model because not all of the parameters we tested are constrained to equality across groups. The re-specified partial invariance multiple group model still did not fit the data well after releasing the equality constraint [LISREL: $\chi^2(6) = 55.629$, $p < .05$; CFI = .956; NNFI/TLI = .912; RMSEA = .218 (90% CI: .168, .272) with $p < .05$; Group 1 SRMR = .030; Group 2 SRMR = .065] and [M*plus*: $\chi^2(6) = 55.998$, $p < .05$; CFI = .969; NNFI/TLI = .938; RMSEA = .218 (90% CI: .168, .272) with $p < .05$; SRMR = .482]. Also, there was a significant difference between the baseline and partial invariance model [LISREL: $\Delta\chi^2 = 55.298$, $\Delta df = 4$, $p < .05$] and [M*plus*: $\Delta\chi^2 = 55.666$, $\Delta df = 4$, $p < .05$]. Thus, there is one (or more) equality constraint that should be released.

The modification indices in LISREL involving equality constraints are provided below for Groups 1 and 2, respectively:

```
GROUP 1: PARTIAL INVARIANCE MULTIPLE GROUP MODEL

Modification Indices and Expected Change

        The Modification Indices Suggest to Add the
Path to      from          Decrease in Chi-Square      New Estimate
   SATISFAC     PROGRESS          24.1                 1.45 IN GROUP 1
   SATISFAC        GRADE          11.8                 0.83 IN GROUP 1
   SATISFAC     SATISFAC          33.1                 0.17 IN GROUP 1

        The Modification Indices Suggest to Add the
Path to      from          Decrease in Chi-Square        New Estimate
   SATISFAC      MOTIVAT          49.8                 1.92 IN GROUP 1

GROUP 2: PARTIAL INVARIANCE MULTIPLE GROUP PATH MODEL

Modification Indices and Expected Change

        The Modification Indices Suggest to Add the
Path to      from          Decrease in Chi-Square        New Estimate
   SATISFAC     PROGRESS          71.7                 1.00 IN GROUP 2
   SATISFAC        GRADE           7.9                 0.67 IN GROUP 2
   SATISFAC     SATISFAC          33.1                -0.17 IN GROUP 2

        The Modification Indices Suggest to Add the
Path to      from          Decrease in Chi-Square        New Estimate
   SATISFAC      MOTIVAT          29.8                 0.89 IN GROUP 2
```

The modification indices in M*plus* involving parameter equality constraints are provided below for both groups:

	M.I.	E.P.C.	Std E.P.C.	StdYX E.P.C.
Group MAJORS				
ON Statements				
PROGRESS ON SATISFAC	4.809	−0.056	−0.056	−0.181
SATISFAC ON PROGRESS	24.258	0.091	0.091	0.028
SATISFAC ON GRADE	11.876	0.062	0.062	0.034
SATISFAC ON SATISFAC	33.288	0.170	0.170	0.170
SATISFAC ON MOTIVATN	50.112	0.652	0.652	0.137
Group NONMAJORS				
ON Statements				
SATISFAC ON PROGRESS	24.258	−0.427	−0.427	−0.098
SATISFAC ON GRADE	11.876	−0.088	−0.088	−0.060
SATISFAC ON SATISFAC	33.288	−0.170	−0.170	0.170
SATISFAC ON MOTIVATN	50.112	−0.235	−0.235	−0.062

Both LISREL and M*plus* show that the largest drop in chi-square (by about 50 points) would be associated with the direct effect from *Motivation* to *Satisfaction*.

The SIMPLIS code below is estimating the partial invariance model with the equality constraint from *Motivation* to *Satisfaction* now released:

```
GROUP 1: PARTIAL INVARIANCE MULTIPLE GROUP MODEL 2
RAW DATA FROM FILE MAJORS.LSF
RELATIONSHIPS:
SATISFAC = PROGRESS MOTIVAT GRADE
PROGRESS = MOTIVAT
GRADE = MOTIVAT
GROUP 2: PARTIAL INVARIANCE MULTIPLE GROUP MODEL 2
RAW DATA FROM FILE NONMAJORS.LSF
RELATIONSHIPS:
PROGRESS = MOTIVAT
SATISFAC = MOTIVAT
SET THE ERROR VARIANCE OF SATISFAC FREE
SET THE ERROR VARIANCE OF PROGRESS FREE
SET THE ERROR VARIANCE OF GRADE FREE
OPTIONS: SC MI
END OF PROBLEM
```

You will notice that the relationship between *Motivation* and *Satisfaction* (SATIFAC = MOTIVAT) is specified in Group 2. This will estimate this direct effect freely across the two groups and release the equality constraint.

The M*plus* code to freely estimate the direct effect from *Motivation* to *Satisfaction* is provided below:

```
TITLE:      PARTIAL INVARIANCE MULTIPLE GROUP PATH MODEL

DATA:       FILE IS MGPATH.DAT;

VARIABLE:   NAMES ARE PROGRESS GRADE SATISFAC MOTIVAT GROUP;
            USEVAR ARE PROGRESS GRADE SATISFAC MOTIVAT;
            GROUPING IS GROUP(1=MAJORS 2=NONMAJORS);

MODEL:      SATISFAC ON PROGRESS (1)
                    MOTIVAT
                    GRADE    (3);
            PROGRESS ON MOTIVAT;
            GRADE    ON MOTIVAT   (5);

OUTPUT:     STDYX MODINDICES(3.84);
```

In M*plus*, you will delete the number 2 in parenthesis [(2)] that was used to constrain the direct effect from *Motivation* to *Satisfaction* equal across groups.

With the exception of the SRMR in M*plus*, the partial invariance multiple group model with 2 equality constraints released fit the data well [LISREL: $\chi^2(5) = 1.704$, $p > .05$; CFI = 1.00; NNFI/TLI = 1.01; RMSEA = .00 (90% CI: .00, .05) with $p > .05$; Group 1 SRMR = .026; Group 2 SRMR = .031] and [M*plus*: $\chi^2(5) = 1.719$, $p > .05$; CFI = 1.00; NNFI/TLI = 1.00; RMSEA = .00 (90% CI: .00, .05) with $p > .05$; SRMR = .266]. Also, there was no longer a significant difference between the baseline and partial invariance model with 2 equality constraints released [LISREL: $\Delta\chi^2 = 1.373$, $\Delta df = 3$, $p > .05$] and [M*plus*: $\Delta\chi^2 = 1.387$, $\Delta df = 3$, $p > .05$]. Moreover, none of the modification indices were associated with significant decreases (greater than 3.84) in the model's chi-square statistic.

Table 9.2 summarizes the model fit for all of the models tested. It is important to note that the sensitivity of the chi-square statistic also applies to the $\Delta\chi^2$ test statistic. Thus, multiple group model testing using the $\Delta\chi^2$ test in large sample sizes may lead to conclusions that model parameters are non-invariant even with a small magnitude of a difference. Cheung and Rensvold (2002) proposed using the ΔCFI between nested models when testing for invariance. They conducted a simulation study and found that the CFI was not impacted by sample size as well as model complexity. Cheung and Rensvold (2002) suggested that a ΔCFI $\leq .01$ between nested multiple group models supports the model with more equality constraints. Using the CFI values from LISREL and M*plus*, respectively, the ΔCFI between the baseline model and the fully constrained model would support non-invariance of one or more constrained model parameters (LISREL: ΔCFI = 1.00 − .86 = .14;

M*plus*: ΔCFI = 1.00 – .91 = .09). The ΔCFI between the baseline model and the partial invariance model #1 would also suggest that one or more of the constrained model parameters are non-invariant (LISREL: ΔCFI = 1.00 – .96 = .04; M*plus*: ΔCFI = 1.00 – .97 = .03). The ΔCFI between the baseline model and the partial invariance model #2 would support the equality constraints in the model (LISREL: ΔCFI = 1.00 – 1.00 = .00; M*plus*: ΔCFI = 1.00 – 1.00 = .00). Hence, the partial invariance model with 2 equality constraints released should be discussed in terms of group differences and similarities.

Table 9.2: Model Fit Information for the Multiple Group Path Models

Software/Model	χ^2	df	CFI	NNFI/ TLI	RMSEA (90% CI) p-value	SRMR (M/ NM[a])
LISREL						
Baseline model	.331	2	1.00	1.01	.00 (.00, .08) p > .05	.005/ .003
Fully constrained model	160.95*	7	.86	.77	.36 (.31, .40) p < .05	.157/ .343
Partial invariance model 1 (M → P freely estimated)	55.63*	6	.96	.91	.22 (.17, .27) p < .05	.030/ .065
Partial invariance model 2 (M → S freely estimated)	1.70	5	1.00	1.01	.00 (.00, .05) p > .05	.026/ .031
M*plus*						
Baseline model	.332	2	1.00	1.00	.00 (.00, .08) p > .05	.011
Fully constrained model	161.98*	7	.91	.84	.36 (.31, .40) p < .05	1.467
Partial invariance model 1 (M → P freely estimated)	55.99*	6	.97	.94	.22 (.17, .27) p < .05	.482
Partial invariance model 2 (M → S freely estimated)	1.72	5	1.00	1.00	.00 (.00, .05) p > .05	.266

Note: * p < .05; [a] = SRMR is reported for groups separately in LISREL; M = Majors; NM = Non-majors; *N* = 200 for Psychology majors; *N* = 150 for Non-majors.

Final Model Interpretation

Selected output from LISREL and M*plus* for the final partial invariance model are provided in Tables 9.3 and 9.4, respectively. Table 9.5 formally presents the relevant model estimates from M*plus* for the final partial invariance multiple group path model. You will notice in Tables 9.3 and 9.4 that the unstandardized estimates for 3 of the direct effects are constrained to equality across groups whereas 2 are estimated freely in each group. Specifically, the unstandardized direct effects from *Progress* to *Satisfaction*, from *Grade* to *Satisfaction*, and from *Motivation* to *Grade* are constrained to equality across groups. This implies that these 3 relationships are similar for Psychology majors and non-majors. However, the unstandardized direct effects from *Motivation* to *Satisfaction* and from *Motivation* to *Progress* are different across the two groups.

Table 9.3: Selected LISREL Output for Final Partial Invariance Path Model

```
GROUP 1: PARTIAL INVARIANCE MULTIPLE GROUP MODEL 2

LISREL Estimates (Maximum Likelihood)

        Structural Equations

PROGRESS = 1.331*MOTIVAT, Errorvar.= 0.236 , R² = 0.816
Standerr  (0.0448)                (0.0237)
Z-values  29.685                   9.975
P-values   0.000                   0.000

GRADE = 1.562*MOTIVAT, Errorvar.= 2.453 , R² = 0.370
Standerr  (0.110)               (0.246)
Z-values  14.234                 9.975
P-values   0.000                 0.000

SATISFAC = 1.072*PROGRESS + 0.775*GRADE + 1.911*MOTIVAT, Errorvar.= 0.484 , R² = 0.966
Standerr  (0.0729)        (0.0238)      (0.122)                  (0.0485)
Z-values  14.692          32.576        15.639                   9.975
P-values   0.000           0.000         0.000                    0.000

Within Group Standardized Solution

        BETA

            PROGRESS     GRADE    SATISFAC
            --------   --------   --------
PROGRESS       - -        - -        - -
   GRADE       - -        - -        - -
SATISFAC     0.319      0.402        - -

        GAMMA

            MOTIVAT
            --------
PROGRESS      0.903
GRADE         0.608
SATISFAC      0.386


GROUP 2: PARTIAL INVARIANCE MULTIPLE GROUP MODEL 2

LISREL Estimates (Maximum Likelihood)

        Structural Equations

PROGRESS = 0.497*MOTIVAT, Errorvar.= 0.259, R² = 0.327
Standerr  (0.0584)                (0.0300)
Z-values  8.505                    8.631
P-values  0.000                    0.000

   GRADE = 1.562*MOTIVAT, Errorvar.= 2.163 , R² = 0.365
Standerr  (0.110)               (0.251)
Z-values  14.234                 8.631
P-values   0.000                 0.000
```

Table 9.3: (*cont.*)

```
SATISFAC = 1.072*PROGRESS + 0.775*GRADE + 1.025*MOTIVAT, Errorvar.= 0.425 , R² = 0.928
Standerr   (0.0729)          (0.0238)      (0.0910)                 (0.0492)
Z-values  14.692            32.576        11.254                   8.631
P-values   0.000             0.000         0.000                    0.000
```

Within Group Standardized Solution

```
        BETA

            PROGRESS      GRADE     SATISFAC
            --------    --------    --------
PROGRESS      - -         - -         - -
   GRADE      - -         - -         - -
SATISFAC     0.273       0.588        - -

        GAMMA

            MOTIVAT
            --------
PROGRESS     0.572
   GRADE     0.604
SATISFAC     0.301
```

Table 9.4: Selected M*plus* Output for Final Partial Invariance Path Model

MODEL RESULTS

	Estimate	S.E.	Est./S.E.	Two-Tailed P-Value
Group MAJORS				
SATISFAC ON				
PROGRESS	1.072	0.073	14.718	0.000
MOTIVAT	1.911	0.122	15.629	0.000
GRADE	0.775	0.024	32.596	0.000
PROGRESS ON				
MOTIVAT	1.331	0.045	29.762	0.000
GRADE ON				
MOTIVAT	1.562	0.110	14.251	0.000
Intercepts				
PROGRESS	3.018	0.139	21.765	0.000
GRADE	69.864	0.347	201.091	0.000
SATISFAC	1.728	1.699	1.017	0.309
Residual Variances				
PROGRESS	0.235	0.023	10.000	0.000
GRADE	2.441	0.244	9.993	0.000
SATISFAC	0.481	0.048	9.990	0.000
Group NONMAJORS				
SATISFAC ON				
PROGRESS	1.072	0.073	14.718	0.000
MOTIVAT	1.024	0.092	11.106	0.000
GRADE	0.775	0.024	32.596	0.000

Table 9.4: (*cont.*)

PROGRESS ON				
MOTIVAT	0.497	0.058	8.533	0.000
GRADE ON				
MOTIVAT	1.562	0.110	14.251	0.000
Intercepts				
PROGRESS	3.210	0.180	17.827	0.000
GRADE	69.771	0.351	198.777	0.000
SATISFAC	1.564	1.691	0.925	0.355
Residual Variances				
PROGRESS	0.258	0.030	8.660	0.000
GRADE	2.149	0.248	8.652	0.000
SATISFAC	0.422	0.049	8.649	0.000

STANDARDIZED MODEL RESULTS

STDYX Standardization

	Estimate	S.E.	Est./S.E.	Two-Tailed P-Value
Group MAJORS				
SATISFAC ON				
PROGRESS	0.319	0.023	13.767	0.000
MOTIVAT	0.386	0.025	15.517	0.000
GRADE	0.403	0.020	20.629	0.000
PROGRESS ON				
MOTIVAT	0.903	0.013	69.346	0.000
GRADE ON				
MOTIVAT	0.608	0.038	16.154	0.000
Intercepts				
PROGRESS	2.672	0.224	11.909	0.000
GRADE	35.490	1.722	20.609	0.000
SATISFAC	0.456	0.451	1.011	0.312
Residual Variances				
PROGRESS	0.184	0.024	7.829	0.000
GRADE	0.630	0.046	13.750	0.000
SATISFAC	0.034	0.005	7.348	0.000
Group NONMAJORS				
SATISFAC ON				
PROGRESS	0.273	0.023	11.716	0.000
MOTIVAT	0.300	0.027	11.334	0.000
GRADE	0.588	0.027	22.044	0.000
PROGRESS ON				
MOTIVAT	0.572	0.055	10.401	0.000

Table 9.4: (*cont.*)

GRADE ON				
MOTIVAT	0.604	0.042	14.406	0.000
Intercepts				
PROGRESS	5.190	0.536	9.685	0.000
GRADE	37.922	1.939	19.552	0.000
SATISFAC	0.644	0.703	0.916	0.359
Residual Variances				
PROGRESS	0.673	0.063	10.712	0.000
GRADE	0.635	0.051	12.519	0.000
SATISFAC	0.072	0.011	6.738	0.000

R-SQUARE

Group MAJORS				
Observed				Two-Tailed
Variable	Estimate	S.E.	Est./S.E.	P-Value
PROGRESS	0.816	0.024	34.673	0.000
GRADE	0.370	0.046	8.077	0.000
SATISFAC	0.966	0.005	211.984	0.000
Group NONMAJORS				
Observed				Two-Tailed
Variable	Estimate	S.E.	Est./S.E.	P-Value
PROGRESS	0.327	0.063	5.200	0.000
GRADE	0.365	0.051	7.203	0.000
SATISFAC	0.928	0.011	87.257	0.000

Table 9.5: Unstandardized Direct Effects, Standard Errors (S.E.), and Standardized Direct Effects for the Partial Invariance Multiple Group Path Model of Motivation

	Psychology Majors			Non-majors		
Association	Unstandardized Estimate	S.E.	Standardized Estimate	Unstandardized Estimate	S.E.	Standardized Estimate
Progress → Satisfaction	1.07*	.07	.32	1.07*	.07	.27
Motivation → Satisfaction[a]	1.91*	.12	.39	1.02*	.09	.30
Grade → Satisfaction	.78*	.02	.40	.78*	.02	.60
Motivation → Progress[a]	1.33*	.05	.90	.50*	.06	.57
Motivation → Grade	1.56*	.11	.61	1.56*	.11	.60

Note: * $p < .05$; [a] = non-invariant model parameter estimate across groups was detected for this relationship.

The unstandardized estimates are commonly used to describe differences for the non-invariant model parameter estimates. You will notice in Tables 9.3–9.5 that the standardized estimates are freely estimated across groups. Again, the unstandardized estimates are constrained to equality in multiple group models in order to account for the variance of the variables in the model. Looking at Table 9.5, the direct effect from *Motivation* to *Satisfaction* is stronger for Psychology majors than for non-majors. Likewise, the direct effect from *Motivation* to *Progress* is stronger for Psychology majors than for non-majors. Perhaps Psychology majors have different motivations when taking a course in their major curriculum than students who are taking the same course that is not in their major curriculum.

Multiple group observed variable path models can highlight important differences across groups with respect to the hypothesized relationships in the model. Multiple group CFA can highlight important differences in the measurement of a latent factor. The next section introduces multiple group CFA.

MULTIPLE GROUP CFA/MEASUREMENT MODEL

Measurement Invariance

Multiple group modeling has been discussed more commonly within the context of *measurement invariance*, which deals with CFA or measurement models. Multiple group modeling with CFA models is largely done to ensure that the factors are being measured similarly across groups to support the construct validity of a scale or survey instrument. Measurement invariance is also tested prior to testing factor mean differences across groups (e.g., with structured means models) in order to make meaningful comparisons.

Measurement invariance pertains to the degree to which the relationships between latent factors and their indicator variables are similar across groups. Measurement invariance is investigated by constraining specific model parameters to be equal across groups. The equality constraints that are imposed across groups when testing measurement invariance are commonly tested with sequentially increasing invariance restrictions. The levels of measurement invariance that are most regularly examined are: 1) configural invariance; 2) metric or weak invariance; 3) strong or scalar invariance; and 4) strict invariance (Meredith, 1993; Widamen & Reise, 1997).

Configural invariance is the least restrictive level in the measurement invariance hierarchy to be tested and refers to the pattern of fixed and free factor loadings across groups (Horn & McArdle, 1992). Thus, the same configuration of the factor model should hold across groups with no equality constraints imposed. *Configural invariance* is supported if the model fits the data acceptably. Once configural invariance is supported, the next level of invariance that is tested is *metric* or *weak invariance*, which requires that the corresponding factor loadings are equal across groups. *Metric invariance* is supported by a non-significant $\Delta\chi^2$ test or a $\Delta\text{CFI} \leq$ .01 between the configural and metric invariance models. Once metric invariance is supported, the next level of invariance that is tested is *strong* or *scalar invariance*, which requires that the corresponding indicator variable intercepts (in addition to factor loadings) are equal across groups. *Strong invariance* is supported by a non-significant $\Delta\chi^2$ test or a $\Delta CFI \leq .01$ between the metric and strong invariance models. Once strong invariance is supported, the next level of invariance that can be tested is *strict invariance*, which is the highest level of invariance in the hierarchy. *Strict invariance* requires that indicator variable error variances (in addition to factor loadings and intercepts) are equal across groups and is supported by a non-significant $\Delta\chi^2$ test or a $\Delta CFI \leq .01$ between the strong and strict invariance models.

If a significant $\Delta\chi^2$ test or a $\Delta CFI > .01$ between configural and metric invariance models is calculated, this would suggest that one or more of the unstandardized factor loadings are not equal across groups. Specifically, metric non-invariance means that a regression slope estimated when regressing the indicator variable on their respective factor is not equal across groups. Hence, a one unit increase in the factor would not result in the same unit change in the responses to the same indicator variable across groups. A significant $\Delta\chi^2$ test or a $\Delta CFI > .01$ between metric and strong invariance models would suggest that one or more of the indicator variable intercepts is not equal across groups. Strong non-invariance means that participants in different groups with the same factor score have different scores on the same indicator variable. A significant $\Delta\chi^2$ test or a $\Delta CFI > .01$ between strong and strict invariance models would suggest that one or more of the indicator variable's error variances is not equal across groups. Strict non-invariance means that the indicator variable is not measuring their respective factor across groups with the same accuracy.

Strict invariance has been suggested as a prerequisite for making meaningful comparisons across groups on the construct(s) of interest (Meredith, 1993; Wu, Li, & Zumbo, 2007). Nonetheless, many contend that strict invariance tests are too restrictive and not generally of interest (Widaman & Reise, 1997; Byrne, 2012). Little, Card, Slegers, and Ledford (2007) reasoned that because error variances are comprised of both indicator-specific variance as well as random measurement error, imposing equality constraints on error variances is not practical. Therefore, strict invariance tests are not generally recommended as a prerequisite for making meaningful group comparisons (Little, 2013; Thompson & Green, 2013). Nonetheless, it is highly recommended that scalar or strong invariance is satisfied for latent mean comparisons across groups (Widaman & Reise, 1997; Little et al., 2007; Little, 2013; Thompson & Green, 2013). Further, partial invariance, when detected, should be allowed in order to make meaningful comparisons across groups. That is, when parameters are detected as different across groups, they should be freely estimated to avoid biased estimates elsewhere in the model when comparing groups (Byrne, Shavelson, & Muthén, 1989; Whittaker, 2013). It must be noted, however, that detecting a large number of non-invariant parameter estimates would reduce the meaningfulness of group comparisons on a latent factor. The multiple group CFA example in this chapter will illustrate how to test for configural, metric, and strong invariance.

An important issue to mention when conducting multiple group CFA is the selection of the reference indicator used to set the metric of the factor(s). The most commonly used method for setting the scale of the factors in multiple group CFA tends to be the reference indicator method where an indicator's loading is set equal to a value of 1 (Cheung & Rensvold, 1999). The reason the reference indicator approach is used most often is because setting the factor variances to values of 1 in multiple group analysis implies that the factor variances are equal across groups. Assuming this equality while testing for invariance can bias results elsewhere in the model. Likewise, using a reference indicator that has a non-invariant or unequal factor loading across groups can lead to wrong conclusions about parameter invariance somewhere else in the multiple group model (Johnson, Meade, & DuVernet, 2009). Thus, it is recommended that the reference indicator's factor loading be invariant or equal across groups.

Cheung and Rensvold (1999) and Yoon and Millsap (2007) have suggested different methods that may be used to determine the most appropriate reference indicator to use with multiple group analysis. Studies have shown that the proposed reference indicator selection methods perform fairly accurately under different conditions (French & Finch, 2008; Johnson et al., 2009; Whittaker & Khojasteh, 2013). A method mentioned previously in Chapter 6 for scaling the factor was suggested by Little, Slegers, and Card (2006), which is effects coding. When using effects coding, the loadings of the indicators on the same factor are constrained so that their average equals the value of 1. The use of effects coding avoids having to use a reference indicator approach or fixing the factor variance equal to a value of 1. This method can also be used in multiple group CFA. In the end, it will be up to the researcher to determine the best factor scaling method to use with their analysis as well as reasonable assumptions for the models and groups that are being compared (e.g., equal factor variances across groups).

Multiple Group CFA Example

The theoretical measurement model we will use as an example when testing for measurement invariance across groups is based on the Holzinger and Swineford (1939) study. We presented this confirmatory factor analysis model of spatial and verbal ability previously in Chapter 6 as a two-factor correlated model. However, this time, the raw data are simulated and will be used to compare the CFA model across two different school groups. The data files for use in LISREL are called

MGHS1.LSF and *MGHS2.LSF* and the data set for use in M*plus* is called *MGHS.DAT*. All the data files are available on the book website. There are 156 participants in the Pasteur school group (in *MGHS1.LSF*) and 145 participants in the Grant-White school group (in *MGHS2.LSF*). In each of the data sets are responses to the six indicator variables (*VISPERC, CUBES, LOZENGES, PARAGRAP, SENTENCE,* and *WORDMEAN*). In the *MGHS.DAT* file, there is also a variable indicating school designation (1 = Pasteur school, 2 = Grant-White school). Figure 9.4 presents the two-factor correlated model that will be compared across the two school groups.

Model Identification in Separate Groups

Again, the first step in multiple group analysis is to test the model fit in each group separately. Identification with multiple group CFA models will need to incorporate the mean structure because we will be testing for strong or scalar invariance where indicator intercepts are constrained to equality across groups. You will notice in Figure 9.4 that the intercepts are included in the model, represented by the triangles containing values of 1 directly affecting each indicator variable. So, we will need to determine the number of distinct elements in the sample covariance matrix as well as the number of observed indicator variable means. You could separate this into the covariance and mean structures. For instance, the number of distinct elements in the sample covariance matrix for each group is $p(p + 1)/2 = 6(7)/2 = 21$. With 6 observed indicator variables, the number of means is 6. Summing the covariance and mean structure elements equals the total pieces of information provided by the data. In this case, we have 27 *pieces of information* provided by the data. Another way to calculate pieces of information provided by your data is to use the following equation: $p(p + 3)/2 = 6(9)/2 = 27$.

The parameters needing to be estimated in the model in each group separately include 6 error variances (one for each indicator variable); 2 factor variances (for spatial and verbal factors); 4 factor loadings (for indicators not serving as reference indicators); 1 factor covariance (between spatial and verbal factors); and 6 intercepts (one for each indicator variable). In total, there are 19 parameters to estimate in the model in each group separately. The *df* for the model in each group is 8 (27 – 19 = 8). You will notice that the means model is just-identified because there are 6 observed means and 6 indicator intercepts being estimated.

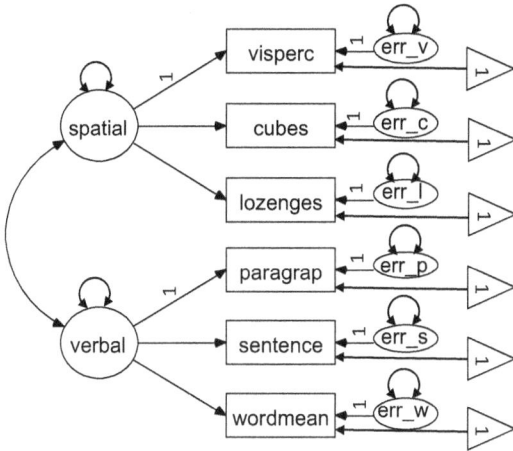

■ **Figure 9.4:** TWO-FACTOR CORRELATED CFA MODEL FOR INVARIANCE TESTING ACROSS SCHOOL GROUPS

Model Estimation in Separate Groups

The SIMPLIS code that tests the model separately in each school group is provided below:

```
CFA MODEL FOR PASTEUR SCHOOL
RAW DATA FROM FILE MGHS1.LSF
LATENT VARIABLES: SPATIAL VERBAL
RELATIONSHIPS:
VISPERC = CONST + 1*SPATIAL
CUBES = CONST + SPATIAL
LOZENGES = CONST + SPATIAL
PARAGRAP = CONST + 1*VERBAL
SENTENCE = CONST + VERBAL
WORDMEAN = CONST + VERBAL
OPTIONS: SC MI
PATH DIAGRAM
END OF PROBLEM

CFA MODEL FOR GRANT-WHITE SCHOOL
RAW DATA FROM FILE MGHS2.LSF
LATENT VARIABLES: SPATIAL VERBAL
RELATIONSHIPS:
VISPERC = CONST + 1*SPATIAL
CUBES = CONST + SPATIAL
LOZENGES = CONST + SPATIAL
```

```
PARAGRAP = CONST + 1*VERBAL
SENTENCE = CONST + VERBAL
WORDMEAN = CONST + VERBAL
OPTIONS: SC MI
PATH DIAGRAM
END OF PROBLEM
```

To incorporate the mean structure in the estimation process in LISREL, the intercepts were included in the RELATIONSHIPS statements by using the term **CONST** for each of the indicator variables. This will estimate an intercept for each of the indicator variables.

The M*plus* code that tests the model separately in each school group is provided below:

```
TITLE:      CFA MODEL FOR PASTEUR SCHOOL

DATA:       FILE IS MGHS.DAT;

VARIABLE:   NAMES ARE VISPERC CUBES LOZENGES
                    PARAGRAP SENTENCE WORDMEAN SCHOOL;

            USEVAR ARE VISPERC CUBES LOZENGES
                    PARAGRAP SENTENCE WORDMEAN;

            USEOBS ARE SCHOOL EQ 1;

MODEL:      SPATIAL BY VISPERC CUBES LOZENGES;
            VERBAL BY PARAGRAP SENTENCE WORDMEAN;
            SPATIAL WITH VERBAL;

OUTPUT:     STDYX MODINDICES(3.84);


TITLE:      CFA MODEL FOR GRANT-WHITE SCHOOL

DATA:       FILE IS MGHS.DAT;

VARIABLE:   NAMES ARE VISPERC CUBES LOZENGES
                    PARAGRAP SENTENCE WORDMEAN SCHOOL;

            USEVAR ARE VISPERC CUBES LOZENGES
                    PARAGRAP SENTENCE WORDMEAN;

            USEOBS ARE SCHOOL EQ 2;

MODEL:      SPATIAL BY VISPERC CUBES LOZENGES;
            VERBAL BY PARAGRAP SENTENCE WORDMEAN;
            SPATIAL WITH VERBAL;

OUTPUT:     STDYX MODINDICES(3.84);
```

The intercepts are estimated by default in M*plus* when raw data are used. ML estimation is used in this example.

Model Testing in Separate Groups

The model fit the data well in each group separately. Table 9.6 provides model fit information from LISREL and M*plus* for the model in each of the groups. If the model did not fit the data well, the modification indices, expected parameter changes, and the residuals could be examined to find potential model misspecification in each of the groups. As described before, you can still proceed with multiple group modeling procedures even if different parameters are estimated in different groups. You would just be limited in terms of testing the invariance of the same model parameters estimated in each group.

Table 9.6: Model Fit Information for the CFA Model in Each School Group

Software/Group	χ^2	df	CFI	NNFI/ TLI	RMSEA (90% CI) p-value	SRMR
LISREL						
Pasteur school	3.25	8	1.00	1.01	.00 (.00, .03) p > .05	.018
Grant-White school	6.45	8	1.00	1.00	.00 (.00, .08) p > .05	.026
M*plus*						
Pasteur school	3.27	8	1.00	1.00	.00 (.00, .03) p > .05	.016
Grant-White school	6.50	8	1.00	1.00	.00 (.00, .09) p > .05	.023

Note: N = 156 for Pasteur school group; N = 145 for Grant-White school group.

Model Identification of the Configural Multiple Group CFA Model – With No Equality Constraints

Since the model fit the data adequately in each of the groups separately, the next step is to estimate a *configural* multiple group model with no parameter equality constraints imposed. This model will also serve as the baseline model which is compared to the metric model with factor loadings constrained to be equal across groups. The configural model estimates model parameters for each group separately, but simultaneously. Model identification for the configural model will require summing the pieces of information that are provided by the data for each group separately. Remember that we calculated 27 pieces of information provided by the data for each group [$p(p+3)/2 = 6(9)/2 = 27$]. Thus, the pieces of information provided by the data for the configural model is 54 (27 + 27 = 54). Also, recall that we estimated 19 parameters in the model for each group separately. Hence, the number of parameters to be estimated in the configural model is 38 (19 + 19 = 38). The *df* for the configural multiple group CFA model is 16 (54 – 38 = 16), which is the sum of the *df* for the CFA model tested in each group separately (8 + 8). The multiple group configural CFA model is an over-identified model with 16 *df*.

Model Estimation of the Configural Multiple Group CFA Model – With No Equality Constraints

The SIMPLIS code to estimate the configural multiple group CFA model is provided below:

```
GROUP 1: CONFIGURAL CFA MODEL-PASTEUR SCHOOL
RAW DATA FROM FILE MGHS1.LSF
LATENT VARIABLES: SPATIAL VERBAL
RELATIONSHIPS:
VISPERC  = CONST + 1*SPATIAL
CUBES    = CONST + SPATIAL
LOZENGES = CONST + SPATIAL
PARAGRAP = CONST + 1*VERBAL
SENTENCE = CONST + VERBAL
WORDMEAN = CONST + VERBAL
GROUP 2: CONFIGURAL CFA MODEL-GRANT-WHITE SCHOOL
RAW DATA FROM FILE MGHS2.LSF
LATENT VARIABLES: SPATIAL VERBAL
RELATIONSHIPS:
VISPERC  = CONST + 1*SPATIAL
CUBES    = CONST + SPATIAL
LOZENGES = CONST + SPATIAL
PARAGRAP = CONST + 1*VERBAL
SENTENCE = CONST + VERBAL
WORDMEAN = CONST + VERBAL
SET THE ERROR VARIANCE OF VISPERC FREE
SET THE ERROR VARIANCE OF CUBES FREE
SET THE ERROR VARIANCE OF LOZENGES FREE
SET THE ERROR VARIANCE OF PARAGRAP FREE
SET THE ERROR VARIANCE OF SENTENCE FREE
SET THE ERROR VARIANCE OF WORDMEAN FREE
SET VARIANCE OF SPATIAL FREE
SET VARIANCE OF VERBAL FREE
SET THE COVARIANCE OF SPATIAL AND VERBAL FREE
OPTIONS: SC
END OF PROBLEM
```

In LISREL, there are default equality constraints that are imposed unless otherwise specified, including error variance, factor variance, and factor covariance constraints. The 9 statements in the code for **GROUP 2** are releasing the 6 error variance constraints for indicator variables (e.g., SET THE ERROR VARIANCE OF WORDMEAN FREE); the variance constraints of the 2 factors (e.g., SET

VARIANCE OF SPATIAL FREE); and the covariance constraint between the two factors (e.g., SET THE COVARIANCE OF SPATIAL AND VERBAL FREE) across groups. Again, relationships specified in Group 2 will be freely estimated and not constrained (e.g., intercepts and factor loadings).

M*plus* has an option to run sequential invariance testing while estimating the configural, metric, and strong/scalar invariance multiple group models. The M*plus* code to run sequential invariance testing is provided below:

```
TITLE:      SEQUENTIAL TESTING FOR MGCFA

DATA:       FILE IS MGHS.DAT;

VARIABLE:   NAMES ARE VISPERC CUBES LOZENGES
                    PARAGRAP SENTENCE WORDMEAN SCHOOL;

            USEVAR ARE VISPERC CUBES LOZENGES
                    PARAGRAP SENTENCE WORDMEAN;

            GROUPING IS SCHOOL (1 = PASTEUR 2 = GRANT-WHITE);

ANALYSIS:   MODEL IS CONFIGURAL METRIC SCALAR;

MODEL:      SPATIAL BY VISPERC CUBES LOZENGES;
            VERBAL BY PARAGRAP SENTENCE WORDMEAN;
            SPATIAL WITH VERBAL;
```

The sequential invariance testing in M*plus* is invoked in the ANALYSIS command: MODEL IS CONFIGURAL METRIC SCALAR, which will estimate the configural model (with no constraints), the metric model (with factor loading constraints), and the scalar or strong model (with factor loading and intercept constraints).

Model Testing of the Configural Multiple Group CFA Model – With No Equality Constraints

The configural model fit the data well [LISREL: $\chi^2(16) = 9.699$, $p > .05$; CFI = 1.00; NNFI/TLI = 1.01; RMSEA = .00 (90% CI: .00, .04) with $p > .05$; Group 1 SRMR = .018; Group 2 SRMR = .026] and [M*plus*: $\chi^2(16) = 9.766$, $p > .05$; CFI = 1.00; NNFI/TLI = 1.00; RMSEA = .00 (90% CI: .00, .04) with $p > .05$; SRMR = .019]. Again, LISREL does not provide an overall SRMR, but provides one for each group whereas M*plus* calculates an overall SRMR.

Model Identification of the Metric Multiple Group CFA Model

After estimating the configural multiple group CFA model, the next step is to constrain the factor loadings to equality across the groups, which is the metric invariance model. The pieces of information provided by the data is still 54, but the number of parameters to be estimated will be reduced by the number of factor loadings that are constrained to equality across groups. Recall that *VISPERC* and *PARAGRAP* are serving as reference indicators across groups, which implies that their loadings are constrained to equality already. The 4 remaining factor loadings will be constrained to equality across groups. Hence the number of parameter estimates will be reduced by 4. Specifically, there will be 12 indicator error variances estimated (6 per group); 4 factor variances estimated (2 per group); 2 factor covariances estimated (1 per group); 12 indicator variable intercepts estimated (6 per group); and 4 factor loadings estimated (which are constrained to equality across groups). The total number of parameters to estimate in the metric invariance model is 34. The *df* for the metric invariance model is 20 (54 − 34 = 20). Notice that the *df* increased in the metric model by 4 as compared to the configural model (*df* = 16) because we are estimating 4 fewer parameters.

Model Estimation of the Metric Multiple Group CFA Model

The SIMPLIS code to estimate the metric multiple group CFA model is provided below:

```
GROUP 1: METRIC CFA MODEL-PASTEUR SCHOOL
RAW DATA FROM FILE MGHS1.LSF
LATENT VARIABLES: SPATIAL VERBAL
RELATIONSHIPS:
VISPERC  = CONST + 1*SPATIAL
CUBES    = CONST + SPATIAL
LOZENGES = CONST + SPATIAL
PARAGRAP = CONST + 1*VERBAL
SENTENCE = CONST + VERBAL
WORDMEAN = CONST + VERBAL
GROUP 2: METRIC CFA MODEL-GRANT-WHITE SCHOOL
RAW DATA FROM FILE MGHS2.LSF
LATENT VARIABLES: SPATIAL VERBAL
RELATIONSHIPS:
VISPERC  = CONST + 1*SPATIAL
CUBES    = CONST
LOZENGES = CONST
PARAGRAP = CONST + 1*VERBAL
SENTENCE = CONST
WORDMEAN = CONST
SET THE ERROR VARIANCE OF VISPERC FREE
SET THE ERROR VARIANCE OF CUBES FREE
SET THE ERROR VARIANCE OF LOZENGES FREE
SET THE ERROR VARIANCE OF PARAGRAP FREE
```

```
SET THE ERROR VARIANCE OF SENTENCE FREE
SET THE ERROR VARIANCE OF WORDMEAN FREE
SET VARIANCE OF SPATIAL FREE
SET VARIANCE OF VERBAL FREE
SET THE COVARIANCE OF SPATIAL AND VERBAL FREE
OPTIONS: SC MI
END OF PROBLEM
```

You will notice that the factors are removed from the indicator equations in the RELATIONSHIPS heading for Group 2 and only the indicator intercepts are estimated (with the exception of the reference indicators). This specifies that the 4 factor loadings should be constrained to equality across groups.

The metric invariance model in M*plus* was already tested using the sequential invariance testing option available in the ANALYSIS command, which will be discussed in more detail subsequently. However, the M*plus* code for testing the metric invariance multiple group CFA model is provided below:

```
TITLE:      METRIC INVARIANCE CFA MODEL

DATA:       FILE IS MGHS.DAT;

VARIABLE:   NAMES ARE VISPERC CUBES LOZENGES
                    PARAGRAP SENTENCE WORDMEAN SCHOOL;

            USEVAR ARE VISPERC CUBES LOZENGES
                    PARAGRAP SENTENCE WORDMEAN;

            GROUPING IS SCHOOL (1 = PASTEUR 2 = GRANT-WHITE);

MODEL:      SPATIAL BY VISPERC CUBES LOZENGES;
            VERBAL BY PARAGRAP SENTENCE WORDMEAN;
            SPATIAL WITH VERBAL;
            [SPATIAL@0 VERBAL@0];

MODEL GRANT-WHITE:
     [VISPERC CUBES LOZENGES
      PARAGRAP SENTENCE WORDMEAN];

OUTPUT:     STDYX MODINDICES(3.84);
```

The default in M*plus* with multiple group CFA models is to constrain factor loadings *and* indicator intercepts to equality across groups. For the metric model, only the factor loadings should be constrained to equality across groups. Thus, in the M*plus* code, you will notice that a second MODEL command is used for the Grant-White school group (**MODEL GRANT-WHITE:**). You will notice under this MODEL command are the indicator variable names contained in brackets. M*plus* model parameters contained in brackets are parameters for the means model or mean structure. In this case, because the indicator variables are endogenous continuous observed variables, the intercepts associated with the indicator variables in the brackets will be estimated freely in the Grant-White school group. This overrides the default of imposing equality constraints on the indicator variable intercepts across groups.

It must be noted that M*plus* will fix factor means equal to zero in group 1 while estimating the factor means in group 2 by default with multiple groups when not using the sequential invariance testing option. Thus, you will notice that the means for the SPATIAL and VERBAL factors were set equal to zero to avoid this estimation [SPATIAL@0 VERBAL@0]. Estimating the factor means is important when conducting structured means analysis, which examines factor mean differences across groups. When testing measurement invariance in general, however, examining factor mean differences is not typically a question of interest (Byrne, 2012). Still, the invariance testing procedures are important to conduct prior to testing for factor mean differences. Thus, the steps we are taking are prerequisite steps for comparing factor means across groups. Moreover, estimating the factor means without intercept equality constraints will result in an under-identified means model or mean structure (see Hancock, 1997 for more information about testing latent means across groups).

It is also important to note that the error variance for *LOZENGES* in the Grant-White school group became negative when imposing the factor loading constraints. The following error message was provided by LISREL:

```
LOZENGES =    0.314    + 2.672*SPATIAL, Errorvar.=-0.454, R² = 1.123
Standerr   (0.160)   (0.124)                       (0.141)
Z-values    1.957    21.586                        -3.223
P-values    0.050     0.000                          0.001

W_A_R_N_I_N_G: Error variance is negative.
```

Likewise, M*plus* provided the following error message:

```
WARNING: THE RESIDUAL COVARIANCE MATRIX (THETA) IN GROUP GRANT-
WHITE IS NOT POSITIVE DEFINITE. THIS COULD INDICATE A NEGATIVE
VARIANCE/RESIDUAL VARIANCE FOR AN OBSERVED VARIABLE, A CORRELATION
GREATER OR EQUAL TO ONE BETWEEN TWO OBSERVED VARIABLES, OR A LINEAR
DEPENDENCY AMONG MORE THAN TWO OBSERVED VARIABLES. CHECK THE
RESULTS SECTION FOR MORE INFORMATION. PROBLEM INVOLVING VARIABLE
LOZENGES.
```

This would already suggest a potential model misspecification with equal factor loadings constraints imposed across groups.

Model Testing and Modification of the Metric Multiple Group CFA Model

The metric invariance model did not fit the data well [LISREL: $\chi^2(20) = 70.891$, $p < .05$; CFI = .967; NNFI/TLI = .951; RMSEA = .130 (90% CI: .098, .164) with $p < .05$; Group 1 SRMR = .034; Group 2 SRMR = .169] and [M*plus*: $\chi^2(20) = 71.390$, $p < .05$; CFI = .975; NNFI/TLI = .963; RMSEA = .131 (90% CI: .099, .164) with $p < .05$; SRMR = .065]. In addition to the error message about a Heywood case, the poor model fit suggests that one or more of the unstandardized factor loadings are not equal across groups. When using the sequential invariance testing option in M*plus*, the chi-square statistics for the configural, metric, and scalar models as well as $\Delta\chi^2$ tests between the different invariance models are provided:

```
Invariance Testing

                       Number of            Degrees of
Model                  Parameters  Chi-Square   Freedom   P-Value
Configural                 38        9.766        16      0.8786
Metric                     34       71.390        20      0.0000
Scalar                     30      150.425        24      0.0000

                                                Degrees of
Models Compared                    Chi-Square     Freedom   P-Value
Metric against Configural             61.624         4      0.0000
Scalar against Configural            140.659         8      0.0000
Scalar against Metric                 79.034         4      0.0000
```

As you can see from the M*plus* output, there is a significant $\Delta\chi^2$ test between the configural and metric multiple group CFA models [$\Delta\chi^2 = 61.624$, $\Delta df = 4$, $p < .05$]. Using LISREL chi-square statistics also indicates a significant $\Delta\chi^2$ test between the configural and metric multiple group CFA models [$\Delta\chi^2 = 61.192$, $\Delta df = 4$, $p < .05$]. The ΔCFI between metric and configural models was also greater than .01, indicating that one or more of the factor loading constraints across groups should be released.

The modification indices in LISREL that involve factor loading equality constraints across the school groups are provided below, respectively:

```
GROUP 1: METRIC CFA MODEL-PASTEUR SCHOOL

Modification Indices and Expected Change

        The Modification Indices Suggest to Add the
Path to        from         Decrease in Chi-Square      New Estimate
        CUBES      SPATIAL            35.3              2.66 IN GROUP 1
        LOZENGES   SPATIAL            54.7              2.92 IN GROUP 1
```

```
GROUP 2: METRIC CFA MODEL-GRANT-WHITE SCHOOL

Modification Indices and Expected Change

        The Modification Indices Suggest to Add the
Path to        from              Decrease in Chi-Square      New Estimate
     CUBES       SPATIAL              58.8                4.02 IN GROUP 2
     LOZENGES    SPATIAL              91.8                1.29 IN GROUP 2
```

Modification indices are not provided when using the sequential invariance testing option in M*plus*. Hence, the metric invariance model will need to be estimated separately in order to obtain modification indices. The modification indices in M*plus* that involve equality constraints across school groups are provided below:

```
                        M.I.    E.P.C.    Std E.P.C.    stdYX E.P.C.

Group PASTEUR

BY Statements

SPATIAL    BY CUBES     35.586   -0.190      0.235          0.066
SPATIAL    BY LOZENGES  55.063    0.246      0.304          0.089

Group GRANT-WHITE

BY Statements

SPATIAL    BY CUBES     35.589    1.097      0.834          0.311
SPATIAL    BY LOZENGES  55.064   -1.288     -0.979         -0.511
VERBAL     BY CUBES      6.090    0.272      0.290          0.108
VERBAL     BY LOZENGES   4.299   -0.171     -0.182         -0.095
```

Again, modification indices can be associated with all fixed and constrained parameters, which may not make sense and could lead to convergence problems. Thus, it is important that you examine the modification indices associated with parameter equality constraints with multiple group models. Thus, the modification indices associated with factor loadings should be examined (i.e., the "BY Statements" in M*plus*). Both LISREL and M*plus* suggest that the largest chi-square decrease (by approximately 55 points) would result when releasing the equality constraint of the factor loading for *LOZENGES* on the SPATIAL factor.

The SIMPLIS code with this factor loading freely estimated in both school groups is provided below:

```
GROUP 1: PARTIAL METRIC CFA MODEL-PASTEUR SCHOOL
RAW DATA FROM FILE MGHS1.LSF
LATENT VARIABLES: SPATIAL VERBAL
RELATIONSHIPS:
```

```
VISPERC = CONST + 1*SPATIAL
CUBES = CONST + SPATIAL
LOZENGES = CONST + SPATIAL
PARAGRAP = CONST + 1*VERBAL
SENTENCE = CONST + VERBAL
WORDMEAN = CONST + VERBAL
GROUP 2: PARTIAL METRIC CFA MODEL-GRANT-WHITE SCHOOL
RAW DATA FROM FILE MGHS2.LSF
LATENT VARIABLES: SPATIAL VERBAL
RELATIONSHIPS:
VISPERC = CONST + 1*SPATIAL
CUBES = CONST
LOZENGES = CONST + SPATIAL
PARAGRAP = CONST + 1*VERBAL
SENTENCE = CONST
WORDMEAN = CONST
SET THE ERROR VARIANCE OF VISPERC FREE
SET THE ERROR VARIANCE OF CUBES FREE
SET THE ERROR VARIANCE OF LOZENGES FREE
SET THE ERROR VARIANCE OF PARAGRAP FREE
SET THE ERROR VARIANCE OF SENTENCE FREE
SET THE ERROR VARIANCE OF WORDMEAN FREE
SET VARIANCE OF SPATIAL FREE
SET VARIANCE OF VERBAL FREE
SET THE COVARIANCE OF SPATIAL AND VERBAL FREE
OPTIONS: SC MI
END OF PROBLEM
```

You will notice that the **SPATIAL** factor for *LOZENGES* is now specified in Group 2. This will estimate this factor loading freely across the two groups, releasing the equality constraint.

The M*plus* code to freely estimate the factor loading of *LOZENGES* on the SPATIAL factor is provided below:

```
TITLE:      PARTIAL METRIC INVARIANCE CFA MODEL

DATA:       FILE IS MGHS.DAT;

VARIABLE:   NAMES ARE VISPERC CUBES LOZENGES
                    PARAGRAP SENTENCE WORDMEAN SCHOOL;

            USEVAR ARE VISPERC CUBES LOZENGES
                    PARAGRAP SENTENCE WORDMEAN;
```

```
                    GROUPING IS SCHOOL (1 = PASTEUR 2 = GRANT-WHITE);

MODEL:        SPATIAL BY VISPERC CUBES LOZENGES;
              VERBAL BY PARAGRAP SENTENCE WORDMEAN;
              SPATIAL WITH VERBAL;
              [SPATIAL@0 VERBAL@0];

MODEL GRANT-WHITE:
    SPATIAL BY LOZENGES;
    [VISPERC CUBES LOZENGES
    PARAGRAP SENTENCE WORDMEAN];

OUTPUT:       STDYX MODINDICES(3.84);
```

In M*plus*, the following statement: SPATIAL BY LOZENGES in the MODEL GRANT-WHITE: command will freely estimate the factor loading across groups.

Because we released a parameter equality constraint, we are estimating one additional model parameter for the re-specified *partial metric invariance* multiple group model. Thus, the *df* for the partial metric invariance multiple group model decreased by 1, making *df* for this model equal to 19. We are calling this model a partial metric invariance model because not all of the factor loadings are constrained to equality across groups. The error variance for *LOZENGES* (which was negative when constraining factor loadings) is now positive and no error messages were provided in the output.

The re-specified partial metric invariance multiple group model fit the data well after releasing the equality constraint [LISREL: $\chi^2(19) = 10.480$, $p > .05$; CFI = 1.00; NNFI/TLI = 1.01; RMSEA = .00 (90% CI: .000, .015) with $p > .05$; Group 1 SRMR = .022; Group 2 SRMR = .029] and [M*plus*: $\chi^2(19) = 10.553$, $p > .05$; CFI = 1.00; NNFI/TLI = 1.00; RMSEA = .00 (90% CI: .000, .017) with $p > .05$; SRMR = .021]. The $\Delta\chi^2$ test between the configural and partial metric invariance models was not significant [LISREL: $\Delta\chi^2 = .781$, $\Delta df = 3$, $p > .05$] and [M*plus*: $\Delta\chi^2 = .787$, $\Delta df = 3$, $p > .05$]. Moreover, the ΔCFI between the configural and partial metric invariance models was less than .01. Thus, the scalar or strong invariance model can be tested next.

Model Identification of the Strong/Scalar Multiple Group CFA Model

After estimating the partial metric multiple group CFA model, the next step is to constrain the indicator variable intercepts to equality across the groups, which is the scalar or strong invariance model. The pieces of information are still 54, but the number of parameters to be estimated will be reduced by the number of indicator intercepts that are constrained to equality across groups. That is, the 6 indicator intercepts will be constrained to equality across groups, reducing the number of parameter estimates by 6. Specifically, there will be 12 indicator error variances estimated (6 per group); 4 factor variances estimated (2 per group); 2 factor covariances estimated (1 per group); 6 indicator variable intercepts estimated (which are constrained across groups); and 5 factor loadings estimated (with all but the LOZENGES loading constrained to equality across groups). The total number of parameters to estimate in the scalar invariance model is 29. The *df* for the scalar invariance model is 25 (54 – 29 = 25). Notice that the *df* increased in the scalar model by 6 as compared to the partial metric model (*df* = 19) because we are estimating 6 fewer parameters.

Model Estimation of the Scalar Multiple Group CFA Model

The SIMPLIS code to estimate the scalar multiple group CFA model is provided below:

```
GROUP 1: SCALAR CFA MODEL-PASTEUR SCHOOL
RAW DATA FROM FILE MGHS1.LSF
LATENT VARIABLES: SPATIAL VERBAL
RELATIONSHIPS:
VISPERC = CONST + 1*SPATIAL
CUBES = CONST + SPATIAL
LOZENGES = CONST + SPATIAL
PARAGRAP = CONST + 1*VERBAL
SENTENCE = CONST + VERBAL
WORDMEAN = CONST + VERBAL
GROUP 2: SCALAR CFA MODEL-GRANT-WHITE SCHOOL
RAW DATA FROM FILE MGHS2.LSF
LATENT VARIABLES: SPATIAL VERBAL
RELATIONSHIPS:
VISPERC = 1*SPATIAL
LOZENGES = SPATIAL
PARAGRAP = 1*VERBAL
SET THE ERROR VARIANCE OF VISPERC FREE
SET THE ERROR VARIANCE OF CUBES FREE
SET THE ERROR VARIANCE OF LOZENGES FREE
SET THE ERROR VARIANCE OF PARAGRAP FREE
SET THE ERROR VARIANCE OF SENTENCE FREE
SET THE ERROR VARIANCE OF WORDMEAN FREE
SET VARIANCE OF SPATIAL FREE
SET VARIANCE OF VERBAL FREE
SET THE COVARIANCE OF SPATIAL AND VERBAL FREE
OPTIONS: SC MI
END OF PROBLEM
```

You will notice that the intercepts (CONST) are removed from the equations in the RELATIONSHIPS heading for Group 2. This specifies that the intercepts (constants) should be constrained to equality across groups.

The M*plus* code for testing the scalar invariance multiple group CFA model is provided below:

```
TITLE:      SCALAR INVARIANCE CFA MODEL

DATA:       FILE IS MGHS.DAT;

VARIABLE:   NAMES ARE VISPERC CUBES LOZENGES
                  PARAGRAP SENTENCE WORDMEAN SCHOOL;

            USEVAR ARE VISPERC CUBES LOZENGES
                  PARAGRAP SENTENCE WORDMEAN;

            GROUPING IS SCHOOL (1 = PASTEUR 2 = GRANT-WHITE);

MODEL:      SPATIAL BY VISPERC CUBES LOZENGES;
            VERBAL BY PARAGRAP SENTENCE WORDMEAN;
            SPATIAL WITH VERBAL;
            [SPATIAL@0 VERBAL@0];

MODEL GRANT-WHITE:
      SPATIAL BY LOZENGES;

OUTPUT:     STDYX MODINDICES(3.84);
```

You will notice that the indicator variables contained in brackets are not specified for the Grant-White school as they were in the metric invariance model.

Model Testing and Modification of the Scalar Multiple Group CFA Model

The scalar invariance model did not fit the data well [LISREL: $\chi^2(25) = 100.858$, $p < .05$; CFI = .951; NNFI/TLI = .942; RMSEA = .142 (90% CI: .114, .172) with $p < .05$; Group 1 SRMR = .021; Group 2 SRMR = .033] and [M*plus*: $\chi^2(25) = 102.064$, $p < .05$; CFI = .963; NNFI/TLI = .956; RMSEA = .143 (90% CI: .115, .173) with $p < .05$; SRMR = .114]. The poor model fit suggests that one or more of the indicator intercepts are not equal across groups. There was a significant $\Delta\chi^2$ test between the partial metric and scalar multiple group CFA models [LISREL: $\Delta\chi^2 = 90.378$, $\Delta df = 4$, $p < .05$] and [M*plus*: $\Delta\chi^2 = 91.511$, $\Delta df = 6$, $p < .05$]. The ΔCFI between partial metric and configural models was also greater than .01, indicating that one or more of the indicator intercept constraints across groups should be released.

The modification indices in LISREL that involve indicator intercept equality constraints across the school groups are provided below, respectively:

GROUP 1: SCALAR CFA MODEL-PASTEUR SCHOOL

Modification Indices and Expected Change

The Modification Indices Suggest to Add the

Path to	from	Decrease in Chi-Square	New Estimate
CUBES	CONST	56.8	-0.48 IN GROUP 1
LOZENGES	CONST	73.0	1.45 IN GROUP 1

GROUP 2: SCALAR CFA MODEL-GRANT-WHITE SCHOOL

Modification Indices and Expected Change

The Modification Indices Suggest to Add the

Path to	from	Decrease in Chi-Square	New Estimate
CUBES	CONST	56.8	0.64 IN GROUP 2
LOZENGES	CONST	73.0	0.29 IN GROUP 2

The modification indices in M*plus* that involve intercept equality constraints across school groups are provided below:

	M.I.	E.P.C.	Std E.P.C.	StdYX E.P.C.
Group PASTEUR				
Means/Intercepts/Thresholds				
[CUBES]	57.500	0.145	0.145	0.040
[LOZENGES]	73.850	1.482	1.482	0.390
[SPATIAL]	22.926	0.510	0.418	0.418
Group GRANT-WHITE				
BY Statements				
Means/Intercepts/Thresholds				
[CUBES]	57.493	1.263	1.263	0.380
[LOZENGES]	73.843	0.322	0.322	0.154
[SPATIAL]	12.660	0.410	0.381	0.381

Both LISREL and M*plus* suggest that the largest chi-square decrease (by approximately 73 points) would result when releasing the equality constraint of the intercept for *LOZENGES*.

The SIMPLIS code with this intercept freely estimated in both school groups is provided below:

```
GROUP 1: PARTIAL SCALAR CFA MODEL-PASTEUR SCHOOL
RAW DATA FROM FILE MGHS1.LSF
LATENT VARIABLES: SPATIAL VERBAL
RELATIONSHIPS:
```

```
VISPERC  = CONST + 1*SPATIAL
CUBES    = CONST + SPATIAL
LOZENGES = CONST + SPATIAL
PARAGRAP = CONST + 1*VERBAL
SENTENCE = CONST + VERBAL
WORDMEAN = CONST + VERBAL
GROUP 2: PARTIAL SCALAR CFA MODEL-GRANT-WHITE SCHOOL
RAW DATA FROM FILE MGHS2.LSF
LATENT VARIABLES: SPATIAL VERBAL
RELATIONSHIPS:
VISPERC  = 1*SPATIAL
LOZENGES = CONST + SPATIAL
PARAGRAP = 1*VERBAL
SET THE ERROR VARIANCE OF VISPERC FREE
SET THE ERROR VARIANCE OF CUBES FREE
SET THE ERROR VARIANCE OF LOZENGES FREE
SET THE ERROR VARIANCE OF PARAGRAP FREE
SET THE ERROR VARIANCE OF SENTENCE FREE
SET THE ERROR VARIANCE OF WORDMEAN FREE
SET VARIANCE OF SPATIAL FREE
SET VARIANCE OF VERBAL FREE
SET THE COVARIANCE OF SPATIAL AND VERBAL FREE
OPTIONS: SC MI
END OF PROBLEM
```

You will notice that the **CONST** for *LOZENGES* is now specified for Group 2. This will estimate this intercept freely across the two groups.

The M*plus* code to freely estimate the intercept for *LOZENGES* is provided below:

```
TITLE:      PARTIAL SCALAR INVARIANCE CFA MODEL

DATA:       FILE IS MGHS.DAT;

VARIABLE:   NAMES ARE VISPERC CUBES LOZENGES
                    PARAGRAP SENTENCE WORDMEAN SCHOOL;
            USEVAR ARE VISPERC CUBES LOZENGES
                    PARAGRAP SENTENCE WORDMEAN;
            GROUPING IS SCHOOL (1 = PASTEUR 2 = GRANT-WHITE);

MODEL:      SPATIAL BY VISPERC CUBES LOZENGES;
            VERBAL BY PARAGRAP SENTENCE WORDMEAN;
            SPATIAL WITH VERBAL;
            [SPATIAL@0 VERBAL@0];
```

```
MODEL GRANT-WHITE:
     SPATIAL BY LOZENGES;
     [LOZENGES];

OUTPUT:     STDYX MODINDICES(3.84);
```

In M*plus*, the following statement: [LOZENGES] in the MODEL GRANT-WHITE: command will freely estimate the indicator's intercept across groups.

Because we released a parameter equality constraint, we are estimating one additional model parameter for the re-specified *partial scalar invariance* multiple group model. Thus, the *df* for the partial scalar invariance multiple group model decreased by 1, making *df* for this model equal to 24. We are calling this model a partial scalar invariance model because not all of the indicator intercepts are constrained to equality across groups. The re-specified partial scalar invariance multiple group model fit the data well after releasing the equality constraint [LISREL: $\chi^2(24) = 16.471$, $p > .05$; CFI = 1.00; NNFI/TLI = 1.01; RMSEA = .00 (90% CI: .000, .035) with $p > .05$; Group 1 SRMR = .022; Group 2 SRMR = .030] and [M*plus*: $\chi^2(24) = 16.625$, $p > .05$; CFI = 1.00; NNFI/TLI = 1.00; RMSEA = .00 (90% CI: .000, .036) with $p > .05$; SRMR = .031]. The $\Delta\chi^2$ test between the partial metric and partial scalar invariance models was not significant [LISREL: $\Delta\chi^2 = 6.145$, $\Delta df = 5$, $p > .05$] and [M*plus*: $\Delta\chi^2 = 6.072$, $\Delta df = 5$, $p > .05$]. Also, the ΔCFI between the partial metric and partial scalar invariance models was less than .01. Table 9.7 summarizes the model fit for all of the models tested.

Table 9.7: Model Fit Information for the Multiple Group CFA Models

Software/Model	χ^2	df	CFI	NNFI/ TLI	RMSEA (90% CI) p-value	SRMR (P/ GW[a])
LISREL						
Configural model	9.70	16	1.00	1.01	.00 (.00, .04) p > .05	.018/ .026
Metric model	70.89*	7	.97	.95	.13 (.10, .16) p < .05	.034/ .169
Partial metric model	10.48	19	1.00	1.01	.00 (.00, .02) p > .05	.022/ .029
Scalar model	100.86*	25	.95	.94	.14 (.11, .17) p < .05	.021/ .033
Partial scalar model	16.47	24	1.00	1.01	.00 (.00, .04) p > .05	.022/ .030
M*plus*						
Configural model	9.77	16	1.00	1.00	.00 (.00, .04) p > .05	.019
Metric model	71.39*	20	.98	.96	.13 (.10, .16) p < .05	.065

Table 9.7: (*cont.*)

Software/Model	χ^2	df	CFI	NNFI/ TLI	RMSEA (90% CI) p-value	SRMR (P/ GW[a])
Partial metric model	10.55	19	1.00	1.00	.00 (.00, .02) p > .05	.021
Scalar model	102.06*	25	.96	.96	.14 (.12, .17) p < .05	.114
Partial scalar model	16.63	24	1.00	1.00	.00 (.00, .04) p > .05	.031

Note: *p < .05; [a] = SRMR is reported for groups separately in LISREL; P = Pasteur school; GW = Grant-White school; *N* = 156 for Pasteur students; *N* = 145 for Grant-White students.

Final Model Interpretation

Selected output from LISREL and M*plus* for the final partial scalar invariance model are provided in Tables 9.8 and 9.9, respectively. Table 9.10 formally presents the relevant model estimates from LISREL for the final partial scalar invariance multiple group path model. You will notice in Tables 9.8 and 9.9 that the unstandardized estimates for 3 of the factor loadings are constrained to equality as well as 5 of the indicator intercepts. More specifically, the unstandardized factor loading and the intercept for *LOZENGES* are not constrained to equality across groups. This implies that the relationship between *LOZENGES* and the SPATIAL factor and the scores on LOZENGES are different at the two schools. It is important to note that it is not uncommon to find a non-invariant indicator intercept for an indicator that has a non-invariant loading, which is the case here.

Table 9.8: Selected LISREL Output for the Final Partial Scalar Invariance CFA Model

```
GROUP 1: PARTIAL SCALAR CFA MODEL-PASTEUR SCHOOL

LISREL Estimates (Maximum Likelihood)

      Measurement Equations
 VISPERC = 0.0733 + 1.000*SPATIAL, Errorvar.= 0.530 , R² = 0.722
Standerr   (0.0750)                            (0.0632)
Z-values   0.977                               8.389
P-values   0.329                               0.000

   CUBES = 0.453 +    2.965*SPATIAL, Errorvar.= 0.444 , R² = 0.965
Standerr  (0.192)    (0.120)                   (0.170)
Z-values   2.353     24.665                     2.605
P-values   0.019      0.000                     0.009

LOZENGES = 1.457 +    2.959*SPATIAL, Errorvar.= 0.466 , R² = 0.963
Standerr  (0.199)    (0.129)                   (0.171)
Z-values   7.302     22.898                     2.727
P-values   0.000      0.000                     0.006
```

Table 9.8: (*cont.*)

```
PARAGRAP = - 0.0398 + 1.000*VERBAL, Errorvar.= 0.499, R² = 0.697
Standerr    (0.0750)                          (0.0602)
Z-values    -0.530                            8.283
P-values    0.596                             0.000

  SENTENCE = 0.133    + 2.782*VERBAL, Errorvar.= 0.452 , R² = 0.952
Standerr    (0.176)    (0.119)                 (0.152)
Z-values    0.759     23.317                   2.979
P-values    0.448      0.000                   0.003

  WORDMEAN = 0.148    + 2.768*VERBAL, Errorvar.= 0.505 , R² = 0.946
Standerr    (0.176)    (0.119)                 (0.153)
Z-values    0.840     23.207                   3.307
P-values    0.401      0.000                   0.001
```

```
        Covariance Matrix of Independent Variables
                SPATIAL     VERBAL
                --------    --------
SPATIAL          1.378
                (0.191)
                 7.224
VERBAL           0.530       1.149
                (0.115)     (0.161)
                 4.597       7.143
```

Within Group Completely Standardized Solution

```
        LAMBDA-X
                SPATIAL     VERBAL
                --------    --------
VISPERC          0.850       - -
 CUBES           0.982       - -
LOZENGES         0.981       - -
PARAGRAP          - -       0.835
SENTENCE          - -       0.976
WORDMEAN          - -       0.973
```

```
        PHI
                SPATIAL     VERBAL
                --------    --------
SPATIAL          1.000
VERBAL           0.421       1.000
```

```
        THETA-DELTA

                VISPERC    CUBES  LOZENGES  PARAGRAP  SENTENCE  WORDMEAN
                --------  -------- -------- -------- -------- --------
                 0.278     0.035    0.037    0.303     0.048     0.054
```

Table 9.8: (*cont.*)

GROUP 2: PARTIAL SCALAR CFA MODEL-GRANT-WHITE SCHOOL

LISREL Estimates (Maximum Likelihood)

Measurement Equations

VISPERC = 0.0733 + 1.000*SPATIAL, Errorvar.= 0.462, R² = 0.692
Standerr (0.0750)   (0.0595)
Z-values 0.977   7.771
P-values 0.329   0.000

   CUBES = 0.453   + 2.965*SPATIAL, Errorvar.= 0.754 , R² = 0.924
Standerr (0.192)   (0.120)   (0.225)
Z-values 2.353   24.665   3.350
P-values 0.019   0.000   0.001

LOZENGES = 0.286   + 1.959*SPATIAL, Errorvar.= 0.313, R² = 0.927
Standerr (0.134)   (0.0974)   (0.0981)
Z-values 2.132   20.112   3.190
P-values 0.033   0.000   0.001

PARAGRAP = - 0.0398 + 1.000*VERBAL, Errorvar.= 0.604, R² = 0.656
Standerr   (0.0750)   (0.0745)
Z-values   -0.530   8.101
P-values   0.596   0.000

SENTENCE = 0.133   + 2.782*VERBAL, Errorvar.= 0.382 , R² = 0.959
Standerr (0.176)   (0.119)   (0.162)
Z-values 0.759   23.317   2.364
P-values 0.448   0.000   0.018

WORDMEAN = 0.148   + 2.768*VERBAL, Errorvar.= 0.499 , R² = 0.946
Standerr (0.176)   (0.119)   (0.165)
Z-values 0.840   23.207   3.031
P-values 0.401   0.000   0.002

Covariance Matrix of Independent Variables

                 SPATIAL   VERBAL
                 --------   --------
   SPATIAL   1.039
                  (0.149)
                   6.978
    VERBAL   0.307   1.150
                  (0.099)   (0.166)
                   3.113   6.941

Within Group Completely Standardized Solution

        LAMBDA-X

                 SPATIAL   VERBAL
                 --------   --------
   VISPERC   0.832   - -
     CUBES   0.961   - -

Table 9.8: (*cont.*)

```
LOZENGES     0.963         - -
PARAGRAP      - -          0.810
SENTENCE      - -          0.979
WORDMEAN      - -          0.973

        PHI

             SPATIAL      VERBAL
             --------     --------
SPATIAL      1.000
  VERBAL     0.281        1.000

        THETA-DELTA
             VISPERC    CUBES  LOZENGES  PARAGRAP  SENTENCE  WORDMEAN
             --------  --------  --------  --------  --------  --------
             0.308     0.076     0.073     0.344     0.041     0.054
```

Table 9.9: Selected M*plus* Output for the Final Partial Scalar Invariance CFA Model

MODEL RESULTS

	Estimate	S.E.	Est./S.E.	Two-Tailed P-Value
Group PASTEUR				
SPATIAL BY				
VISPERC	1.000	0.000	999.000	999.000
CUBES	2.965	0.120	24.659	0.000
LOZENGES	2.959	0.129	22.858	0.000
VERBAL BY				
PARAGRAP	1.000	0.000	999.000	999.000
SENTENCE	2.782	0.119	23.371	0.000
WORDMEAN	2.768	0.119	23.279	0.000
SPATIAL WITH				
VERBAL	0.526	0.114	4.599	0.000
Means				
SPATIAL	0.000	0.000	999.000	999.000
VERBAL	0.000	0.000	999.000	999.000
Intercepts				
VISPERC	0.073	0.075	0.977	0.328
CUBES	0.453	0.192	2.363	0.018
LOZENGES	1.457	0.199	7.334	0.000
PARAGRAP	−0.040	0.075	−0.532	0.595
SENTENCE	0.133	0.175	0.762	0.446
WORDMEAN	0.148	0.175	0.843	0.399
Variances				
SPATIAL	1.369	0.190	7.221	0.000
VERBAL	1.142	0.160	7.142	0.000

Table 9.9: (*cont.*)

Residual Variances				
VISPERC	0.527	0.063	8.341	0.000
CUBES	0.441	0.172	2.558	0.011
LOZENGES	0.463	0.173	2.675	0.007
PARAGRAP	0.496	0.060	8.283	0.000
SENTENCE	0.449	0.148	3.034	0.002
WORDMEAN	0.501	0.150	3.348	0.001
Group GRANT-WHITE				
SPATIAL BY				
VISPERC	1.000	0.000	999.000	999.000
CUBES	2.965	0.120	24.659	0.000
LOZENGES	1.959	0.097	20.120	0.000
VERBAL BY				
PARAGRAP	1.000	0.000	999.000	999.000
SENTENCE	2.782	0.119	23.371	0.000
WORDMEAN	2.768	0.119	23.279	0.000
SPATIAL WITH				
VERBAL	0.305	0.098	3.115	0.002
Means				
SPATIAL	0.000	0.000	999.000	999.000
VERBAL	0.000	0.000	999.000	999.000
Intercepts				
VISPERC	0.073	0.075	0.977	0.328
CUBES	0.453	0.192	2.363	0.018
LOZENGES	0.286	0.134	2.142	0.032
PARAGRAP	−0.040	0.075	−0.532	0.595
SENTENCE	0.133	0.175	0.762	0.446
WORDMEAN	0.148	0.175	0.843	0.399
Variances				
SPATIAL	1.032	0.148	6.993	0.000
VERBAL	1.142	0.164	6.945	0.000
Residual Variances				
VISPERC	0.459	0.059	7.749	0.000
CUBES	0.749	0.230	3.261	0.001
LOZENGES	0.311	0.099	3.137	0.002
PARAGRAP	0.599	0.074	8.127	0.000
SENTENCE	0.380	0.158	2.403	0.016
WORDMEAN	0.496	0.160	3.104	0.002

Table 9.9: (*cont.*)

STANDARDIZED MODEL RESULTS

STDYX Standardization

	Estimate	S.E.	Est./S.E.	Two-Tailed P-Value
Group PASTEUR				
SPATIAL BY				
VISPERC	0.850	0.022	38.547	0.000
CUBES	0.982	0.007	135.063	0.000
LOZENGES	0.981	0.007	133.107	0.000
VERBAL BY				
PARAGRAP	0.835	0.024	35.053	0.000
SENTENCE	0.976	0.008	115.910	0.000
WORDMEAN	0.973	0.009	111.380	0.000
SPATIAL WITH				
VERBAL	0.421	0.068	6.216	0.000
Means				
SPATIAL	0.000	0.000	999.000	999.000
VERBAL	0.000	0.000	999.000	999.000
Intercepts				
VISPERC	0.053	0.055	0.976	0.329
CUBES	0.128	0.055	2.349	0.019
LOZENGES	0.413	0.061	6.821	0.000
PARAGRAP	−0.031	0.058	−0.531	0.596
SENTENCE	0.044	0.057	0.763	0.445
WORDMEAN	0.049	0.057	0.844	0.398
Variances				
SPATIAL	1.000	0.000	999.000	999.000
VERBAL	1.000	0.000	999.000	999.000
Residual Variances				
VISPERC	0.278	0.037	7.413	0.000
CUBES	0.035	0.014	2.475	0.013
LOZENGES	0.037	0.014	2.568	0.010
PARAGRAP	0.303	0.040	7.606	0.000
SENTENCE	0.048	0.016	2.945	0.003
WORDMEAN	0.054	0.017	3.192	0.001
Group GRANT-WHITE				
SPATIAL BY				
VISPERC	0.832	0.025	33.560	0.000
CUBES	0.961	0.013	74.499	0.000
LOZENGES	0.963	0.013	75.734	0.000
VERBAL BY				
PARAGRAP	0.810	0.026	30.648	0.000
SENTENCE	0.979	0.009	108.773	0.000
WORDMEAN	0.973	0.009	106.248	0.000

Table 9.9: (*cont.*)

SPATIAL WITH				
VERBAL	0.281	0.079	3.548	0.000
Means				
SPATIAL	0.000	0.000	999.000	999.000
VERBAL	0.000	0.000	999.000	999.000
Intercepts				
VISPERC	0.060	0.061	0.977	0.329
CUBES	0.145	0.062	2.337	0.019
LOZENGES	0.139	0.065	2.121	0.034
PARAGRAP	−0.030	0.057	−0.532	0.595
SENTENCE	0.044	0.058	0.759	0.448
WORDMEAN	0.049	0.058	0.840	0.401
Variances				
SPATIAL	1.000	0.000	999.000	999.000
VERBAL	1.000	0.000	999.000	999.000
Residual Variances				
VISPERC	0.308	0.041	7.463	0.000
CUBES	0.076	0.025	3.075	0.002
LOZENGES	0.073	0.024	2.970	0.003
PARAGRAP	0.344	0.043	8.044	0.000
SENTENCE	0.041	0.018	2.336	0.019
WORDMEAN	0.054	0.018	3.011	0.003

R-SQUARE

Group PASTEUR

Observed Variable	Estimate	S.E.	Est./S.E.	Two-Tailed P-Value
VISPERC	0.722	0.037	19.273	0.000
CUBES	0.965	0.014	67.532	0.000
LOZENGES	0.963	0.014	66.553	0.000
PARAGRAP	0.697	0.040	17.526	0.000
SENTENCE	0.952	0.016	57.955	0.000
WORDMEAN	0.946	0.017	55.690	0.000

Group GRANT-WHITE

Observed Variable	Estimate	S.E.	Est./S.E.	Two-Tailed P-Value
VISPERC	0.692	0.041	16.780	0.000
CUBES	0.924	0.025	37.249	0.000
LOZENGES	0.927	0.024	37.867	0.000
PARAGRAP	0.656	0.043	15.324	0.000
SENTENCE	0.959	0.018	54.386	0.000
WORDMEAN	0.946	0.018	53.124	0.000

Table 9.10: Unstandardized Estimates, Standard Errors (S.E.), and Standardized Estimates for the Partial Scalar Invariance Multiple Group CFA Model

Parameter	Pasteur School			Grant-White School		
	Unstandardized Estimate	S.E.	Standardized Estimate	Unstandardized Estimate	S.E.	Standardized Estimate
Spatial						
Visperc	1.00	–	.85	1.00	–	.83
Cubes	2.97*	.12	.98	2.97*	.12	.96
Lozenges[a]	2.96*	.13	.98	1.96*	.10	.96
Verbal						
Paragrap	1.00	–	.84	1.00	–	.81
Sentence	2.78*	.12	.98	2.78*	.12	.98
Wordmean	2.77*	.12	.97	2.77*	.12	.97
Intercepts						
Visperc	.07	.08	.28	.07	.08	.31
Cubes	.45*	.19	.04	.45*	.19	.08
Lozenges[a]	1.46*	.20	.04	.29*	.13	.07
Paragrap	−.04	.08	.30	−.04	.08	.34
Sentence	.13	.18	.05	.13	.18	.04
Wordmean	.15	.18	.05	.15	.18	.05
Spatial variance	1.38*	.19	1.00	1.04*	.15	1.00
Verbal variance	1.15*	.16	1.00	1.15*	.17	1.00
Spatial and verbal covariance	.53*	.12	.42	.31*	.10	.28

Note: * $p < .05$; [a] = non-invariant model parameter estimate across groups was detected for this relationship. – = S.E. not estimated for reference indicators set equal to 1.

The unstandardized estimates are commonly used to describe differences for the non-invariant model parameter estimates. You will notice in Tables 9.9–9.10 that the standardized estimates are freely estimated across groups. Again, the unstandardized estimates are constrained to equality in multiple group models in order to account for the variance of the variables in the model. Looking at Table 9.10, the relationship between LOZENGES and the Spatial factor is stronger for students at the Pasteur school than for students at the Grant-White school. Moreover, the intercept for LOZENGES is larger for students at the Pasteur school than for students at the Grant-White school. Perhaps spatial relations are included in the curriculum at the Pasteur school, but not at the Grant-White school? When detecting differences with multiple group modeling, it is important to then determine potential reasons for the differences, noting that potential explanations for the differences should be evaluated in future research studies.

STRICT INVARIANCE TESTING

We did not test for strict invariance for the multiple group CFA model example. If strict invariance was to be tested, the SIMPLIS code below would constrain error variances to equality across groups:

```
GROUP 1: STRICT INVARIANCE CFA MODEL-PASTEUR SCHOOL
RAW DATA FROM FILE MGHS1.LSF
LATENT VARIABLES: SPATIAL VERBAL
RELATIONSHIPS:
VISPERC = CONST + 1*SPATIAL
CUBES = CONST + SPATIAL
LOZENGES = CONST + SPATIAL
PARAGRAP = CONST + 1*VERBAL
SENTENCE = CONST + VERBAL
WORDMEAN = CONST + VERBAL
GROUP 2: STRICT INVARIANCE CFA MODEL-GRANT-WHITE SCHOOL
RAW DATA FROM FILE MGHS2.LSF
LATENT VARIABLES: SPATIAL VERBAL
RELATIONSHIPS:
VISPERC = 1*SPATIAL
LOZENGES = CONST + SPATIAL
PARAGRAP = 1*VERBAL
SET VARIANCE OF SPATIAL FREE
SET VARIANCE OF VERBAL FREE
SET THE COVARIANCE OF SPATIAL AND VERBAL FREE
OPTIONS: SC MI
END OF PROBLEM
```

You will notice that the six statements in Group 2 to set error variances for the indicators free are deleted in the SIMPLIS code.

The M*plus* code provided below would constrain error variances for the indicators to equality across groups:

```
TITLE:     STRICT INVARIANCE CFA MODEL

DATA:      FILE IS MGHS.DAT;

VARIABLE:  NAMES ARE VISPERC CUBES LOZENGES
                PARAGRAP SENTENCE WORDMEAN SCHOOL;

           USEVAR ARE VISPERC CUBES LOZENGES
                PARAGRAP SENTENCE WORDMEAN;

           GROUPING IS SCHOOL (1 = PASTEUR 2 = GRANT-WHITE);
```

```
MODEL:      SPATIAL BY VISPERC CUBES LOZENGES;
            VERBAL BY PARAGRAP SENTENCE WORDMEAN;
            SPATIAL WITH VERBAL;
            [SPATIAL@0 VERBAL@0];
            VISPERC  (1);
            CUBES    (2);
            LOZENGES (3);
            PARAGRAP (4);
            SENTENCE (5);
            WORDMEAN (6);
MODEL GRANT-WHITE:
            SPATIAL BY LOZENGES;
            [LOZENGES];

OUTPUT:     STDYX MODINDICES(3.84);
```

You will notice that the indicators are included in the MODEL command with numbers in parenthesis. When endogenous variables are listed in the MODEL command, they represent error variances. The numbers in parenthesis will constrain the error variance for each indicator to be equal across groups. Invariance tests could proceed with strict equality constraints as they have for the previous metric and scalar invariance tests with the use of $\Delta\chi^2$ tests and the ΔCFI.

STRUCTURAL MODEL GROUP DIFFERENCES

The multiple group method can also test for group differences in a structural model. The invariance testing would proceed as previously described where sequentially more stringent invariance constraints are imposed across groups in the following order: 1) conduct the two-step SEM model testing separately in each group; 2) test the configural model with no equality constraints; 3) test the metric model with factor loading constraints; 4) test the scalar model with intercept constraints; and 5) test the structural invariance model with structural equality constraints imposed across groups. The $\Delta\chi^2$ test and/or the ΔCFI between the sequentially constrained, nested models can be used to determine whether the equality constraints that were imposed resulted in a significant decrease in fit. If factor mean comparisons across groups are not of interest in the structural multiple group model, they can be set equal to zero during the invariance testing sequence (Byrne, 2012).

MULTIPLE GROUP MODELS WITH ORDINAL INDICATORS

The indicators for the multiple group CFA example presented in this chapter were continuous variables, for which maximum likelihood (ML) is an appropriate estimator. When the indicators are ordinal, alternative estimators are recommended in order to avoid biased parameter estimates. For instance, robust weighted least squares (DWLS in LISREL; WLSMV in M*plus*) estimation is recommended when the indicators are ordered categorical data, especially when Likert scales with less than 5 option responses are used. Invariance testing would proceed as previously described, beginning with testing models separately in each group through testing the scalar or strong invariance model. Recall that thresholds are estimated with ordinal indicators instead of intercepts (see Chapter 6). Thus, when testing scalar or strong invariance with ordinal indicators, the thresholds would be constrained to equality across groups. In addition, the $\Delta\chi^2$ test would need to be adjusted when using DWLS or WLSMV estimators. As mentioned in Chapter 5, a $\Delta\chi^2$ test has been proposed by Asparouhov and Muthén (2010) with DWLS or WLSMV, which can be conducted using the **DIFFTEST** procedure in M*plus*. More research about invariance testing of multiple group CFA models with ordinal indicators is warranted, especially in terms of using the difference between goodness-of-fit indices (e.g., ΔCFI; see Sass, Schmitt, & Marsh, 2014). Svetina, Rutkowski, and Rutkowski (2020) provide a tutorial on invariance testing in multiple group CFA with ordinal indicators.

CAUTIONS ABOUT INVARIANCE TESTING

When conducting multiple group modeling, it is important to note that equality constraints imposed on non-invariant parameters can result in detecting non-invariance falsely elsewhere in the model. As previously mentioned, the reference indicator selected for a factor has important implications since its loading is assumed invariant across groups by setting it equal to 1. If the loading of the reference indicator is non-invariant across groups, non-invariance can show up in other parameters constrained to equality in the model due to the misspecification. There are methods that can help researchers when first selecting a reference indicator for a multiple group analysis (Cheung & Rensvold, 1999; Yoon & Millsap, 2007). Alternatively, effects coding can be used to set the scale of the factors in a multiple group CFA (Little et al., 2006). Although more research in this area is needed, once the factors are scaled adequately, it has been suggested that invariance tests be conducted one parameter at a time (e.g., Hancock, Stapleton, & Arnold-Berkovits, 2009). This would help to prevent false detections of non-invariance in invariant model parameters when constraining non-invariant parameters to equality.

The same concerns discussed in Chapter 5 about model modifications also pertain to multiple group testing. Specifically, when testing for the invariance of many parameters, the $\Delta\chi^2$ test and/or the ΔCFI may falsely indicate non-invariance simply by chance. With large sample sizes, the $\Delta\chi^2$ test will be more likely to detect small magnitudes between model parameters across groups as statistically significant. The use of Δgoodness-of-fit indices for invariance testing, particularly with large sample sizes, has been promising. The ΔCFI cutoff we have been using in this chapter (ΔCFI $\leq$ .01) between nested, invariance constrained models to provide support for invariance was suggested by Cheung and Rensvold (2002) and is commonly used. Others have suggested different ΔCFI cutoff values for different invariance tests as well as different Δgoodness-of-fit indices to be used (e.g., ΔRMSEA). For instance, Chen (2007) recommended using a ΔCFI $\leq$ .01 combined with a ΔRMSEA $\leq$ .015 for both metric and scalar invariance tests. Meade, Johnson, and Braddy (2008) recommended a smaller cutoff of ΔCFI $\leq$ .002 to provide support for invariance. Rutkowski and Svetina (2014) proposed using a ΔCFI $\leq$ .02 combined with a ΔRMSEA $\leq$ .03 for metric invariance and a ΔCFI $\leq$ .01 combined with a ΔRMSEA $\leq$ .01 for scalar invariance with large numbers of groups being compared (i.e., 10 and 20). Thus, there is not one Δgoodness-of-fit index that works optimally in all scenarios.

Provided some of the cautions about multiple group modeling and testing, it is important that decisions about factor scaling and which parameters to constrain and/or freely estimate be soundly based on theoretical foundations. Once a final multiple group model is established, it is important to cross-validate the model in subsequent samples. If possible, we recommend randomly splitting the data set to be used for multiple group testing in order to conduct invariance tests independently in each of the two data sets. This may be difficult to do in practice, but it is important to be more assured of the results found during invariance testing.

SUMMARY

The multiple group approach in SEM is useful for testing whether group differences exist for a theoretically specified model. The group comparisons can be conducted with multiple regression models, observed variable path models, confirmatory factor models, and structural models. When using multiple group methods in SEM, theoretically meaningful differences can be highlighted and be subsequently investigated in future research studies. Multiple group methods can also provide support for measurement invariance, which provides support for construct validity. However, a researcher can hypothesize that group differences exist in confirmatory factor analysis models based on previous literature and theory.

It is important to consider factor scaling methods and invariance testing methods when conducting multiple group modeling. Although statistical tests and Δgoodness-of-fit indices (e.g., $\Delta\chi^2$ test and ΔCFI) are largely used to determine invariance, theoretical considerations are an important part of the process. Moreover, it is strongly recommended that the final multiple group model be cross-validated.

This chapter illustrated multiple group testing procedures with an observed variable path model and a CFA model. As previously mentioned, multiple group methods can be extended to other models. It is hoped that the examples in this chapter will allow readers to be able to conduct multiple group modeling in their own research.

EXERCISES

A researcher was interested in whether nursing students' "School Self-Concept" is different depending upon semester in school. During the same academic year, data were collected from students in their first semester ($n = 160$) and from students in their second semester ($n = 150$) of nursing school. Participants responded to the following five items:

1. I enjoy my accomplishments in school.
2. I am able to manage time for school.
3. I have a hard time understanding material presented in class.
4. My teachers like me.
5. I do well on homework assignments.

Ratings were on a 5-point Likert scale (1 = strongly disagree to 5 = strongly agree). Item 3 was reverse-coded. The items were hypothesized to load on a single "School Self-Concept" factor. The data files for LISREL are *SSC1.LSF* and *SSC2.LSF* for first- and second-semester students, respectively. The data file for M*plus* is in *SSC.DAT*. All the data files are on the book website. For this exercise, test the measurement invariance of the single-factor across the two student semester groups. Theory supports using item 1 as the reference indicator in the multiple group model.

For this exercise, be sure to do the following:

a. Test the one-factor model in each group separately and document model fit.
b. Test the configural model with no equality constraints imposed and document model fit.
c. Test the metric model with factor loading constraints imposed and document model fit.

d. Compare the metric and configural models to determine if any of the loadings should be freely estimated across groups. Freely estimate factor loadings if necessary.

e. Test the scalar or strong model with item intercept constraints imposed and document model fit.

f. Compare the scalar/strong and the model you retained from part d to determine if any of the intercepts should be freely estimated across groups. Freely estimate item intercepts if necessary.

g. Include a table of results for each model tested and describe any differences that you discovered during the invariance testing process.

REFERENCES

Aparouhov, T., & Muthén, B. (2010). *Simple second-order chi-square correction.* Retrieved from www.statmodel.com/download/WLSMV_new_chi21.pdf.

Arbuckle, J. L., & Wothke, W. (2003). *Amos 5.0 user's guide.* Chicago, IL: Smallwaters Corporation.

Byrne, B. M. (2012). *Structural equation modeling with Mplus: Basic concepts, applications, and programming.* New York: Routledge/Taylor & Francis Group.

Byrne, B. M., Shavelson, R. J., & Muthén, B. (1989). Testing for the equivalence of factor covariance and mean structures: The issue of partial measurement invariance. *Psychological Bulletin*, 105(3), 456–466.

Byrne, B. & Sunita, M. S. (2006). The MACS approach to testing for multigroup invariance of a second-order structure – A walk through the process. *Structural Equation Modeling: A Multidisciplinary Journal*, 13(2), 287–321.

Chen, F. F. (2007). Sensitivity of goodness of fit indexes to lack of measurement invariance. *Structural Equation Modeling*, 14(3), 464–504.

Cheung, G. W., & Rensvold, R. B. (1999). Testing factorial invariance across groups: Areconceptualization and proposed new method. *Journal of Management*, 25, 1–27.

Cheung, G. W., & Rensvold, R. B. (2002). Evaluating the goodness-of-fit indexes for testing measurement invariance. *Structural Equation Modeling*, 9(2), 233–255.

French, B. F., & Finch, W. H. (2008). Multigroup confirmatory factor analysis: Locating the invariant referent sets. *Structural Equation Modeling*, 15(1), 96–113.

Hancock, G. R. (1997). Structural equation modeling methods of hypothesis testing of latent variable means. *Measurement and Evaluation in Counseling and Development*, 30, 91–105.

Hancock, G. R., Stapleton, L. M., & Arnold-Berkovits, I. (2009). The tenuousness of invariance tests within multisample covariance and mean structure models. In T. Teo & M. S. Khine (Eds.), *Structural equation modeling: Concepts and applications in educational research* (pp. 137–174). Rotterdam, Netherlands: Sense Publishers.

Holzinger, K. J., & Swineford, F. A. (1939). *A study in factor analysis: The stability of a bi-factor solution.* (Supplementary Educational Monographs, No. 48). Chicago, IL: University of Chicago, Department of Education.

Horn, J. L., & McArdle, J. J. (1992). A practical and theoretical guide to measurement invariance in aging research. *Experimental Aging Research*, 18(3–4), 117–144.

Howard, G. S., & Maxwell, S. E. (1982). Do grades contaminate student evaluations of instruction? *Research in Higher Education*, 16, 175–188.

Johnson, E. C., Meade, A. W., & DuVernet, A. M. (2009). The role of referent indicators in tests of measurement invariance. *Structural Equation Modeling*, 16, 642–657.

Jöreskog, K. (1971). Simultaneous factor analysis in several populations. *Psychometrika*, 36, 409–426.

Jöreskog, K., & Sörbom, D. (1993). *LISREL 8: Structural equation modeling with the SIMPLIS command language.* Chicago, IL: Scientific Software International.

Little, T. D. (2013). *Longitudinal structural equation modeling.* New York: Guilford Press.

Little, T. D., Card, N. A., Slegers, D. W., & Ledford, E. C. (2007). Representing contextual effects in multiple-group MACS models. In T. D. Little, J. A. Bovaird, & N. A. Card (Eds.), *Modeling contextual effects in longitudinal studies* (pp. 121–147). New York: Lawrence Erlbaum Associates Publishers.

Little, T. D., Slegers, D. W., & Card, N. A. (2006). A non-arbitrary method of identifying and scaling latent variables in SEM and MACS models. *Structural Equation Modeling*, 13(1), 59–72

Marcoulides, G., & Schumacker, R. E. (Eds.). (2001). *New developments and techniques in structural equation modeling: Issues and techniques.* Mahwah, NJ: Lawrence Erlbaum.

Meade, A. W., Johnson, E. C., & Braddy, P. W. (2008). Power and sensitivity of alternative fit indices in tests of measurement invariance. *Journal of Applied Psychology*, 93(3), 568–592.

Meredith, W. (1993). Measurement invariance, factor analysis and factorial invariance. *Psychometrika*, 58, 525–543.

Rutkowski, L., & Svetina, D. (2014). Assessing the hypothesis of measurement invariance in the context of large-scale international surveys. *Educational and Psychological Measurement*, 74(1), 31–57.

Sass, D. A., Schmitt, T. A., & Marsh, H. W. (2014). Evaluating model fit with ordered categorical data within a measurement invariance framework: A comparison of estimators. *Structural Equation Modeling*, 21(2), 167–180.

Svetina, D., Rutkowski, L., & Rutkowski, D. (2020). Multiple-group invariance with categorical outcomes using updated guidelines: An illustration using M*plus* and the lavaan/semTools packages. *Structural Equation Modeling*, 27(1), 111–130.

Thompson, M. S., & Green, S. B. (2013). Evaluating between-group differences in latent variable means. In G. R. Hancock & R. O. Mueller (Eds.), *Structural equation modeling: A second course* (pp. 163–218). Charlotte, NC: IAP Information Age Publishing.

Vandenberg, R. J., & Lance, C. E. (2000). A review and synthesis of the measurement invariance literature: Suggestions, practices, and recommendations for organizational research. *Organizational Research Methods*, 3(1), 4–70.

Whittaker, T. A. (2013). The impact of noninvariant intercepts in latent means models. *Structural Equation Modeling*, 20, 108–130.

Whittaker, T. A., & Khojasteh, J. (2013). A comparison of methods to detect invariant reference indicators in structural equation modelling. *International Journal of Quantitative Research in Education*, 1(4), 426–442.

Widaman, K. F., & Reise, S. P. (1997). Exploring the measurement invariance of psychological instruments: Applications in the substance use domain. In K. J. Bryant, M. Windle, & S. G. West (Eds.), *The science of prevention: Methodological advances from alcohol and substance abuse research* (pp. 281–324). Washington, DC: American Psychological Association.

Wu, A. D., Li, Z., & Zumbo, B. D. (2007). Decoding the meaning of factorial invariance and updating the practice of multi-group confirmatory factor analysis: A demonstration with TIMSS data. *Practical Assessment, Research, and Evaluation*, 12(3), 1–26.

Yoon, M., & Millsap, R. E. (2007). Detecting violations of factorial invariance using data-based specification searches: A Monte Carlo study. *Structural Equation Modeling*, 14(3), 435–463.

Chapter 10

SEM CONSIDERATIONS

BEST PRACTICES IN SEM

Many authors in the field of SEM have suggested important guidelines for researchers when conducting SEM and reporting the results. For instance, Thompson (2000) provided guidance for conducting structural equation modeling by citing key issues and including the following list of ten commandments for good structural equation modeling analysis and reporting: (1) avoid conclusions that a model is the only one to fit the data; (2) cross-validate any modified model with split-sample or new data; (3) test multiple competing models; (4) evaluate measurement models first, then structural models; (5) evaluate models by fit, theory, and practical concerns; (6) report multiple model fit indices; (7) meet multivariate normality assumptions; (8) pursue parsimonious models; (9) consider variable scales of measurement and distributions; and (10) do not use small samples. Boomsma (2000) discussed how to write a research paper when structural equation models were used in empirical research and how to decide what information to report. His basic premise was that all information necessary for someone else to replicate the analysis should be reported. He provided a flowchart of the SEM process and discussed key features, including the theoretical framework, plausible models, model characteristics, data characteristics, model estimation, model evaluation and selection, and conclusions.

MacCallum and Austin (2000) provided an excellent survey of problems in applications of SEM in psychological research. McDonald and Ho (2002) examined 41 articles meeting specific criteria that were published from 1995 to 1997 across 13 different psychology journals. They advised that SEM researchers should provide a detailed rationale of the SEM model tested together with alternative models, describe model identification, address non-normality and missing data issues, include all parameter estimates with standard errors, provide the correlation matrix with standard deviations (and perhaps residuals), and report model

DOI: 10.4324/9781003044017-10

fit indices. Shah and Goldstein (2006) extensively reviewed 93 articles meeting specific criteria that were published in four different operations management journals from the earliest application of SEM in the journal to 2003. Their summary of studies (see Table 1 in their article) provides valuable information for methodological researchers in SEM in terms of considerations for manipulated factors (e.g., sample size, indicators per factor, number of latent variables) in Monte Carlo simulation studies. They also provide recommendations for improving the use of SEM in operations management research. They suggested that researchers should use alternative models, discuss missing data and non-normality issues, include the input data matrix in the paper for other researchers to be able to use, and discuss issues related to sample size and model identification.

Mueller and Hancock (2008) presented illustrations for each of the SEM stages (i.e., model conceptualization, identification and estimation, model fit assessment, and model modification) and provided recommendations for best practices when writing manuscript sections (i.e., introduction, method section, results section, and discussion) describing SEM analysis and results. Hoyle and Isherwood (2013) developed a questionnaire for studies reporting SEM findings that supplements the research report standards proposed by the Publication and Communications Board Working Group on Journal Article Reporting Standards (JARS) of the American Psychological Association (APA Publication and Communications Board Working Group on Journal Article Reporting Standards, 2008).

Recently, Zhang, Dawson, and Kline (2021) conducted a review of SEM studies published in top tier organizational and management journals using an assessment system based on published works about best practices in SEM. Their assessment system used nine criteria when rating the studies selected for their review, including: 1) justification for using SEM provided; 2) hypothesis tested in the model provided; 3) sample size justified and statistical power tested; 4) distributional assumptions examined and met; 5) missing data reported and handled appropriately; 6) reliability information provided; 7) two-step SEM is conducted; 8) the chi-square test, TLI, RMSEA, and SRMR are reported; and 9) residuals reported (for local fit assessment). Their evaluation of published studies revealed that a majority of the studies reviewed did not include a strong justification for using SEM, information about missing data, information about sample size or statistical power, RMSEA fit information, and information about the residuals. Zhang et al. (2021) provided the following practical suggestions for researchers conducting and reporting SEM analysis: provide a research and analytic plan; describe re-specifications to the original model; cross-validate the model; and do not retain a model that is unlikely to cross-validate.

Based on the previously mentioned references and many preceding references (e.g., Cliff, 1983; Breckler, 1990; Raykov, Tomer, & Nesselroade, 1991; Hoyle & Panter, 1995; Maxwell & Cole, 1995), we find the following *general* suggestions to be valuable when publishing SEM research:

1. Provide a review of literature that supports your theoretical model.
2. Provide information about the software program used along with the version.
3. Indicate the type of SEM model analysis.
4. Include correlation matrix, sample size, means, and standard deviations of variables.
5. Include a diagram of your theoretical model.
6. Describe issues concerning normality and missing data.
7. For interpretation of results, indicate estimation procedure used and why; describe fit indices used and why; include power and sample size determination.
8. Provide unstandardized parameter estimates with corresponding standard errors as well as standardized parameter estimates.

Our suggestions are important because the SEM software, model, data, and program will be archived in the journal. The power, sample size, and estimates provided will permit future use in research synthesis studies. Providing this research information will also permit future cross-cultural research, multi-sample or multi-group comparisons, replication, or validation by others in the research community because the analysis can be further examined.

We have made many of these same suggestions in our previous chapters, so our intentions in this chapter are to succinctly summarize guidelines and recommendations for SEM researchers. A brief summary checklist should help to remind us of the issues and analysis considerations a researcher makes:

CHECKLIST FOR SEM

Basic Issues

1. Is sample size sufficient (power, effect size)?
2. Have you addressed missing data (MCAR, MAR, etc.)?
3. Have you addressed normality, outliers, linearity, restriction of range?
4. Are you using the correct covariance matrix?
5. Have you selected the correct estimation method?
6. Is the theoretical model identified ($df = 0$ or greater)?

Analysis Issues

1. Have you reported the correct fit indices?
2. Have you provided unstandardized estimates (with corresponding standard errors) and standardized estimates?
3. Have you scaled the latent factors appropriately?
4. Have you justified any model modifications (e.g., adding error covariances)?
5. Have you cross-validated the model (assuming sufficient sample size)?
6. Have you diagrammed the model and/or provided estimates in the diagram?

We will follow through on these checklists of items in more detail as we discuss the modeling steps a researcher takes in conducting structural equation modeling.

MODEL SPECIFICATION

A researcher should begin an SEM research study with a rationale and purpose for the study, followed by a sound theoretical foundation of the path model, measurement model, and/or the structural model. *Model specification* involves determining every relationship and parameter in the model that is of interest to the researcher. This includes a discussion of the latent variables and how they are defined in the measurement model. The hypothesis should involve the testing of the hypothesized model and/or a difference between alternative models. Moreover, the goal of the researcher is to determine, as best as possible, the theoretical model that explains the observed relationships among the data in the sample. If the theoretical model is misspecified, it could yield biased parameter estimates. We do not typically know the true population model, so bias in parameter estimates can be attributed to specification error. The model should be developed from the available theory and research in the substantive area, which should be the main purpose of the literature review. A set of recommendations for model specification is given in Table 10.1.

Table 10.1: Model Specification Recommendations

1. Did you provide a rationale and purpose for your study, including why SEM rather than another statistical analysis approach was required?
2. Did you describe your latent variables, thus providing a substantive background to how they are measured?
3. Did you establish a sound theoretical basis for your path models, measurement models, and/or structural models?
4. Did you theoretically justify alternative models for comparison (e.g., nested models)?
5. Did you clearly state the hypotheses for testing the parameters and/or models?
6. Did you discuss the expected magnitude and direction of expected parameter estimates?
7. Did you include a figure or diagram of your path, measurement, and/or structural models?

MODEL IDENTIFICATION

In structural equation modeling, it is crucial that the researcher resolve the *identification problem* prior to the estimation of parameters in observed variable path models, measurement models, and structural models. In the identification problem, we ask the following question: On the basis of the sample data contained in the sample covariance matrix (**S**) and the theoretical model implied by the population covariance matrix ($\widehat{\Sigma}$), can a unique set of parameters be estimated? A quick check on model identification is whether the degrees of freedom (*df*) of the theoretical model are equal to or greater than 0. Models with *df* = 0 are just-identified and will fit the data perfectly. Thus, if the goal is to test the fit of the theoretical model to the data, the *df* of the model should be greater than 0. Also, it is important to make sure that latent factors are scaled appropriately in the model. A set of recommendations for model identification is given in Table 10.2.

Table 10.2: Model Identification Recommendations

1. Did you specify the number of distinct values in your sample covariance matrix?
2. Did you indicate the number of free parameters to be estimated?
3. Did you report the number of degrees of freedom and thereby the level of identification of the model?
4. How did you scale the latent variables (i.e., fix either one factor loading per latent variable or the latent variable variances to 1)?
5. Did you avoid non-recursive models until identification has been assured?
6. Did you utilize parsimonious models to assist with identification?

DATA PREPARATION

In most research applications today, the raw data will be used and read into SEM software programs. It is important for the researcher to provide information about the appropriateness of the data for the analysis to be conducted. As described in Chapter 2, it is crucial for the data to be inspected for outliers, normality, linearity, and missing data. It is also important for researchers to report psychometric information about the data collected for the SEM analysis, such as reliability and validity information provided in previous research as well as for the new sample of data collected.

It is also crucial to consider issues about adequate sample size and power. We recommend that sample size and power be considered prior to data collection. SEM requires large sample sizes to prevent convergence issues as well as unstable model parameter estimates. Offering a recommendation on sample size requirements in SEM is difficult because it depends upon model complexity, the strength of the relationships among variables and factors, normality of the variables, missing data, and the estimator, to name a few factors. It has been suggested that appropriate sample size be determined using the $N{:}t$ ratio (called the $N{:}q$ ratio in other studies) or the number of participants (N) to number of parameters to freely estimate (t) ratio. Bentler and Chou (1987) proposed that this ratio be no less than 5:1, but ideally it should be 10:1. Jackson's (2003) study resulted in suggesting a ratio of 20:1 for more confidence in the SEM results. We recommend that researchers attempt to meet the 20:1 ratio if achievable, but not fall below the 10:1 ratio. Hence, 20 participants per parameter to freely estimate in the model is our general recommendation.

A related issue to sample size is having adequate power for the SEM analysis to detect significant parameters when they exist in the population. Power can be determined for the overall model or for a particular model parameter. MacCallum, Browne, and Sugawara (1996) provided an approach to testing the power of a model using the root mean square error of approximation (RMSEA). Hancock and French (2013) presented an in-depth discussion of the null and alternative model testing process in SEM, especially in regard to the non-centrality parameter (NCP; λ) and the root mean square error of approximation (RMSEA; ε). They provided sample size tables for power and values of the RMSEA, which can also be produced using Schoemann, Preacher, and Coffman's (2010) online calculator.

A better way to compute power for an overall SEM model as well as specific model parameter estimates is to perform a Monte Carlo study. This method was proposed by Muthén and Muthén (2002) who demonstrated how to conduct a Monte Carlo study for an SEM model to determine appropriate sample size and statistical power. This can be done with the MONTECARLO feature in M*plus*. They provide examples using different models in their article that researchers can implement.

In addition to the previously mentioned considerations, it is recommended that researchers provide the correlations with standard deviations in their article or with supplementary materials. It is important for this information to be provided for other researchers to replicate the results as reported and for research synthesis purposes. A set of recommendations for data preparation is given in Table 10.3.

Table 10.3: Data Preparation Recommendations

1. Have you adequately described the population from which the sample data were drawn?
2. Did you use a reasonable sample size for adequate power? What is the ratio of sample size to number of parameters to estimate? Did you conduct a power analysis?
3. Did you report the measurement level and psychometric properties (i.e., reliability and validity) of your variables?
4. Did you report the descriptive statistics for your variables?
5. Did you create a table with correlations, means, and standard deviations?
6. Did you consider and treat any missing data? What was the sample size both before and after treating the missing data?
7. Did you consider and treat any outliers?
8. Did you consider the range of values obtained for variables, as restricted range of one or more variables can reduce the magnitude of correlations?
9. Did you consider and treat any non-normality of the data (e.g., skewness and kurtosis, data transformations)?
10. Did you consider and treat any multicollinearity among the variables?
11. Did you consider whether variables are linearly related, which can reduce the magnitude of correlations?
12. Did you take the measurement scale of the variables into account when computing statistics such as means, standard deviations, and correlations?
13. When using the correlation matrix, did you include standard deviations of the variables in order to obtain correct estimates of standard errors for the parameter estimates?
14. How can others access your data and SEM program (e.g., appendix, website, email)?

MODEL ESTIMATION

With *model estimation*, we need to decide which estimation technique is most appropriate for estimating the parameters in our models based on the data to be used in the analysis. For example, we might choose the maximum likelihood estimation technique because we meet the multivariate normality assumption (acceptable skewness and kurtosis); there are no missing data, no outliers, and have continuous variable data. If the observed variables are interval scaled and multivariate normal, then the ML estimates, standard errors, and chi-square test are appropriate. When we have ordered categorical data, particularly with four or fewer response categories (e.g., with Likert ratings), then *robust* weighted least squares (WLS) estimation is recommended.

Even when estimation procedures are correctly selected, there could still be other factors that can affect parameter estimation in general. Missing data, outliers, multicollinearity, and non-normality of data distributions can seriously affect the estimation process and often result in error messages pertaining to Heywood cases (correlations greater than 1 or negative variance), non-positive definite matrices (determinant of the matrix is zero), or failure to reach convergence (unable to compute a final set of parameter estimates). Thus, iterations may need to be increased or starting values used to help the estimation process converge. It is essential that these problems be reported in SEM applications to inform readers of potential problems with the collected data. We have included some recommendations for model estimation in Table 10.4.

Table 10.4: Model Estimation Recommendations

1. Did you identify the estimation technique based on the type of data matrix?
2. What estimation technique is appropriate for the distribution of the sample data (ML for multivariate normal data; DWLS or WLSMV for categorical data)?
3. Did you specify the type of matrix used in the analysis (e.g., covariance, correlation (Pearson, polychoric, polyserial), augmented moment, or asymptotic matrices)?
4. Did you encounter Heywood cases (correlation greater than 1 or negative variance), multi-collinearity, or non-positive definite matrices?
5. Did you encounter and resolve any convergence problems or inadmissible solution problems by using start values, setting the admissibility check off (in LISREL), using a larger sample size, or using a different method of estimation?
6. Which SEM program and version did you use?
7. Did you include the code used in your study?

MODEL TESTING

Model testing allows researchers to evaluate the appropriateness of the theoretical model against the sample data. Model testing is largely done with model fit indices. However, there are other components involved in model testing (see Chapter 5). Specifically, model testing involves global model fit, parameter fit, and model comparisons, which represent different components of model fit. In addition, it has recently become a suggestion to provide residuals (e.g., standardized or normalized) for the final model in order to allow readers to assess model fit at the local level (Zhang et al., 2021). Residuals are primarily used to detect potential model misspecification during the model modification process. However, they can also highlight relationships that are not explained well by the model, even if global model fit supports adequate fit. We provide a set of recommendations for model testing in Table 10.5.

Table 10.5: Model Testing Recommendations

1. Did you compare alternative models or equivalent models?
2. Did you report several model-fit indices (e.g., for a single model: chi-square, df, CFI, NNFI/TLI, RMSEA, SRMR; for a nested model: $\Delta\chi^2$ test, AIC, BIC)?
3. Did you specify separate measurement models and structural models?
4. Did you provide a table of unstandardized and standardized parameter estimates, standard errors?
5. Do parameter estimates have the expected magnitude and direction?
6. Did you report R^2 values or residual/error variances for endogenous variables?
7. Did you provide the residuals for the final model (could be provided as supplementary information or in an Appendix)?

MODEL MODIFICATION/RE-SPECIFICATION

If the fit of a theoretical model is not acceptable, which is sometimes the case with an initial model, the next step is *model modification* and subsequent evaluation of the new, re-specified model. Modification indices, expected parameter change values, and residuals can be used to help during the modification process. The model modification process should always be detailed in a study once the initial model does not fit the data well. Modified models are less likely to cross-validate due to capitalizations on chance during the modification process. Thus, it is recommended that enough data be gathered to randomly split the sample and run the analysis on both sets of data or gather new data to cross-validate the model. This may be difficult to accomplish in practice, but it is important. At the very least, the researchers should note limitations with model modifications and plan to cross-validate the model in a future study. A set of recommendations for model modification is given in Table 10.6.

Table 10.6: Model Modification Recommendations

---

1. Did you clearly indicate how you modified the initial model?
2. Did you provide a theoretical justification for the modified model?
3. Did you add or delete one parameter at a time? What parameters were deleted and what parameters were added?
4. Did you provide parameter estimates and model-fit indices for both the initial model and the modified models?
5. Did you report statistical significance of free parameters, modification indices, and expected change statistics of fixed parameters, and residual information for all models?
6. How did you evaluate and select the best model?
7. Did you cross-validate your SEM model?

NON-RECURSIVE MODELS

It is important to briefly introduce a general classification used in SEM. This classification applies to both observed variable and latent variable path models. There are two different types of models in SEM: recursive and non-recursive. Recursive models tend to be less complex than non-recursive models whereas non-recursive models may have residual covariances or what are named feedback loops. The models presented in previous chapters in the book are all examples of recursive models. An example of a non-recursive model is provided in Figure 10.1. The model is non-recursive because the direct effect from Y2 to Y1 that is modeled results in a reciprocal pattern of causation between variables Y2 and Y1 given that the errors of Y1 and Y2 covary (i.e., Y2 → Y1 ↔ Y2).

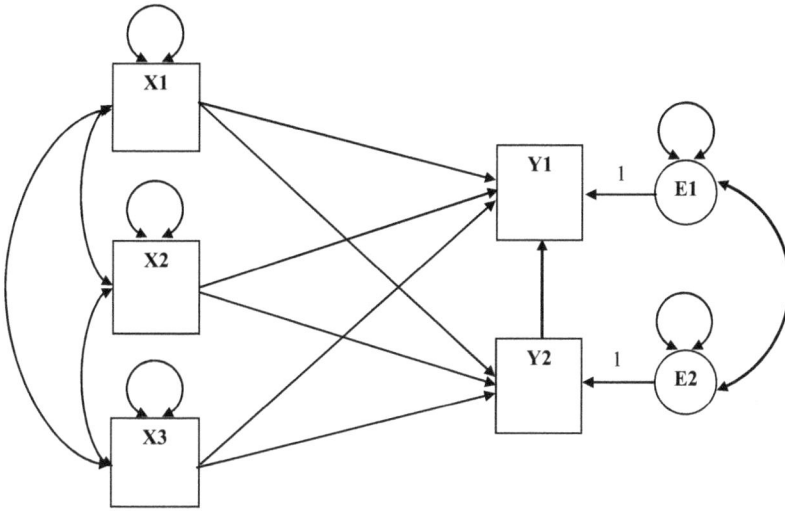

■ **Figure 10.1:** NON-RECURSIVE PATH MODEL WITH A RECIPROCAL PATTERN OF CAUSATION

The model in Figure 10.2 is a non-recursive model with a direct feedback loop. In Figure 10.2, Y1 directly affects Y2 which, in turn, directly affects Y1. When direct feedback loops are modeled, they occur in studies with data collected from a single occasion because the two variables (e.g., Y1 and Y2) are reciprocally related. Hence, temporal precedence is not a concern.

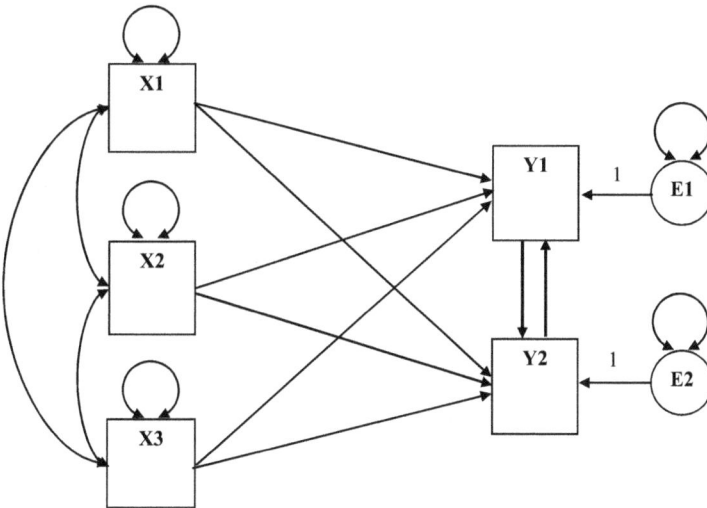

■ **Figure 10.2:** NON-RECURSIVE PATH MODEL WITH DIRECT FEEDBACK LOOP

A non-recursive model with an indirect feedback loop is presented in Figure 10.3. These types of feedback loops involve three or more variables, creating a reciprocal pattern of causation. For example, Y2 directly affects Y1 which, in turn, directly affects Y3 which, in turn, directly affects Y2.

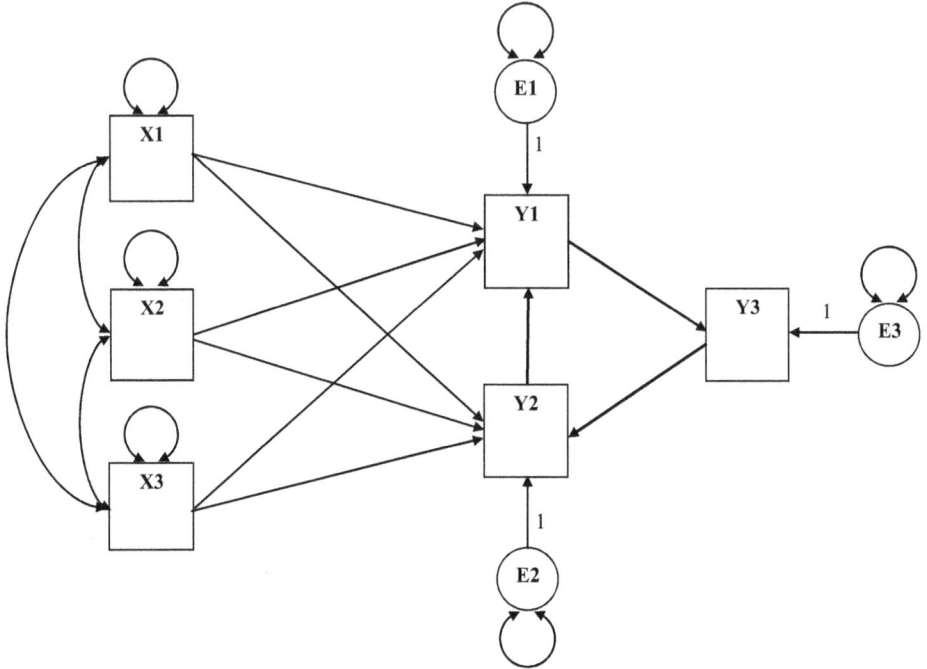

▦ **Figure 10.3:** NON-RECURSIVE PATH MODEL WITH INDIRECT FEEDBACK LOOPS

The main differences with respect to recursive and non-recursive models are identification and estimation. A recursive model is generally an identified model for which model parameters may be able to be estimated. Identification of a non-recursive model is more difficult and could lead to estimation problems (e.g., non-convergence, improper solutions such as Heywood cases) or no model parameter estimation if the model is under-identified. As with any model, theoretical justifications should guide the specification of non-recursive models. Also, researchers should determine whether the non-recursive model is identified prior to model estimation. See Bollen (1989) for more information concerning non-recursive models.

EQUIVALENT MODELS

Another matter in SEM that must be considered is equivalent models. Equivalent models are models where the causal relationships among variables may be specified differently, but they mathematically result in the same implied covariance matrices. Thus, equivalent models result in the same model fit information because they fit the data identically. Equivalent models can be observed with any of the models discussed in this book (e.g., observed variable path models, CFA/measurement models, and full SEM models).

The four models illustrated in Figure 10.4 are all equivalent path models. You will notice that the models represent very different theoretical relationships among the variables. For instance, the covariance between X1 and X2 in Figures 10.4a and 10.4c represents a different relationship than that illustrated in Figures 10.4b and 10.4d. The direct effects between X1 and Y2 and between X2 and Y1 also change direction causally in the presented models. Because equivalent models cannot be distinguished by model fit, relevant theory is needed to make decisions about the direction of causal relationships to be modeled. Also, the parameter estimates in the equivalent models should be evaluated to make sure they are reasonable.

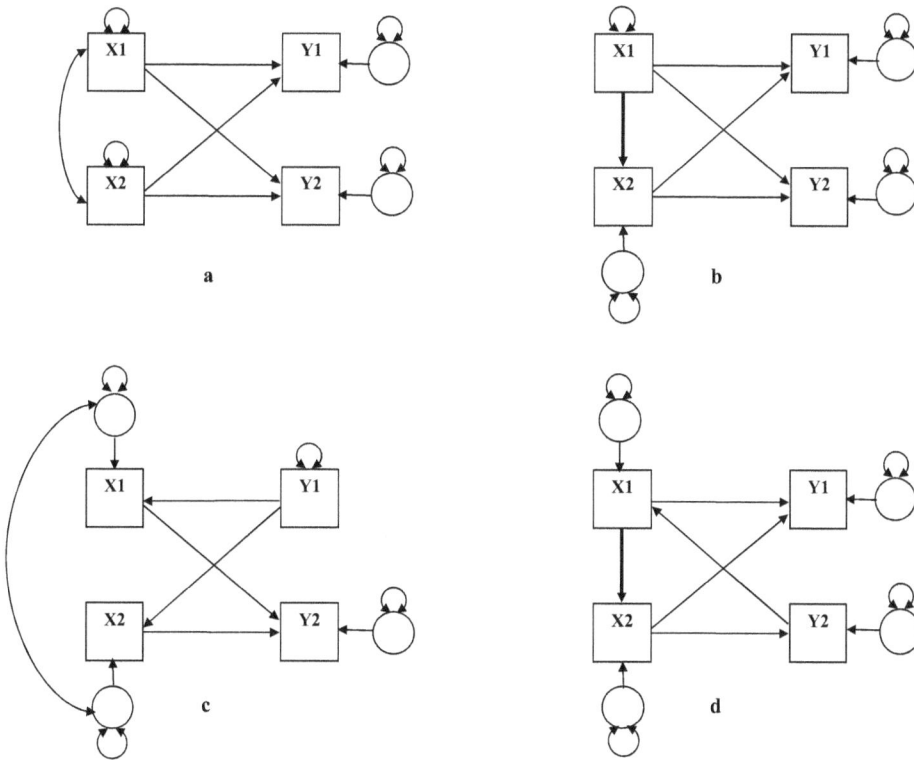

■ **Figure 10.4:** EQUIVALENT PATH MODELS

Researchers do not frequently evaluate equivalent models, but they should be considered in research applications where causal directions are not clearly prescribed. In a study about equivalent models, MacCallum, Wegener, Uchino, and Fabrigar (1993) concluded the following:

> A tempting way to manage the problem with equivalent models is to argue that the original model is more compelling and defensible than the equivalent models because of its a priori status. That is, because the original model was developed from theory or prior research or both and was found to yield an interpretable solution and adequate fit to the data, it is inherently more justifiable as a meaningful explanation of the data than the alternative equivalent models. We believe, however, that such a defense would often be the product of wishful thinking. This argument implies that no other equally good explanation of the data is plausible as the researcher's a priori model simply because the researcher did not generate the alternatives a priori. Such a position ignores the possibilities that prior research and theoretical development might not have been adequate or complete, that the researcher might not have been aware of alternative theoretical views, or that the researcher simply did not think of reasonable alternative models of the relationships among the variables under study.
>
> *p. 197*

SUMMARY

In structural equation modeling, the researcher follows the steps of model specification, identification, estimation, testing, and modification. We advise the researcher to base path models, measurement models, and structural models on sound theory. When analyzing full SEM models, we recommend utilizing the two-step approach to establish measurement model fit before model testing of the latent variable relationships in the structural model. We also recommend that theoretical models need to be cross-validated to determine the stability of the parameter estimates. Finally, we stated that researchers should include their SEM program, data, and path diagram of their model in any article. This permits a replication of the analysis and verification of the results. We do not advocate using specification searches to find the best fitting model without having a theoretically justified reason for modifying the initial model. We further advocate using another sample of data to help validate that the modified model is a meaningful and substantive theoretical structural model. We have provided several recommendations for the five steps when doing structural equation modeling. We also introduced non-recursive models and recommended that equivalent models be considered during SEM analysis.

SEM is a flexible and popular method that may be used to answer a variety of research questions. However, it is a mathematical method that is commonly used in non-experimental designs. It is the SEM user's job to make sure that the appropriate assumptions are met and the statistical conclusions are theoretically reasonable. When finding support for a model during the model testing process, there may be alternative and/or equivalent models that fit just as well or identically. Again, the importance of theoretical knowledge in SEM analysis cannot be overstressed. Theory is what directs the model specification and re-specification processes and it warns us about the acceptability of model parameter estimates.

We wish you well in your SEM endeavors. We would like to end this chapter with words written by Wolfe (1985) about analyzing causal models. He wrote that "the easy part is mathematical. The hard part is constructing causal models that are consistent with sound theory. In short, causal models are no better than the ideas that go with them" (Wolfe, 1985, p. 385).

REFERENCES

APA Publication and Communications Board Working Group on Journal Article Reporting Standards. (2008). Reporting standards for research in psychology: Why do we need them? What might they be? *American Psychologist*, 63, 839–851

Bentler, P. M., & Chou, C-P. (1987). Practical issues in structural modeling. *Sociological Methods & Research*, 16(1), 78–117.

Bollen, K. A. (1989). *Structural equations with latent variables*. New York: John Wiley & Sons.

Boomsma, A. (2000). Reporting analyses of covariance structure. *Structural Equation Modeling*, 7, 461–483.

Breckler, S. J. (1990). Applications of covariance structure modeling in psychology: Cause for concern? *Psychological Bulletin*, 107, 260–273.

Cliff, N. (1983). Some cautions concerning the application of causal modeling methods. *Multivariate Behavioral Research,* 18, 115–126.

Hancock, G. R. & French, B. (2013). Power analysis in structural equation modeling. In G. R. Hancock & R. O. Mueller (Eds.), *Structural equation modeling: A second course* (2nd ed., pp. 117–159). Greenwich, CT: Information Age.

Hoyle, R. H., & Isherwood, J. C. (2013). Reporting results from structural equation modeling analyses in *Archives of Scientific Psychology*. *Archives of Scientific Psychology*, 1, 14–22.

Hoyle, R. H., & Panter, A. T. (1995). Writing about structural equation models. In R. H. Hoyle (Ed.), *Structural equation modeling: Concepts, issues, and applications* (pp. 158–176). Thousand Oaks, CA: Sage.

Jackson, D. L. (2003). Revisiting sample size and number of parameter estimates: Some support for the N:q hypothesis. *Structural Equation Modeling*, 10(1), 128–141.

MacCallum, R. C., & Austin, J. T. (2000). Applications of structural equation modeling in psychological research. *Annual Review of Psychology*, 51, 201–226.

MacCallum, R. C., Browne, M. W., & Sugawara, H. M. (1996). Power analysis and determination of sample size for covariance structure modeling. *Psychological Methods*, 1, 130–149.

MacCallum, R. C., Wegener, D. T., Uchino, B. N., & Fabrigar, L. R. (1993). The problem of equivalent models in applications of covariance structure analysis. *Psychological Bulletin*, 114(1), 185–199.

Maxwell, S. E., & Cole, D. A. (1995). Tips for writing (and reading) methodological articles. *Psychological Bulletin*, 118, 193–198.

McDonald, R. P., & Ho, R. M. (2002). Principles and practice in reporting structural equation analyses. *Psychological Methods*, 7, 64–82.

Mueller, R. O., & Hancock, G. R. (2008). Best practices in structural equation modeling. In J. W. Osborne (Ed.), *Best practices in quantitative methods* (pp. 488–508). Thousand Oaks, CA: Sage.

Muthén, L. K., & Muthén, B. O. (2002). How to use a Monte Carlo study to decide on sample size and determine power. *Structural Equation Modeling*, 9(4), 599–620.

Raykov, T., Tomer, A., & Nesselroade, J. R. (1991). Reporting structural equation modeling results in *Psychology and Aging*: Some proposed guidelines. *Psychology and Aging*, 6, 499–533.

Schoemann, A. M., Preacher, K. J., & Coffman, D. L. (2010, April). Plotting power curves for RMSEA [Computer software]. Available from http://quantpsy.org/.

Shah, R., & Goldstein, S. M. (2006). Use of structural equation modeling in operations management research: Looking back and forward. *Journal of Operations Management*, 24, 148–169.

Thompson, B. (2000). Ten commandments of structural equation modeling. In L. Grimm & P. Yarnold (Eds.), *Reading and understanding more multivariate statistics* (pp. 261–284). Washington, DC: American Psychological Association.

Wolfe, L. M. (1985). Applications of causal models in higher education. In J. C. Smart (Ed.), *Higher education: Handbook of theory and research* (Vol. 1, pp. 381–413). New York: Agathon.

Zhang, M. F., Dawson, J. F., & Kline, R. B. (2021). Evaluating the use of covariance-based structural equation modelling with reflective measurement in organizational and management research: A review and recommendations for best practice. *British Journal of Management*, 32, 257–272.

INTRODUCTION TO MATRIX OPERATIONS

Structural equation modeling performs calculations using several different matrices. The matrix operations to perform the calculations involve addition, subtraction, multiplication, and division of elements in the different matrices. We present these basic matrix operations, followed by a simple multiple regression example.

Matrix Definition

A matrix is indicated by a capital letter, e.g., A, B, or R, and takes the form:

$$A_{22} = \begin{bmatrix} 3 & 5 \\ 5 & 6 \end{bmatrix}$$

The matrix can be rectangular or square shaped and contains an array of numbers. A correlation matrix is a square matrix with the value of 1.0 in the diagonal and variable correlations in the off-diagonal. A correlation matrix is symmetrical because the correlation coefficients in the lower half of the matrix are the same as the correlation coefficients in the upper half of the matrix. We usually only report the diagonal values and the correlations in the lower half of the correlation matrix. For example:

$$R_{33} = \begin{bmatrix} 1.0 & .30 & .50 \\ .30 & 1.0 & .60 \\ .50 & .60 & 1.0 \end{bmatrix}$$

but we report the following as a correlation matrix:

```
1.0
 .30 1.0
 .50 .60 1.0
```

Matrices have a certain number of rows and columns. The A matrix above has 2 rows and 2 columns. The **order** of a matrix is the **size** of the matrix, or number of rows times the number of columns. The order of the A matrix is 22, and shown as subscripts, where the first subscript is the number of rows and second subscript is the number of columns.

When we refer to elements in the matrix, we use row and column designations to identify the location of the element in the matrix. The location of an element has a subscript using the row number first, followed by the column number. For example, the correlation $r = .30$ is in the R_{21} matrix location or row 2, column 1.

Matrix Addition and Substraction

Matrix addition adds corresponding elements in two matrices, while matrix subtraction subtracts corresponding elements in two matrices. Consequently, the two matrices must have the same order (number of rows and columns), so we can add $A_{32} + B_{32}$ or subtract $A_{32} - B_{32}$. In the following example, Matrix A elements are added to Matrix B elements:

$$A + B = \begin{bmatrix} 3 & 5 & 2 \\ 1 & 6 & 0 \\ 9 & 1 & 2 \end{bmatrix} + \begin{bmatrix} 1 & -3 & 5 \\ 2 & 1 & 3 \\ 0 & 7 & -3 \end{bmatrix} = \begin{bmatrix} 4 & 2 & 7 \\ 3 & 7 & 3 \\ 9 & 8 & -1 \end{bmatrix}$$

Matrix Multiplication

Matrix multiplication is not as straightforward as matrix addition and subtraction. For a product of matrices we write $A \cdot B$ or AB. If A is a **m x n** matrix and B is a **n x p** matrix, then AB is a **m x p** matrix of rows and columns. The number of columns in the first matrix must match the number of rows in the second matrix to be compatible and permit multiplication of the elements of the matrices. The following example will illustrate how the row elements in the first matrix (A) are multiplied times the column elements in the second matrix (B) to yield the elements in the third matrix C.

$$c_{11} = 1 \times 2 + 2 \times 1 = 2 + 2 = 4$$

$$c_{12} = 1 \times 4 + 2 \times 8 = 4 + 16 = 20$$

$$c_{13} = 1 \times 6 + 2 \times 7 = 6 + 14 = 20$$

$$c_{21} = 3 \times 2 + 5 \times 1 = 6 + 5 = 11$$

$$c_{22} = 3 \times 4 + 5 \times 8 = 12 + 40 = 52$$

$$c_{23} = 3 \times 6 + 5 \times 7 = 18 + 35 = 53$$

$$A \bullet B = \begin{bmatrix} 1 & 2 \\ 3 & 5 \end{bmatrix} \bullet \begin{bmatrix} 2 & 4 & 6 \\ 1 & 8 & 7 \end{bmatrix} = \begin{bmatrix} 4 & 20 & 20 \\ 11 & 52 & 53 \end{bmatrix}$$

Matrix C is:

$$C = \begin{bmatrix} 4 & 20 & 20 \\ 11 & 52 & 53 \end{bmatrix}$$

It is important to note that matrix multiplication is **noncommutative**, i.e., $AB \neq BA$. The order of operation in multiplying elements of the matrices is therefore very important. Matrix multiplication, however, is associative, i.e., $A\,(BC) = (AB)\,C$, because the order of matrix multiplication is maintained.

A special matrix multiplication is possible when a single number is multiplied by the elements in a matrix. The single number is called a **scalar**. The scalar is simply multiplied by each of the elements in the matrix. For example,

$$D = 2 \begin{bmatrix} 3 & 4 \\ 4 & 6 \end{bmatrix} = \begin{bmatrix} 6 & 8 \\ 8 & 12 \end{bmatrix}$$

Matrix Division

Matrix division is similar to matrix multiplication with a little twist. In regular division, we divide the numerator by the denominator. However, we can also multiply the numerator by the inverse of the denominator. For example, in regular division, 4 is divided by 2 to give 2, however we get the same result if we multiply 4 by $\frac{1}{2}$. Therefore, matrix division is simply $A\,/\,B$ or $A \bullet 1/B = AB^{-1}$. The special designation of the B^{-1} matrix is called the **inverse** of the B matrix.

Matrix division thus requires finding the inverse of a matrix, which involves computing the **determinant** of a matrix, the **matrix of minors**, and the **matrix of**

cofactors. We then create a **transposed matrix** and an **inverse matrix**, which when multiplied yield an **identity matrix**. We now turn our attention to finding these values and matrices involved in matrix division.

Determinant of a Matrix

The determinant of a matrix is a unique number (not a matrix) that uses all the elements in the matrix for its calculation, and is a generalized variance for that matrix. For our illustration we will compute the determinant of a 2 by 2 matrix; leaving higher order matrix determinant computations for high speed computers. The determinant is computed by cross multiplying the elements of the matrix if:

$$A = \begin{bmatrix} a & b \\ c & d \end{bmatrix}$$

then the determinant of $A = ad - cb$.

For example, when

$$A = \begin{bmatrix} 2 & 5 \\ 3 & 6 \end{bmatrix}$$

the determinant of $A = 2 \times 6 - 3 \times 5 = -3$.

Matrix of Minors

Each element in a matrix has a **minor**. To find the minor of each element, simply draw a vertical and a horizontal line through that element to form a matrix with one less row and column. We next calculate the determinants of these minor matrices, and then place them in a **matrix of minors**. The matrix of minors has the same number of rows and columns as the original matrix.

The **matrix of minors** for the following 3 by 3 matrix is computed as follows:

$$A = \begin{bmatrix} 1 & 6 & -3 \\ -2 & 7 & 1 \\ 3 & -1 & 4 \end{bmatrix}$$

$$M_{11} = \begin{bmatrix} 7 & 1 \\ -1 & 4 \end{bmatrix} = (7)(4) - (1)(-1) = 29$$

$$M_{12} = \begin{bmatrix} -2 & 1 \\ 3 & 4 \end{bmatrix} = (-2)(4) - (1)(3) = -11$$

$$M_{13} = \begin{bmatrix} -2 & 7 \\ 3 & -1 \end{bmatrix} = (-2)(-1) - (7)(3) = -19$$

$$M_{21} = \begin{bmatrix} 6 & -3 \\ -1 & 4 \end{bmatrix} = (6)(4) - (-1)(-3) = 21$$

$$M_{22} = \begin{bmatrix} 1 & -3 \\ 3 & 4 \end{bmatrix} = (1)(4) - (-3)(3) = 13$$

$$M_{23} = \begin{bmatrix} 1 & 6 \\ 3 & -1 \end{bmatrix} = (1)(-1) - (6)(3) = -19$$

$$M_{31} = \begin{bmatrix} 6 & -3 \\ 7 & 1 \end{bmatrix} = (6)(1) - (-3)(7) = 27$$

$$M_{32} = \begin{bmatrix} 1 & -3 \\ -2 & 1 \end{bmatrix} = (1)(1) - (-3)(-2) = -5$$

$$M_{33} = \begin{bmatrix} 1 & 6 \\ -2 & 7 \end{bmatrix} = (1)(7) - (6)(-2) = 19$$

$$A_{Minors} = \begin{bmatrix} 29 & -11 & -19 \\ 21 & 13 & -19 \\ 27 & -5 & 19 \end{bmatrix}$$

Matrix of Cofactors

A **matrix of cofactors** is created by multiplying each element of the **matrix of minors** by (–1) for $i + j$ elements, where i = row number of the element and j = column number of the element. These values form a new matrix, called a **matrix of cofactors**.

An easy way to remember this multiplication rule is to observe the matrix below. Start with the first row and multiply the first entry by (+), second entry by (–), third by (+), and so on to the end of the row. For the second row start multiplying by (–), then (+), then (–), and so on. All odd rows begin with a – sign and all even rows begin with a + sign.

$$
\begin{array}{ccc}
+ & - & + \\
- & + & - \\
+ & - & + \\
- & + & -
\end{array}
$$

We now proceed by multiplying elements in the **matrix of minors** by –1 for the $i + j$ elements.

$$
A_{Minors} = \begin{bmatrix} +1 & -1 & +1 \\ -1 & +1 & -1 \\ +1 & -1 & +1 \end{bmatrix} \begin{bmatrix} 29 & -11 & -19 \\ 21 & 13 & -19 \\ 27 & -5 & 19 \end{bmatrix}
$$

So for the matrix above weobtain the **matrix of cofactors**:

$$
C_{Cofactors} = \begin{bmatrix} 29 & 11 & -19 \\ -21 & 13 & 19 \\ 27 & 5 & 19 \end{bmatrix}
$$

Determinant of Matrix Revisited

The matrix of cofactors makes finding the determinant of any size matrix easy. We multiply elements in any row or column of our original A matrix, by any one corresponding row or column in the **matrix of cofactors** to compute the determinant of the matrix. We can compute the determinant using any row or column, so rows with zeroes make the calculation of the determinant easier. The determinant of our original 3 by 3 matrix (A) using the 3 by 3 matrix of cofactors would be: $\det A = a_{11}c_{11} + a_{12}c_{12} + a_{13}c_{13}$

Recall that matrix A was:

$$
A = \begin{bmatrix} 1 & 6 & -3 \\ -2 & 7 & 1 \\ 3 & -1 & 4 \end{bmatrix}
$$

The matrix of cofactors was:

$$
C_{Cofactors} = \begin{bmatrix} 29 & 11 & -19 \\ -21 & 13 & 19 \\ 27 & 5 & 19 \end{bmatrix}
$$

So, the determinant of matrix A, using the first row of both matrices is.

$$\det A = (1)(29) + (6)(11) + (-3)(-19) = \quad 152$$

We could have also used the second columns of both matrices and obtained the same determinant value:

$$\det A = (6)(11) + (7)(13) + (-1)(5) = \quad 152$$

Two special matrices that we have already mentioned, also have determinants: a **diagonal matrix** and a **triangular matrix**. A **diagonal matrix** is a matrix which contains zero or non-zero elements on its main diagonal, but zeroes everywhere else. A **triangular matrix** has zeros only either above or below the main diagonal. To calculate the determinants of these matrices, we only need to multiply the elements on the main diagonal. For example, the following triangular matrix K has a determinant of 96.

$$K = \begin{bmatrix} 2 & 0 & 0 & 0 \\ 4 & 1 & 0 & 0 \\ -1 & 5 & 6 & 0 \\ 3 & 9 & -2 & 8 \end{bmatrix}$$

This is computed by multiplying the diagonal values in the matrix:

$$\det K = (2)\,(1)\,(6)\,(8) = \quad 96.$$

Transpose of a Matrix

The transpose of a matrix is created by taking the *rows* of an original matrix C and placing them into corresponding *columns* of the transpose matrix, C'. For example:

$$C = \begin{bmatrix} 29 & 11 & -19 \\ -21 & 13 & 19 \\ 27 & 5 & 19 \end{bmatrix}$$

$$C' = \begin{bmatrix} 29 & -21 & 27 \\ 11 & 13 & 5 \\ -19 & 19 & 19 \end{bmatrix}$$

The **transposed matrix** of the **matrix of cofactors** is called the **adjoint matrix**, designated as Adj(A). The **adjoint matrix** is important because we use it to create the inverse of a matrix, our final step in matrix division operations.

Inverse of a Matrix

The general formula for finding an inverse of a matrix is one over the determinant of the matrix times the adjoint of the matrix:

$$A^{-1} = [1/ \det A] \text{ ADJ}(A)$$

Since we have already found the determinant and adjoint of A, we find the inverse of A as follows:

$$A^{-1} = \left(\frac{1}{152}\right)\begin{bmatrix} 29 & -21 & 27 \\ 11 & 13 & 5 \\ -19 & 19 & 19 \end{bmatrix} = \begin{bmatrix} .191 & -.138 & .178 \\ .072 & .086 & .033 \\ -.125 & .125 & .125 \end{bmatrix}$$

An important property of the inverse of a matrix is that if we multiply its elements by the elements in our original matrix, we should obtain an **identity matrix**. An identity matrix will have 1.0 in the diagonal and zeroes in the off-diagonal. The identity matrix is computed as:

$$A A^{-1} = I$$

Since we have the original matrix of A and the inverse of matrix A, we multiply elements of the matrices to obtain the **identity** matrix, I:

$$AA^{-1} = \begin{bmatrix} 1 & 6 & -3 \\ -2 & 7 & 1 \\ 3 & -1 & 4 \end{bmatrix} * \begin{bmatrix} .191 & -.138 & .178 \\ .072 & .086 & .033 \\ -.125 & .125 & .125 \end{bmatrix} = \begin{bmatrix} 1 & 0 & 0 \\ 0 & 1 & 0 \\ 0 & 0 & 1 \end{bmatrix}$$

Matrix Operations in Statistics

We now turn our attention to how the matrix operations are used to compute statistics. We will only cover the calculation of the Pearson correlation and provide the matrix approach in multiple regression, leaving more complicated analyses to computer software programs.

Pearson Correlation (Variance–Covariance Matrix)

In the book, we illustrated how to compute the Pearson correlation coefficient from a variance–covariance matrix. Here we demonstrate the matrix approach. An important matrix in computing correlations is the sums of squares and cross-products matrix (SSCP). We will use the following pairs of scores to create the SSCP matrix.

X1	X2
5	1
4	3
6	5

The mean of X1 is 5 and the mean of X2 is 3. We use these mean values to compute deviation scores from each mean. We first create a matrix of deviation scores, D:

$$D = \begin{bmatrix} 5 & 1 \\ 4 & 3 \\ 6 & 5 \end{bmatrix} - \begin{bmatrix} 5 & 3 \\ 5 & 3 \\ 5 & 3 \end{bmatrix} = \begin{bmatrix} 0 & -2 \\ -1 & 0 \\ 1 & 2 \end{bmatrix}$$

Next, we create the transpose of matrix D, D':

$$D' = \begin{bmatrix} 0 & -1 & 1 \\ -2 & 0 & 2 \end{bmatrix}$$

Finally, we multiply the transpose of matrix D by the matrix of deviation scores to compute the sums of squares and cross-products matrix:

$$\text{SSCP} = D' * D$$

$$SSCP = \begin{bmatrix} 0 & -1 & 1 \\ -2 & 0 & 2 \end{bmatrix} * \begin{bmatrix} 0 & -2 \\ -1 & 0 \\ 1 & 2 \end{bmatrix} = \begin{bmatrix} 2 & 2 \\ 2 & 8 \end{bmatrix}$$

The sums of squares are along the diagonal of the matrix, and the sum of squares cross-products are on the off-diagonal. The matrix multiplications are provided below for the interested reader.

$(0)(0) + (-1)(-1) + (1)(1) = 2$ [sums of squares $= (0^2 + -1^2 + 1^2)$]

$(-2)(0) + (0)(-1) + (2)(1) = 2$ [sum of squares cross product]

$(0)(-2) + (-1)(0) + (1)(2) = 2$ [sum of squares cross product]

$(-2)(-2) + (0)(0) + (2)(2) = 8$[sums of squares $= (-2^2 + 0^2 + 2^2)$]

$SSCP = \begin{bmatrix} 2 & 2 \\ 2 & 8 \end{bmatrix}$ Sum of squares in diagonal of matrix

Variance–Covariance Matrix

Structural equation modeling uses a sample variance–covariance matrix in its calculations. The SSCP matrix is used to create the variance–covariance matrix, S:

$$S = \frac{SSCP}{n-1}$$

In matrix notation this becomes ½ times the matrix elements:

$$S = \frac{1}{2} * \begin{bmatrix} 2 & 2 \\ 2 & 8 \end{bmatrix} = \begin{bmatrix} 1 & 1 \\ 1 & 4 \end{bmatrix},$$

with variances of variables in the diagonal of the matrix and covariance terms in the off-diagonal of the matrix.

We can now calculate the Pearson correlation coefficient using the basic formula of covariance divided by the square root of the product of the variances.

$$r = \frac{CovarianceX1X2}{\sqrt{VarianceX1 * VarianceX2}} = \frac{1}{\sqrt{1*4}} = \frac{1}{2} = .50$$

Multiple Regression

The multiple linear regression equation with two predictor variables is:

$$y = \beta_0 + \beta_1 X_1 + \beta_2 X_2 + e_i$$

where y is the dependent variable, X_1 and X_2 are the two predictor variables,

β_0 is the regression constant or y - intercept,

β_1 and β_2 are the regression weights to be estimated,

and e is the error of prediction.

Given the data below, we can use matrix algebra to estimate the regression weights:

Y	X_1	X_2
3	2	1
2	3	5
4	5	3
5	7	6
8	8	7

We model each subject's y score as a linear function of the betas:

$$y_1 = 3 = 1\beta_0 + 2\,\beta_1 + 1\,\beta_2 + e_1$$
$$y_2 = 2 = 1\beta_0 + 3\,\beta_1 + 5\,\beta_2 + e_2$$
$$y_3 = 4 = 1\beta_0 + 5\,\beta_1 + 3\,\beta_2 + e_3$$
$$y_4 = 5 = 1\beta_0 + 7\,\beta_1 + 6\,\beta_2 + e_4$$
$$y_5 = 8 = 1\beta_0 + 8\,\beta_1 + 7\,\beta_2 + e_5$$

This series of equations can be expressed as a single matrix equation:

$$y = \qquad X \qquad \beta + \quad e$$

$$y = \begin{bmatrix} 3 \\ 2 \\ 4 \\ 5 \\ 8 \end{bmatrix} = \begin{bmatrix} 1 & 2 & 1 \\ 1 & 3 & 5 \\ 1 & 5 & 3 \\ 1 & 7 & 6 \\ 1 & 8 & 7 \end{bmatrix} \begin{bmatrix} \beta_0 \\ \beta_1 \\ \beta_2 \end{bmatrix} + \begin{bmatrix} e_1 \\ e_2 \\ e_3 \\ e_4 \\ e_5 \end{bmatrix}$$

The first column of matrix X contains 1s, which compute the regression constant. In matrix form, the multiple linear regression equation is: $y = X\beta + e$.

Using calculus, we translate this matrix to solve for the regression weights:

$$\hat{\beta} = (X'X)^{-1} X'y$$

The matrix equation is:

$$\hat{\beta} = \left\{ \begin{bmatrix} 1 & 1 & 1 & 1 & 1 \\ 2 & 3 & 5 & 7 & 8 \\ 1 & 5 & 3 & 6 & 7 \end{bmatrix} \begin{bmatrix} 1 & 2 & 1 \\ 1 & 3 & 5 \\ 1 & 5 & 3 \\ 1 & 7 & 6 \\ 1 & 8 & 7 \end{bmatrix} \right\}^{-1} * \begin{bmatrix} 1 & 1 & 1 & 1 & 1 \\ 2 & 3 & 5 & 7 & 8 \\ 1 & 5 & 3 & 6 & 7 \end{bmatrix} \begin{bmatrix} 3 \\ 2 \\ 4 \\ 5 \\ 8 \end{bmatrix}$$

where the braces span X' and X, and the final two matrices are X' and y.

We first compute $X'X$ and then compute $X'y$:

$$X'X = \begin{bmatrix} 5 & 25 & 22 \\ 25 & 151 & 130 \\ 22 & 130 & 120 \end{bmatrix} \quad \text{and} \quad X'y = \begin{bmatrix} 22 \\ 131 \\ 111 \end{bmatrix}$$

Next we create the inverse of $X'X$, where 1016 is the determinant of $X'X$.

$$(X'X)^{-1} = \frac{1}{1016} \begin{bmatrix} 1220 & -140 & -72 \\ -140 & 116 & -100 \\ -72 & -100 & 130 \end{bmatrix}$$

Finally, we solve for the X_1 and X_2 regression weights:

$$\hat{\beta} = \frac{1}{1016} \begin{bmatrix} 1220 & -140 & -72 \\ -140 & 116 & -100 \\ -72 & -100 & 130 \end{bmatrix} \begin{bmatrix} 22 \\ 131 \\ 111 \end{bmatrix} = \begin{bmatrix} .50 \\ 1 \\ -.25 \end{bmatrix}$$

The multiple regression equation is:

$$\hat{y}_i = .50 + 1X_1 - .25X_2$$

We use the multiple regression equation to compute predicted scores and then compare the predicted values to the original y values to compute the error of prediction values, e. For example, the first y score was 3 with $X_1 = 2$ and $X_2 = 1$. We substitute the X_1 and X_2 values in the regression equation and compute a predicted y score of 2.25. The error of prediction is computed as y – this predicted y score or $3 - 2.25 = .75$. These computations are listed below and are repeated for the remaining y values.

$$\hat{y}_1 = .50 + 1(2) - .25\,(1)$$
$$\hat{y}_1 = 2.25$$
$$\hat{e}_1 = 3 - 2.25 = .75$$

$$\hat{y}_2 = .50 + 1(3) - .25\,(5)$$
$$\hat{y}_2 = 2.25$$
$$\hat{e}_2 = 2 - 2.25 = -.25$$

$$\hat{y}_3 = .50 + 1(5) - .25\,(3)$$
$$\hat{y}_3 = 4.75$$
$$\hat{e}_3 = 4 - 4.75 = -.75$$

$$\hat{y}_4 = .50 + 1(7) - .25\,(6)$$
$$\hat{y}_4 = 6.00$$
$$\hat{e}_4 = 5 - 6 = -1.00$$

$$\hat{y}_5 = .50 + 1(8) - .25\,(7)$$
$$\hat{y}_5 = 6.75$$
$$\hat{e}_5 = 8 - 6.75 = 1.25$$

The regression equation is: $\hat{y}_i = .50 + 1.0X_1 - .25X_2$

We can now place the Y values, X values, regression weights, and error terms back into the matrices to yield a complete solution for the Y values. Notice that the error term vector should sum to zero (0.0). Also notice that each y value is uniquely composed of an intercept term (.50), a regression weight (1.0) times an X_1 value, a regression weight (−.25) times an X_2 value, and a residual error, e.g., the first y value of $3 = .5 + 1.0(2) - .25\,(1) + .75$.

$$
\begin{bmatrix} 3 \\ 2 \\ 4 \\ 5 \\ 8 \end{bmatrix}
= .5 + 1.0
\begin{bmatrix} 2 \\ 3 \\ 5 \\ 7 \\ 8 \end{bmatrix}
- .25
\begin{bmatrix} 1 \\ 5 \\ 3 \\ 6 \\ 7 \end{bmatrix}
+
\begin{bmatrix} .75 \\ -.25 \\ -.75 \\ -1.00 \\ 1.25 \end{bmatrix}
$$

Appendix 2

STATISTICAL TABLES

Table A.1: Areas Under the Normal Curve (z-scores)

Second decimal place in z

z	.00	.01	.02	.03	.04	.05	.06	.07	.08	.09
.0	.0000	.0040	.0080	.0120	.0160	.0199	.0239	.0279	.0319	.0359
.1	.0398	.0438	.0478	.0517	.0557	.0596	.0636	.0675	.0714	.0753
.2	.0793	.0832	.0871	.0910	.0948	.0987	.1026	.1064	.1103	.1141
.3	.1179	.1217	.1255	.1293	.1331	.1368	.1406	.1443	.1480	.1517
.4	.1554	.1591	.1628	.1664	.1700	.1736	.1772	.1808	.1844	.1879
.5	.1915	.1950	.1985	.2019	.2054	.2088	.2123	.2157	.2190	.2224
.6	.2257	.2291	.2324	.2357	.2389	.2422	.2454	.2486	.2517	.2549
.7	.2580	.2611	.2642	.2673	.2704	.2734	.2764	.2794	.2823	.2852
.8	.2881	.2910	.2939	.2967	.2995	.3023	.3051	.3078	.3106	.3133
.9	.3159	.3186	.3212	.3238	.3264	.3289	.3315	.3340	.3365	.3389
1.0	.3413	.3438	.3461	.3485	.3508	.3531	.3554	.3577	.3599	.3621
1.1	.3643	.3665	.3686	.3708	.3729	.3749	.3770	.3790	.3810	.3830
1.2	.3849	.3869	.3888	.3907	.3925	.3944	.3962	.3980	.3997	.4015
1.3	.4032	.4049	.4066	.4082	.4099	.4115	.4131	.4147	.4162	.4177
1.4	.4192	.4207	.4222	.4236	.4251	.4265	.4279	.4292	.4306	.4319
1.5	.4332	.4345	.4357	.4793	.4382	.4394	.4406	.4418	.4429	.4441
1.6	.4452	.4463	.4474	.4484	.4495	.4505	.4515	.4525	.4535	.4545
1.7	.4554	.4564	.4573	.4582	.4591	.4599	.4608	.4616	.4625	.4633
1.8	.4641	.4649	.4656	.4664	.4671	.4678	.4686	.4693	.4699	.4706
1.9	.4713	.4719	.4726	.4732	.4738	.4744	.4750	.4756	.4761	.4767
2.0	.4772	.4778	.4783	.4788	.4793	.4798	.4803	.4808	.4812	.4817
2.1	.4821	.4826	.4830	.4834	.4838	.4842	.4846	.4850	.4854	.4857
2.2	.4861	.4826	.4868	.4871	.4875	.4878	.4881	.4884	.4887	.4890
2.3	.4893	.4896	.4898	.4901	.4904	.4906	.4909	.4911	.4913	.4916
2.4	.4918	.4920	.4922	.4925	.4927	.4929	.4931	.4932	.4934	.4936
2.5	.4938	.4940	.4941	.4943	.4945	.4946	.4948	.4949	.4951	.4952
2.6	.4953	.4955	.4956	.4957	.4959	.4960	.4961	.4962	.4963	.4964
2.7	.4965	.4966	.4967	.4968	.4969	.4970	.4971	.4972	.4973	.4974
2.8	.4974	.4975	.4976	.4977	.4977	.4978	.4979	.4979	.4980	.4981
2.9	.4981	.4982	.4982	.4983	.4984	.4984	.4985	.4985	.4986	.4986
3.0	.4987	.4987	.4987	.4988	.4988	.4989	.4989	.4989	.4990	.4990
3.1	.4990	.4991	.4991	.4991	.4992	.4922	.4992	.4992	.4993	.4993
3.2	.4993	.4993	.4994	.4994	.4994	.4994	.4994	.4995	.4995	.4995
3.3	.4995	.4995	.4995	.4996	.4996	.4996	.4996	.4996	.4996	.4997
3.4	.4997	.4997	.4997	.4997	.4997	.4997	.4997	.4997	.4997	.4998
3.5	.4998									
4.0	.49997									
4.5	.499997									
5.0	.4999997									

Table A.2: Distribution of *t* for Given Probability Levels

	Level of significance for one-tailed test					
	.10	.05	.025	.01	.005	.0005
	Level of significance for two-tailed test					
df	.20	.10	.05	.02	.01	.001
1	3.078	6.314	12.706	31.821	63.657	636.619
2	1.886	2.920	4.303	6.965	9.925	31.598
3	1.638	2.353	3.182	4.541	5.841	12.941
4	1.533	2.132	2.776	3.747	4.604	8.610
5	1.476	2.015	2.571	3.365	4.032	6.859
6	1.440	1.943	2.447	3.143	3.707	5.959
7	1.415	1.895	2.365	2.998	3.499	5.405
8	1.397	1.860	2.306	2.896	3.355	5.041
9	1.383	1.833	2.262	2.821	3.250	4.781
10	1.372	1.812	2.228	2.764	3.169	4.587
11	1.363	1.796	2.201	2.718	3.106	4.437
12	1.356	1.782	2.179	2.681	3.055	4.318
13	1.350	1.771	2.160	2.650	3.012	4.221
14	1.345	1.761	2.145	2.624	2.977	4.140
15	1.341	1.753	2.131	2.602	2.947	4.073
16	1.337	1.746	2.120	2.583	2.921	4.015
17	1.333	1.740	2.110	2.567	2.898	3.965
18	1.330	1.734	2.101	2.552	2.878	3.992
19	1.328	1.729	2.093	2.539	2.861	3.883
20	1.325	1.725	2.086	2.528	2.845	3.850
21	1.323	1.721	2.080	2.518	2.831	3.819
22	1.321	1.717	2.074	2.508	2.819	3.792
23	1.319	1.714	2.069	2.500	2.807	3.767
24	1.318	1.711	2.064	2.492	2.797	3.745
25	1.316	1.708	2.060	2.485	2.787	3.725
26	1.315	1.706	2.056	2.479	2.779	3.707
27	1.314	1.703	2.052	2.473	2.771	3.690
28	1.313	1.701	2.048	2.467	2.763	3.674
29	1.311	1.699	2.045	2.462	2.756	3.659
30	1.310	1.697	2.042	2.457	2.750	3.646
40	1.303	1.684	2.021	2.423	2.704	3.551
60	1.296	1.671	2.000	2.390	2.660	3.460
120	1.289	1.658	1.980	2.358	2.617	3.373
∞	1.282	1.645	1.960	2.326	2.576	3.291

Table A.3: Distribution of *r* for Given Probability Levels

df	Level of significance for one-tailed test			
	.05	.025	.01	.005
	Level of significance for two-tailed test			
	.10	.05	.02	.01
1	.988	.997	.9995	.9999
2	.900	.950	.980	.990
3	.805	.878	.934	.959
4	.729	.811	.882	.917
5	.669	.754	.833	.874
6	.622	.707	.789	.834
7	.582	.666	.750	.798
8	.540	.632	.716	.765
9	.521	.602	.685	.735
10	.497	.576	.658	.708
11	.576	.553	.634	.684
12	.458	.532	.612	.661
13	.441	.514	.592	.641
14	.426	.497	.574	.623
15	.412	.482	.558	.606
16	.400	.468	.542	.590
17	.389	.456	.528	.575
18	.378	.444	.516	.561
19	.369	.433	.503	.549
20	.360	.423	.492	.537
21	.352	.413	.482	.526
22	.344	.404	.472	.515
23	.337	.396	.462	.505
24	.330	.388	.453	.496
25	.323	.381	.445	.487
26	.317	.374	.437	.479
27	.311	.367	.430	.471
28	.306	.361	.423	.463
29	.301	.355	.416	.486
30	.296	.349	.409	.449
35	.275	.325	.381	.418
40	.257	.304	.358	.393
45	.243	.288	.338	.372
50	.231	.273	.322	.354
60	.211	.250	.295	.325
70	.195	.232	.274	.303
80	.183	.217	.256	.283
90	.173	.205	.242	.267
100	.164	.195	.230	.254

Table A.4: Distribution of Chi-square for Given Probability Levels

Probability df	.99	.98	.95	.90	.80	.70	.50	.30	.20	.10	.05	.02	.01	.001
1	.00016	.00663	.00393	.0158	.0642	.148	.455	1.074	1.642	2.706	3.841	5.412	6.635	10.827
2	.0201	.0404	.103	.211	.446	.713	1.386	2.408	3.219	4.605	5.991	7.824	9.210	13.815
3	.115	.185	.352	.584	1.005	1.424	2.366	3.665	4.642	6.251	7.815	9.837	11.345	16.266
4	.297	.429	.711	1.064	1.649	2.195	3.357	4.878	5.989	7.779	9.488	11.668	13.277	18.467
5	.554	.752	1.145	1.610	2.343	3.000	4.351	6.064	7.289	9.236	11.070	13.388	15.086	20.515
6	.872	1.134	1.635	2.204	3.070	3.828	5.348	7.231	8.558	10.645	12.592	15.033	16.812	22.457
7	1.239	1.564	2.167	2.833	3.822	4.671	6.346	8.383	9.803	12.017	14.067	16.622	18.475	24.322
8	1.646	2.032	2.733	3.490	4.594	5.527	7.344	9.524	11.030	13.362	15.507	18.168	20.090	26.125
9	2.088	2.532	3.325	4.168	5.380	6.393	8.343	10.656	12.242	14.684	16.919	19.679	21.666	27.877
10	2.558	3.059	3.940	4.865	6.179	7.267	9.342	11.781	13.442	15.987	18.307	21.161	23.209	29.588
11	3.053	3.609	4.575	5.578	6.989	8.148	10.341	12.899	14.631	17.275	19.675	22.618	24.725	31.264
12	3.571	4.178	5.226	6.304	7.807	9.034	11.340	14.011	15.812	18.549	21.026	24.054	26.217	32.909
13	4.107	4.765	5.892	7.042	8.634	9.926	12.340	15.119	16.985	19.812	22.362	25.472	27.688	34.528
14	4.660	5.368	6.571	7.790	9.467	10.821	13.339	16.222	18.151	21.064	23.685	26.873	29.141	36.123
15	5.229	5.985	7.261	8.547	10.307	11.721	14.339	17.322	19.311	22.307	24.996	28.259	30.578	37.697
16	5.812	6.614	7.962	9.312	11.152	12.624	15.338	18.418	20.465	23.542	26.296	29.633	32.000	39.252
17	6.408	7.255	8.672	10.085	12.002	13.531	16.338	19.511	21.615	24.769	27.587	30.995	33.409	40.790
18	7.015	7.906	9.390	10.865	12.857	14.440	17.338	20.601	22.760	25.989	28.869	32.346	34.805	42.312
19	7.633	8.567	10.117	11.651	13.716	15.352	18.338	21.689	23.900	27.204	30.144	33.687	36.191	43.820
20	8.260	9.237	10.851	12.443	14.578	16.266	19.337	22.775	25.038	28.412	31.410	35.020	37.566	45.315
21	8.897	9.915	11.591	13.240	15.445	17.182	20.337	23.858	26.171	29.615	32.671	36.343	38.932	46.797
22	9.542	10.600	12.338	14.041	16.314	18.101	21.337	24.939	27.301	30.813	33.924	37.659	40.289	48.268
23	10.196	11.293	13.091	14.848	17.187	19.021	22.337	26.018	28.429	32.007	35.172	38.968	41.638	49.728
24	10.856	11.992	13.848	15.659	18.062	19.943	23.337	27.096	29.553	33.196	36.415	40.270	42.980	51.179
25	11.524	12.697	14.611	16.473	18.940	20.867	24.337	28.172	30.675	34.382	37.652	41.566	44.314	52.620

Table A.4: (cont.)

df	Probability														
	.99	.98	.95	.90	.80	.70	.50	.30	.20	.10	.05	.02	.01	.001
26	12.198	13.409	15.379	17.292	19.820	21.792	25.336	29.246	31.795	35.563	38.885	42.856	45.642	54.052
27	12.879	14.125	16.151	18.114	20.703	22.719	26.336	30.319	32.912	36.741	40.113	44.140	46.963	55.476
28	13.565	14.847	16.928	18.939	21.588	23.647	27.336	31.391	34.027	37.916	41.337	45.419	48.278	56.893
29	14.256	15.574	17.708	19.768	22.475	24.577	28.336	32.461	35.139	39.087	42.557	46.693	49.588	58.302
30	14.953	16.306	18.493	20.599	23.364	25.508	29.336	33.530	36.250	40.256	43.773	47.962	50.892	59.703
32	16.362	17.783	20.072	22.271	25.148	27.373	31.336	35.665	38.466	42.585	46.194	50.487	53.486	62.487
34	17.789	19.275	21.664	23.952	26.938	29.242	33.336	37.795	40.676	44.903	48.602	52.995	56.061	65.247
36	19.233	20.783	23.269	25.643	28.735	31.115	35.336	39.922	42.879	47.212	50.999	55.489	58.619	67.985
38	20.691	22.304	24.884	27.343	30.537	32.992	37.335	42.045	45.076	49.513	53.384	57.969	61.162	70.703
40	22.164	23.838	26.509	29.051	32.345	34.872	39.335	44.165	47.269	51.805	55.759	60.436	63.691	73.402
42	23.650	25.383	28.144	30.765	34.147	36.755	41.335	46.282	49.456	54.090	58.124	62.892	66.206	76.084
44	25.148	26.939	29.787	32.487	35.974	38.641	43.335	48.396	51.639	56.369	60.481	65.337	68.710	78.750
46	26.657	28.504	31.439	34.215	37.795	40.529	45.335	50.507	53.818	58.641	62.830	67.771	71.201	81.400
48	28.177	30.080	33.098	35.949	39.621	42.420	47.335	52.616	55.993	60.907	65.171	70.197	73.683	84.037
50	29.707	31.664	34.764	37.689	41.449	44.313	49.335	54.723	58.164	63.167	67.505	72.613	76.154	86.661
52	31.246	33.256	36.437	39.433	43.281	46.209	51.335	56.827	60.332	65.422	69.832	75.021	78.616	89.272
54	32.793	34.856	38.116	41.183	45.117	48.106	53.335	58.930	62.496	67.673	72.153	77.422	81.069	91.872
56	34.350	36.464	39.801	42.937	46.955	50.005	55.335	61.031	64.658	69.919	74.468	79.815	83.513	94.461
58	35.913	38.078	41.492	44.696	48.797	51.906	57.335	63.129	66.816	72.160	76.778	82.201	85.950	97.039
60	37.485	39.699	43.188	46.459	50.641	53.809	59.335	65.227	68.972	74.397	79.082	84.580	88.379	99.607
62	39.063	41.327	44.889	48.226	52.487	55.714	61.335	67.322	71.125	76.630	81.381	86.953	90.802	102.166
64	40.649	42.960	46.595	49.996	54.336	57.620	63.335	69.416	73.276	78.860	83.675	89.320	93.217	104.716
66	42.240	44.599	48.305	51.770	56.188	59.527	65.335	71.508	75.424	81.085	85.965	91.681	95.626	107.258
68	43.838	46.244	50.020	53.548	58.042	61.436	67.335	73.600	77.571	83.308	88.250	94.037	98.028	109.791
70	45.442	47.893	51.739	55.329	59.898	63.346	69.335	75.689	79.715	85.527	90.531	96.388	100.425	112.317

Note: For larger values of df, the expression $\sqrt{(X^2)^2} - \sqrt{2df} - 1$ may be used as a normal deviate with unit variance, remembering that the probability for X^2 corresponds with that of a single tail of the normal curve.

Table A.5: Distribution of F for Given Probability Levels (.05 Level)

df_1, df_2	1	2	3	4	5	6	7	8	9	10	12	15	20	24	30	40	60	120	∞
1	161.4	199.5	215.7	224.6	230.2	234.0	236.8	238.9	240.5	241.9	243.9	245.9	248.0	249.1	250.1	251.1	252.2	253.3	254.3
2	18.51	19.00	19.16	19.25	19.30	19.33	19.35	19.37	19.38	19.40	19.41	19.43	19.45	19.45	19.46	19.47	19.48	19.49	19.50
3	10.13	9.55	9.28	9.12	9.01	8.94	8.89	8.85	8.81	8.79	8.74	8.70	8.66	8.64	8.62	8.59	8.57	8.55	8.53
4	7.71	6.94	6.59	6.39	6.26	6.16	6.09	6.04	6.00	5.96	5.91	5.86	5.80	5.77	5.75	5.72	5.69	5.66	5.63
5	6.61	5.79	5.41	5.19	5.05	4.95	4.88	4.82	4.77	4.74	4.68	4.62	4.56	4.53	4.50	4.46	4.43	4.40	4.36
6	5.99	5.14	4.76	4.53	4.39	4.28	4.21	4.15	4.10	4.06	4.00	3.94	3.87	3.84	3.81	3.77	3.74	3.70	3.67
7	5.59	4.74	4.35	4.12	3.97	3.87	3.79	3.73	3.68	3.64	3.57	3.51	3.44	3.41	3.38	3.34	3.30	3.27	3.23
8	5.32	4.46	4.07	3.84	3.69	3.58	3.50	3.44	3.39	3.35	3.28	3.22	3.15	3.12	3.08	3.04	3.01	2.97	2.93
9	5.12	4.26	3.86	3.63	3.48	3.37	3.29	3.23	3.18	3.14	3.07	3.01	2.94	2.90	2.86	2.83	2.79	2.75	2.71
10	4.96	4.10	3.71	3.48	3.33	3.22	3.14	3.07	3.02	2.98	2.91	2.85	2.77	2.74	2.70	2.66	2.62	2.58	2.54
11	4.84	3.98	3.59	3.36	3.20	3.09	3.01	2.95	2.90	2.85	2.79	2.72	2.65	2.61	2.57	2.53	2.49	2.45	2.40
12	4.75	3.89	3.49	3.26	3.11	3.00	2.91	2.85	2.80	2.75	2.69	2.62	2.54	2.51	2.47	2.43	2.38	2.34	2.30
13	4.67	3.81	3.41	3.18	3.03	2.92	2.83	2.77	2.71	2.67	2.60	2.53	2.46	2.42	2.38	2.34	2.30	2.25	2.21
14	4.60	3.74	3.34	3.11	2.96	2.85	2.76	2.70	2.65	2.60	2.53	2.46	2.39	2.35	2.31	2.27	2.22	2.18	2.13
15	4.54	3.68	3.29	3.06	2.90	2.79	2.71	2.64	2.59	2.54	2.48	2.40	2.33	2.29	2.25	2.20	2.16	2.11	2.07
16	4.49	3.63	3.24	3.01	2.85	2.74	2.66	2.59	2.54	2.49	2.42	2.35	2.28	2.24	2.19	2.15	2.11	2.06	2.01
17	4.45	3.59	3.20	2.96	2.81	2.70	2.61	2.55	2.49	2.45	2.38	2.31	2.23	2.19	2.15	2.10	2.06	2.01	1.96
18	4.41	3.55	3.16	2.93	2.77	2.66	2.58	2.51	2.46	2.41	2.34	2.27	2.19	2.15	2.11	2.06	2.02	1.97	1.92
19	4.38	3.52	3.13	2.90	2.74	2.63	2.54	2.48	2.42	2.38	2.31	2.23	2.16	2.11	2.07	2.03	1.98	1.93	1.88
20	4.35	3.49	3.10	2.87	2.71	2.60	2.51	2.45	2.39	2.35	2.28	2.20	2.12	2.08	2.04	1.99	1.95	1.90	1.84
21	4.32	3.47	3.07	2.84	2.68	2.57	2.49	2.42	2.37	2.32	2.25	2.18	2.10	2.05	2.01	1.96	1.92	1.87	1.81
22	4.30	3.44	3.05	2.82	2.66	2.55	2.46	2.40	2.34	2.30	2.23	2.15	2.07	2.03	1.98	1.94	1.89	1.84	1.78
23	4.28	3.42	3.03	2.80	2.64	2.53	2.44	2.37	2.32	2.27	2.20	2.13	2.05	2.01	1.96	1.91	1.86	1.81	1.76
24	4.26	3.40	3.01	2.78	2.62	2.51	2.42	2.36	2.30	2.25	2.18	2.11	2.03	1.98	1.94	1.89	1.84	1.79	1.73
25	4.24	3.39	2.99	2.76	2.60	2.49	2.40	2.34	2.28	2.24	2.16	2.09	2.01	1.96	1.92	1.87	1.82	1.77	1.71
26	4.23	3.37	2.98	2.74	2.59	2.47	2.39	2.32	2.27	2.22	2.15	2.07	1.99	1.95	1.90	1.85	1.80	1.75	1.69
27	4.21	3.35	2.96	2.73	2.57	2.46	2.37	2.31	2.25	2.20	2.13	2.06	1.97	1.93	1.88	1.84	1.79	1.73	1.67
28	4.20	3.34	2.95	2.71	2.56	2.45	2.36	2.29	2.24	2.19	2.12	2.04	1.96	1.91	1.87	1.82	1.77	1.71	1.65
29	4.18	3.33	2.93	2.70	2.55	2.43	2.35	2.28	2.22	2.18	2.10	2.03	1.94	1.90	1.85	1.81	1.75	1.70	1.64

Table A.5: (*cont.*)

df_1 df_2	1	2	3	4	5	6	7	8	9	10	12	15	20	24	30	40	60	120	∞
30	4.17	3.32	2.92	2.69	2.53	2.42	2.33	2.27	2.21	2.16	2.09	2.01	1.93	1.89	1.84	1.79	1.74	1.68	1.62
40	4.08	3.23	2.84	2.61	2.45	2.34	2.25	2.18	2.12	2.08	2.00	1.92	1.84	1.79	1.74	1.69	1.64	1.58	1.51
60	4.00	3.15	2.76	2.53	2.37	2.25	2.17	2.10	2.04	1.99	1.92	1.84	1.75	1.70	1.65	1.59	1.53	1.47	1.39
120	3.92	3.07	2.68	2.45	2.29	2.17	2.09	2.02	1.96	1.91	1.83	1.75	1.66	1.61	1.55	1.50	1.43	1.35	1.25
∞	3.84	3.00	2.60	2.37	2.21	2.10	2.01	1.94	1.88	1.83	1.75	1.67	1.57	1.52	1.46	1.39	1.32	1.22	1.00

Table A.6: Distribution of *F* for Given Probability Levels (.01 Level)

df_2 \ df_1	1	2	3	4	5	6	7	8	9	10	12	15	20	24	30	40	60	120	∞
1	4052	4999.5	5403	5625	5764	5859	5928	5982	6022	6056	6106	6157	6209	6235	6261	6287	6313	6339	6366
2	98.5	99.00	99.17	99.25	99.30	99.33	99.36	99.37	99.39	99.40	99.42	99.43	99.45	99.46	99.47	99.47	99.48	99.49	99.50
3	34.12	30.82	29.46	28.71	28.24	27.91	27.67	27.49	27.35	27.23	27.05	26.87	26.69	26.60	26.50	26.41	26.32	26.22	26.13
4	21.20	18.00	16.69	15.98	15.52	15.21	14.98	14.80	14.66	14.55	14.37	14.20	14.02	13.93	13.84	13.75	13.65	13.56	13.46
5	16.26	13.27	12.06	11.39	10.97	10.67	10.46	10.29	10.16	10.05	9.89	9.72	9.55	9.47	9.38	9.29	9.20	9.11	9.02
6	13.75	10.92	9.78	9.15	8.75	8.47	8.26	8.10	7.98	7.87	7.72	7.56	7.40	7.31	7.23	7.14	7.06	6.97	6.88
7	12.25	9.55	8.45	7.85	7.46	7.19	6.99	6.84	6.72	6.62	6.47	6.31	6.16	6.07	5.99	5.91	5.82	5.74	5.65
8	11.26	8.65	7.59	7.01	6.63	6.37	6.18	6.03	5.91	5.81	5.67	5.52	5.36	5.28	5.20	5.12	5.03	4.95	4.86
9	10.56	8.02	6.99	6.42	6.06	5.80	5.61	5.47	5.35	5.26	5.11	4.96	4.81	4.73	4.65	4.57	4.48	4.40	4.31
10	10.04	7.56	6.55	5.99	5.64	5.39	5.20	5.06	4.94	4.85	4.71	4.56	4.41	4.33	4.25	4.17	4.08	4.00	3.91
11	9.65	7.21	6.22	5.67	5.32	5.07	4.89	4.74	4.63	4.54	4.40	4.25	4.10	4.02	3.94	3.86	3.78	3.69	3.60
12	9.33	6.93	5.95	5.41	5.06	4.82	4.64	4.50	4.39	4.30	4.16	4.01	3.86	3.78	3.70	3.62	3.54	3.45	3.36
13	9.07	6.70	5.74	5.21	4.86	4.62	4.44	4.30	4.19	4.10	3.96	3.82	3.66	3.59	3.51	3.43	3.34	3.25	3.17
14	8.86	6.51	5.56	5.04	4.69	4.46	4.28	4.14	4.03	3.94	3.80	3.66	3.51	3.43	3.35	3.27	3.18	3.09	3.00
15	8.68	6.36	5.42	4.89	4.56	4.32	4.14	4.00	3.89	3.80	3.67	3.52	3.37	3.29	3.21	3.13	3.05	2.96	2.87
16	8.53	6.23	5.29	4.77	4.44	4.20	4.03	3.89	3.78	3.69	3.55	3.41	3.26	3.18	3.10	3.02	2.93	2.84	2.75
17	8.40	6.11	5.18	4.67	4.34	4.10	3.93	3.79	3.68	3.59	3.46	3.31	3.16	3.08	3.00	2.92	2.83	2.75	2.65
18	8.29	6.01	5.09	4.58	4.25	4.01	3.84	3.71	3.60	3.51	3.37	3.23	3.08	3.00	2.92	2.84	2.75	2.66	2.57
19	8.18	5.93	5.01	4.50	4.17	3.94	3.77	3.63	3.52	3.43	3.30	3.15	3.00	2.92	2.84	2.76	2.67	2.58	2.49
20	8.10	5.85	4.94	4.43	4.10	3.87	3.70	3.56	3.46	3.37	3.23	3.09	2.94	2.86	2.78	2.69	2.61	2.52	2.42
21	8.02	5.78	4.87	4.37	4.04	3.81	3.64	3.51	3.40	3.31	3.17	3.03	2.88	2.80	2.72	2.64	2.55	2.46	2.36
22	7.95	5.72	4.82	4.31	3.9	3.76	3.59	3.45	3.35	3.26	3.12	2.98	2.83	2.75	2.67	2.58	2.50	2.40	2.31
23	7.88	5.66	4.76	4.26	3.94	3.71	3.54	3.41	3.30	3.21	3.07	2.93	2.78	2.70	2.62	2.54	2.45	2.35	2.26
24	7.82	5.61	4.72	4.22	3.90	3.67	3.50	3.36	3.26	3.17	3.03	2.89	2.74	2.66	2.58	2.49	2.40	2.31	2.21
25	7.77	5.57	4.68	4.18	3.85	3.63	3.46	3.32	3.22	3.13	2.99	2.85	2.70	2.62	2.54	2.45	2.36	2.27	2.17
26	7.72	5.53	4.64	4.14	3.82	3.59	3.42	3.29	3.18	3.09	2.96	2.81	2.66	2.58	2.50	2.42	2.33	2.23	2.13
27	7.68	5.49	4.60	4.11	3.78	3.56	3.39	3.26	3.15	3.06	2.93	2.78	2.63	2.55	2.47	2.38	2.29	2.20	2.10
28	7.64	5.45	4.57	4.07	3.75	3.53	3.36	3.23	3.12	3.03	2.90	2.75	2.60	2.52	2.44	2.35	2.26	2.17	2.06

Table A.6: (*cont.*)

df₁ df₂	1	2	3	4	5	6	7	8	9	10	12	15	20	24	30	40	60	120	∞
29	7.60	5.42	4.54	4.04	3.73	3.50	3.33	3.20	3.09	3.00	2.87	2.73	2.57	2.49	2.41	2.33	2.23	2.14	2.03
30	7.56	5.39	4.51	4.02	3.70	3.47	3.30	3.17	3.07	2.98	2.84	2.70	2.55	2.47	2.39	2.30	2.21	2.11	2.01
40	7.31	5.18	4.31	3.83	3.51	3.29	3.12	2.99	2.89	2.80	2.66	2.52	2.37	2.29	2.20	2.11	2.02	1.92	1.80
60	7.08	4.98	4.13	36.5	3.34	3.12	2.95	2.82	2.72	2.63	2.50	2.35	2.20	2.12	2.03	1.94	1.84	1.73	1.60
120	6.85	4.79	3.95	3.48	3.17	2.96	2.79	2.66	2.56	2.47	2.34	2.19	2.03	1.95	1.86	1.76	1.66	1.53	1.38
∞	6.63	4.61	3.78	3.32	3.02	2.80	2.64	2.51	2.41	2.32	2.18	2.04	1.88	1.79	1.70	1.59	1.47	1.32	1.00

NAME INDEX

SUBJECT INDEX

Note: Entries in **bold** denote tables; entries in *italics* denote figures.

For Product Safety Concerns and Information please contact our EU
representative GPSR@taylorandfrancis.com
Taylor & Francis Verlag GmbH, Kaufingerstraße 24, 80331 München, Germany